The Role of Geospatial Technologies in Landslide Hazard Assessment

This book is designed to provide a detailed, methodological framework for landslide hazard assessment. The focus is on various dimensions of landslide hazard assessment, including the terminologies used in landslide hazard analysis and landslide inventory systems used globally and their relevance in generating a complete and reliable landslide database for further analysis, supported by global case studies. It includes an overview of the methodological developments in landslide hazard assessment and the role of geospatial technologies in landslide studies.

Features:

- Helps readers to understand the technical details of geospatial techniques applied in hazard management.
- Deals with the practicalities of how to recognise and classify unstable terrain.
- Covers recent advances in landslide estimation, particularly the automated means of landslide susceptibility estimation.
- Explores methodological frameworks of landslide hazard assessment.
- Illustrates case studies from the United States, Europe, and Asia, including demonstrations of different methodologies of landslide susceptibility zonation.

This book is aimed at researchers, graduate students, and libraries in geotechnical and environmental engineering.

The Role of Geospatial Technologies in Landslide Hazard Assessment

Sudhakar Dhondu Pardeshi,
Sumant Eknath Autade and
Suchitra Sudhakar Pardeshi

CRC Press
Taylor & Francis Group
Boca Raton London New York

CRC Press is an imprint of the
Taylor & Francis Group, an **informa** business

Designed cover image: Sudhakar Dhondu Pardeshi, Sumant Eknath Autade
and Suchitra Sudhakar Pardeshi

First edition published 2025
by CRC Press
2385 NW Executive Center Drive, Suite 320, Boca Raton FL 33431

and by CRC Press
4 Park Square, Milton Park, Abingdon, Oxon, OX14 4RN

CRC Press is an imprint of Taylor & Francis Group, LLC

Library of Congress Cataloging-in-Publication Data
Names: Pardeshi, Sudhakar Dhondu, author. | Autade, Sumant Eknath, author. | Pardeshi, Suchitra Sudhakar, author.
Title: The role of geospatial technologies in landslide hazard assessment / Sudhakar Dhondu Pardeshi, Sumant Eknath Autade, Suchitra Sudhakar Pardeshi.
Description: First edition. | Boca Raton : CRC Press, 2024. | Includes bibliographical references and index.
Identifiers: LCCN 2024035316 (print) | LCCN 2024035317 (ebook) | ISBN 9781032347165 (hardback) | ISBN 9781032347172 (paperback) | ISBN 9781003323488 (ebook)
Subjects: LCSH: Landslide hazard analysis. | Geographic information systems. | Remote sensing.
Classification: LCC QE599.2 .P37 2024 (print) | LCC QE599.2 (ebook) | DDC 551.3/070285—dc23/eng/20241123
LC record available at https://lccn.loc.gov/2024035316
LC ebook record available at https://lccn.loc.gov/2024035317

ISBN: 9781032347165 (hbk)
ISBN: 9781032347172 (pbk)
ISBN: 9781003323488 (ebk)

DOI: 10.1201/9781003323488

Typeset in Times
by Apex CoVantage, LLC

Dedication

*We dedicate this work to our
parents and teachers.*

– Authors

Contents

Foreword

Landslides are one of the most critical geohazards in hilly areas of the world. The frequency and magnitude of landslides have increased drastically due to growing human interventions. Since the 1960s, the population in many parts of the world has been growing. Pressure on land has continuously been increasing to support the needs of such a population, including developmental activities. Hilly areas are not free of these interventions. Reckless construction and unscientific agricultural and mining activities in hilly areas are responsible for reducing the resistance power of the hillsides, thereby causing devastating landslides.

By this time, several books and articles have been published on this devastating geohazard. But the present book, *The Role of Geospatial Technologies in Landslide Hazard Assessment*, authored by Sudhakar Dhondu Pardeshi, Sumant Eknath Autade, and Suchitra Sudhakar Pardeshi, addresses a new area of investigating landslides. The book will provide a comprehensive exploration of the role of geospatial technologies in understanding, assessing, and mitigating landslide hazards.

The book contains seven chapters. In the first part of the book, the authors have elaborated on the fundamental concepts of landslides and different terminologies and classifications of landslides, as well as the socioeconomic implications of landslides, hazard assessment, and mitigation strategies. It is followed by the recent advances in the methods and techniques of landslide investigation, a review of existing works in the world, and some case studies from India. The latter half of the book, comprising Chapters 5–7, discusses risk zonation case studies from some selected areas of the world, risk assessment, and mitigation strategies.

I believe the book will open a new perspective on landslide studies and will be helpful to geoscientists, engineers, environmentalists, and planners.

Prof. Sunil Kumar De
President, International Association of Geomorphologists (IAG)
President, Indian Institute of Geomorphologists (IGI)

Department of Geography
North-Eastern Hill University, Shillong, India

Preface

In recent times, the increasing frequency and severity of landslides have posed significant challenges to communities worldwide, causing immense socioeconomic and environmental impacts. As we witness these events with growing concern, it becomes increasingly evident that effective mitigation strategies are essential to minimise the adverse effects of landslides. We thought to bring such a book for the academicians, researchers, planners, and executives in landslide hazard management.

This book, *The Role of Geospatial Technologies in Landslide Hazard Assessment*, aims to provide a comprehensive exploration of the role of geospatial technologies in understanding, assessing, and managing landslide hazards. Through a blend of theoretical insights, case studies, and practical applications, this book elucidates various facets of landslide hazard assessment and mitigation.

The journey through this book begins with an introduction to the landslide hazard and related aspects, where the authors delve into the fundamental concepts surrounding landslides, delineating different terminologies and classifications. This foundational understanding sets the stage for discussing the socioeconomic implications of landslides, the importance of hazard assessment, and the implementation of mitigation strategies. Moreover, the chapter sheds light on the pivotal role of government agencies like the National Disaster Management Authority (NDMA) in formulating policies and plans for effective disaster management in India.

Subsequently, there is a global exploration of the recent advances in landslide hazard assessment. This part highlights specifically spatial patterns and methodologies that are used. From a review of landslide inventories to examining various susceptibility zonation techniques, this chapter offers valuable insights into the evolving landscape of landslide research and practices worldwide.

The book also explains different techniques for identifying and mapping landslides, emphasising the significance of geospatial data and remote sensing technologies. Through case studies from regions like the Western Ghats in India and the Nepal Himalayas, the authors demonstrate the application of multi-date satellite data and image interpretation techniques in understanding landslide morphology and distribution patterns.

The book elaborates on the methodologies for landslide identification and mapping, emphasising the importance of comprehensive databases and field investigations. Case studies reinforce the utility of satellite data and image interpretation and underscore the critical role of geospatial technologies in characterising landslide geometry and morphometric characteristics.

Besides this, the book also discusses landslide susceptibility zonation, highlighting selected approaches through case studies across different regions. By examining heuristic and multi-criteria decision-making models, this chapter elucidates the process of assigning weights to thematic data layers, thereby enhancing our understanding of landslide susceptibility in diverse landscapes.

As an important part of landslide risk assessment, the next chapter elucidates qualitative and quantitative approaches alongside the burgeoning role of geospatial technologies. Through a case study from the Western Ghats, India, the authors demonstrate methods for identifying vulnerable elements, mapping landslide vulnerability, and calculating specific and total risk.

Finally, the book encapsulates the discourse with an emphasis on landslide hazard management and mitigation strategies. From outlining the need for mitigation to exploring conventional and modern techniques, this book underscores the importance of stakeholder collaboration and the utilisation of information systems for effective landslide hazard management.

As we navigate through the entire contents of this book, it becomes evident that geospatial technologies serve as invaluable tools in our collective efforts to mitigate landslide hazards. Through this book, we aim to equip researchers, practitioners, policymakers, and stakeholders with the knowledge and insights necessary to address the multifaceted challenges posed by landslides in the contemporary world.

Acknowledgements

We take this opportunity to express our gratitude towards many individuals and institutions who helped us bring this book to reality. The journey of this book began with a book proposal from the Indian Society of Remote Sensing (ISRS), Dehradun. We are grateful to the past and present executive council of ISRS, Dehradun. We are also grateful to the proposal review committee for their valuable suggestions in the early stages of writing this book. It is also important to mention the CRC Press, New Delhi, editorial team for their constant support and follow-up. We all appreciate the support received from the CRC press team. We express our sincere thanks to the public works departments (government of Maharashtra) of Palghar, Thane, and Raigad districts; National Highways Authority of India; and *Times of India*, Mumbai, for their cooperation during data acquisition. We are thankful for the assistance received from our friends, colleagues, and students – mainly, Sai Shelar, Priyanka Hingonekar, and Tushar Raut. This book was possible only because of the constant support and encouragement received from our family members. The mention of Preeti Autade is a must here, for her constant support. Our family members, Atharva, Aryan, and Omkar, were also sources of encouragement in this work. We also humbly acknowledge the blessings received from our parents and other family members. Last, thanks to our friends who constantly supported us in carrying out this work.

About the Authors

Sudhakar Dhondu Pardeshi is presently working as a professor of geography at Savitribai Phule Pune University, Pune. Before joining this university, he served as a lecturer in geography at Ahmednagar College, Ahmednagar. His interest areas are geomorphology, landslide studies, water resource management, remote sensing, and GIS applications. He has 29 years of teaching and research experience at both graduate and postgraduate levels. He has published his research work in various national and international journals.

Sumant Eknath Autade is presently working as an associate professor in geography at Swami Vivekanand Night College in Dombivli, affiliated with the University of Mumbai. He has 18 years of teaching and research experience of at both graduate and postgraduate levels. His research interests are in the application of remotely sensed data in natural hazard assessment, particularly landslides. He has published 13 research papers in various journals of national and international repute. He has completed four research projects funded by state and national funding agencies.

Suchitra Sudhakar Pardeshi is presently working as professor of geography in Prof. Ramkrishna More ACS College, Akurdi, Pune. Prior to this, she served in Shivaji University, Kolhapur, and K.S.K.W. ASC College, Nasik. Her interest areas are geomorphology, landslide studies, flood hazard assessment, remote sensing, and GIS applications. She has 28 years of teaching and research experience at both graduate and postgraduate levels. She has published her research work in various national and international journals. She has completed four research projects funded by different national research funding agencies.

Abbreviations

AHP	Analytic Hierarchy Process
AOR	Angle of Reach approach
ASTER DEM	Advanced Spaceborne Thermal Emission and Reflection Radiometer Digital Elevation Model
BIC	Bayesian Information Criterion
BIS	Bureau of Indian Standards
CI	Consistency Index
CR	Consistency Ratio
CNN	Convolutional Neural Network
DInSAR	Differential Interferometry Synthetic Aperture Radar
EO	Earth Observation
ER	Elongation Ratio
EM-DAT	Emergency Events Database
EFLD	Enhanced Durham Fatal Landslide Database
FN Criteria	False Negative Criteria
FR	Frequency Ratio
GIS	Geographical Information System
GFLD	Global Fatal Landslide Database
GNSS	Global Navigation Satellite System
GPS	Global Positioning System
IVM	Information Value Method
IFFI	Italian Landslide Inventory
LHZ	Landslide Hazard Zonation
LRA	Landslide Risk Assessment
LSZ	Landslide Susceptibility Zonation
L/W	Length-to-Width ratio
LHEF	Landslide Hazard Evaluation Factors
LiDAR	Light Detection and Ranging
MSRDC	Maharashtra State Road Development Corporation
MCDM	Multi-criteria Decision-Making Approach
NASA	National Aeronautics and Space Administration
NASA GLC	NASA Global Landslide Catalog
NDMA	National Disaster Management Authority
NDRF	National Disaster Response Force
NHAI	National Highways Authority of India
NLP	Natural Language Processing
NRDMS	Natural Resource Data Management System
PWD	Public Works Department
RI	Random Inconsistency

SPOT	Satellite for Observation of the Earth
SRTM DEM	Shuttle Radar Topography Mission Digital Elevation Model
TEHD	Total Estimated Hazard
WSN	Wireless Sensor Network

1 Introduction

1.1 INTRODUCTION

Landslides are important geological hazards. Slope instability caused by various preparatory and triggering factors leads to landslides in undulated terrain. Landslides cause disruption in transportation lines, loss of agricultural land, damage to man-made structures and settlements, and sometimes, fatalities, if they occur in proximity to human habitations. Landslides also result in geomorphic changes and alter landforms. Landslides very often are considered local environmental issues; however, the effects of landslides are often found to be beyond local jurisdiction. The growing population across the world increases pressure on the availability of land due to the expansion of human settlements and other land use. It is, therefore, the population that is forced to occupy unstable hill slopes. The encroachments of human habitation on unstable hill slopes lead to an increased probability of landslide occurrence and risk to human habitation.

Considering the socioeconomic and geomorphic implications of landslide occurrence, there is a need to develop appropriate measures to mitigate its negative impacts. Landslide hazard assessment is an important step towards landslide hazard and risk management. It involves an inventory of landslides, landslide susceptibility zonation, and landslide risk assessment. The available literature needs to be reviewed to identify the adoption and development of landslide investigation methodologies throughout the world.

This chapter covers the conceptual background of landslide hazards and related terminologies. The fundamentals of landslide hazard assessment and the various processes involved therein are discussed in subsequent sections of this chapter. The chapter also attempts to evaluate the disaster management policy of India in the context of landslide hazards.

1.2 LANDSLIDE TERMINOLOGY

1.2.1 LANDSLIDES

The term 'landslide' is perceived differently by professionals from different disciplines such as geology, geomorphology, engineering, etc. These definitions reflect the complex nature of the processes involved in a landslide. However, all definitions of landslides are associated with specific mechanisms of slope failures, their characteristics, and their properties. There are varieties of terms used to define this phenomenon, such as mass wasting, landslides, slope instability, mass movement, slope failures, and slope movement. Several authors have defined landslides in different ways. Some of the important definitions of landslides are given below.

DOI: 10.1201/9781003323488-1

According to (Monkhouse, 1970), a landslide refers to a fall of the earth and rock material down a slope or mountainside, the result of gravity and rain lubrication. The term 'landslide' is commonly used in the United States, whereas the term 'landslip' is common in Britain. Selby (1982) defines a landslide as one of the forms of mass wasting, referring to the downslope movement of soil or rock material under the influence of gravity without the direct aid of other media such as water, air, or ice. However, often, many landslides are triggered by intense rainfall, seismic activity, and man-made structures along steep slopes. According to Alexander (1993), landslides occur on transport-limited slopes that lead to the removal of loose, weathered material downslope. Transport-limited slopes tend to be more prone to landslides than weathering-limited slopes. Such kinds of mass movement occur when slopes are steeper than the threshold angle of stability.

According to Highland (2008), landslide refers to the downslope movement of rock, soil, and debris under the influence of gravity and also to the landscape developed by these movements. Slope failure, mass movement, and slope instability are other phrases or terms that are used alternatively with the term 'landslides'. Sharpe (1938) defined a landslide as a type of movement that is perceptible and involves a relatively dry mass of earth debris. Varnes (1984) defined landslides as 'almost all varieties of mass movements on the slope, including rock falls, topples and debris flow that involve little or no true sliding'. Brusden (1984) considered landslides a unique form of mass transport and a process that does not require a transportation medium such as water, air, or ice.

Crozier (1986) defined landslides as 'the outward and downward gravitational movement of the earth material without the aid of running water as a transporting agent'. According to Hutchinson (1988), 'A landslide in its strict sense is a relatively rapid mass wasting process that causes the downslope movement of a mass of rock, debris or earth triggered by a variety of external stimulus'.

Cruden (1991) considered it as the movement of rock debris or earth downslope. In the last few decades, the term 'landslide' is being used to define mass movement, but it refers to only a specific type of slope movement with a specific composition, form, and speed. According to Highland and Bobrowsky (2008), a mass movement is a proper term for any form of detachment, a downslope movement of soil and rock with different velocities. They also mentioned that landslides include all forms of mass movement having a speed greater than 1 millimetre per day.

The National Disaster Management Authority of India (2009) has explained the term 'landslide' as a 'a process involving slope forming materials due to the action of gravity that involves movement like falls, flow, topples and creeps'. Pelty (2010) described a landslide as a 'range of processes that result in the downward and outward movement of slope forming material composed of rock, soils and artificial materials'. A recent definition by Couture states that a landslide is a movement of a mass of soil (earth or debris) or rock in a downslope direction. This concept of landslide is broadened concerning the type of material that moves downslope.

Introduction

1.2.2 LANDSLIDE CLASSIFICATION

A landslide is the downwards movement of rock, soil, or debris on the surface of rupture, either curved (rotational) or planar (translational). Understanding the characteristics and processes dominant in slope instability is essential for developing appropriate landslide mitigation strategies. The classification of landslides helps to estimate the amount of displaced material, runout distance, potentially affected areas, the possibility of recurrence, nature, speed of movement of material, etc. Landslides can be classified based on material type (rock, debris, soil) or type of movement (fall, topple, slide, flow, avalanche, spread). This sub-section describes the major classifications of landslides.

Landslides are caused by several factors and conditions. There is a remarkable spatial variation in landslide occurrence and also in the nature and shape of the landslide phenomenon. Landslides can be sub-grouped based on slope-forming material, slope, rate of weathering, landslide geometry, movement type, hydrological conditions, speed of movement, triggering mechanism, recurrence interval, etc.

Several landslide classification schemes are proposed by researchers worldwide. However, the most common factors considered for the classification of landslides are slope-forming materials, landslide geometry, type of movement, and meteorological information.

Sharpe (1938) proposed a landslide classification scheme based on the type of movement, material type, and role of water and ice (Table 1.1).

TABLE 1.1

Sharpe's Classification Scheme of Mass Movement

				Material (Earth or Rock)			
	Kind	Rate	Ice	Earth, rock, and ice	Rock and earth, dry with a minor amount of water	Earth and rock with water	Water
Free side	Flow	Slow to rapid	Glacial transportation	Rock glacial cap/ solifluction	Rock creep Talus creep Soil creep	Solifluction	Fluvial transportation
	Slip	Slow to rapid		Debris avalanche	--	Earth flow Mudflow Debris avalanche	
No free side	Slip or flow	Slow to rapid		Rockslide Rockfall	Slump Debris slide Debris fall		

(*Source*: Sharpe 1938)

TABLE 1.2

Varnes's Classification Scheme of Mass Movement

Type of Movement			Type of Material		
			Bedrock	Engineering Soils	
				Predominantly Coarse	Predominantly Fine
Falls			Rockfall	Debris fall	Earth fall
Topples			Rock topple	Debris topple	Earth topple
Slides	Rotational	Few Units	Rock slump	Debris slump	Earth slump
	Translational	Many Units	Rock block slide	Debris block slide	Earth block slide
			Rockslide	Debris slide	Earth slide
Lateral spreads			Rock spread	Debris spread	Earth spread
Flows			Rock flow (Deep creep)	Debris flow (Soil creep)	Earth flow
Complex combinations of two or more principal types of movement					

(*Source*: Varnes 1978)

Sharpe (1938) emphasised the role of ice and water in the mass movement. However, his scheme does not provide clear guidelines about the rate of movement and has not given a quantitative range to define the rate of movement of slope-forming material.

Hutchinson (1988) used the type of movement and morphology as a basis for his classification scheme. His classification is specifically based on field observations and aerial photo interpretation. His scheme provides descriptive explanations of each landslide type, which are not defined clearly.

Crozier (1986), in his classification scheme, calculated the threshold values of various morphometric parameters to define landslide types.

Varnes (1984) developed a landslide classification scheme based on the type of slope-forming material and type of movement. He considered the textural characteristics of bedrock and engineering soils for the classification of slope instability (Table 1.2).

Varnes's classification scheme of mass movement is one of the most appropriate classification schemes and has been used popularly for landslide studies. The classification of landslides based on their movement, slope-forming material, and textural characteristics is more realistic for further investigations. Landslide classification schemes are largely used in the preparation of complete landslide inventories and hazard zonation mapping.

Some of the important types of slope failures are the following:

1 Fall
2 Topple

3 Landslide (rotational and translational)
4 Soil creep
5 Flow

1.2.3 LANDSLIDE INVENTORY

Preparation of a complete landslide inventory is one of the most important preliminary exercises in the process of landslide hazard assessment. The concept of landslide inventory is defined by Cruden (1991) as 'the simplest form of landslide information which records the location and where it is known, the date of occurrence of a type of landslide that has left traces of the event in an area'. A landslide inventory map also shows slope failure by a single event, or it may show the cumulative effects of several events (Guzzetti, 2000).

1.2.4 LANDSLIDE PREPARATORY AND TRIGGERING FACTORS

Delineation of potential landslide hazard zones is an integral part of landslide hazard assessment. Classification of the land surface depending on the degree of susceptibility to landslides of different magnitudes is based on various causative factors and triggering mechanisms.

The degree of association of landslide magnitude with different classes of causative factors is analysed separately to provide a ranking for the identification of landslide hazard classes. Therefore, the significance of landslide causative factors cannot be overemphasised. When conditions and processes that promote slope instability are identified, then it is often possible to estimate their relative contribution to slope failure and to give them some qualitative or semi-quantitative measures place by place (Varnes, 1984).

Recognition of variables and conditions that cause landslides is of utmost importance in the process of landslide susceptibility zonation (LSZ) mapping. Wu and Sidle (1995) described two different kinds of variables: viz., preparatory variables and triggering variables. First, preparatory variables are those factors that cause slope failure without actually initiating it and thereby tend to place the slope in a marginally stable status. Chen et al. (2012) elaborated the impact of geological settings on landslide geometry, especially landslides induced by earthquakes. These factors generally constitute the integral characteristics of the landscape. These variables include slope gradient, aspect, slope curvature, lithology, lineaments, faults (structural discontinuities), soil characteristics (e.g., depth, cohesiveness, texture, level of weathering), land cover, land use, drainage, etc. Second, triggering variables are those which shift the slope from a marginally stable to an unstable state, thereby initiating failures in an area of given susceptibility. These factors include natural processes such as rainfall, seismicity, storms, volcanic eruptions, and even anthropogenic activities along the landscape susceptible to slope failure, like road cutting, deforestation, farming, and opencast mining.

Selection and assigning weightage to these landslide causative variables is a critical task mainly because it directly affects the reliability of the landslide hazard susceptibility map. Several authors have carried out LSZ mapping based on different causative factors. The research carried out in the recent period in different parts of the world has been reviewed to compare the landslide causative factors considered for LSZ.

Several researchers have attempted to identify specific causative factors for landslide hazard zonation. However, the influence of any causative condition on landslide occurrence varies from place to place due to variations in geomorphological and climatic parameters. A brief description of the causative factors considered for landslide hazard zonation by a few researchers is given in Table 3.1.

Several researchers have adopted different types of ranking framework for landslide preparatory and triggering factors (Naithani 2007; Singh et al. 2011; Pereira et al. 2012). The literature on consideration of landslide preparatory and triggering factors for landslide susceptibility zonation is discussed in 2.5.

1.2.5 THE MAPPING UNIT

The mapping unit refers to a portion of land containing a set of ground conditions that differ from the adjacent unit across the defined geographical area (Hansen, 1984). The selection of a mapping unit for landslide susceptibility zonation is an important task because the spatial mapping unit determines the quality of the hazard map. The homogeneity in the geological and slope conditions of a mapping unit maximises the reliability of landslide susceptibility mapping. These mapping units are grouped into five: viz., grid-cell (pixel), terrain unit, unique condition unit, slope units, and topographic units (Guzzetti et al., 1999).

The majority of the landslide hazard zonation mapping exercises carried out in the recent past have used grid-cell (pixel) as a mapping unit, which is usually preferred in GIS-based landslide hazard zonation methods. GIS-based landslide susceptibility zonation methods such as information value model (IVM), artificial neural network (ANN), maximum likelihood model, discriminant analysis, frequency ratio model, probabilistic hazard mapping, logistic regression model, and multi-criteria decision-making process carry out pixel-based classification for landslide hazard zonation mapping.

Terrain units are based on geomorphological and geological homogeneity, which can also be used for mapping landslide susceptibility. Rowbotham (1998) used the digital elevation model (DEM) based terrain unit for susceptibility classification in Hong Kong. Panikkar and Subramaniyan (1997b) applied terrain units for landslide hazard assessment in Mussourie in the Kumaun Himalayas, India.

Unique condition units (UCUs) for slope stability analysis are used for the classification of individual parameters related to landslides based on homogeneous conditions (Guzzetti et al., 1999). UCU-based classification has proved to be a suitable mapping unit in slope stability assessment.

Slope units are also commonly used as mapping units in landslide hazard zonation mapping. Slope units are homogeneous surfaces comprising the region between drainage and divide lines. Slope units can be subbasins, main slope units, or small slope facets (Guzzetti et al., 1999). Several heuristic and statistical landslide susceptibility assessments have used slope units as a unit for landslide susceptibility mapping.

Eeckhaut et al. (2009) compared landslide susceptibility maps produced by using a multivariate approach for mapping units: viz., grid-cells (pixel), topographical mapping unit (TMU), and slope units. The study revealed that a TMU-based landslide susceptibility map shows a larger susceptible area than the grid-based LSZ map.

Homogeneous susceptibility units (HSUs) are also being used to classify landslide susceptibility maps. Das et al. (2011) carried out multi-resolution segmentation to divide landslide susceptibility maps into HSUs based on shape and scale parameters. Das et al. (2011) derived HSUs automatically from a grid-based landslide susceptibility map using a region-growing algorithm and an optimum size factor for the national highway corridor in the Garhwal Himalayas.

So far, various mapping units have been used for susceptibility zonation in different parts of the world, but still, there is a lack of agreement to determine the best-suited mapping unit universally for LSZ mapping.

1.2.6 Landslide Risk

Understanding concepts about landslide risk is an important step towards effective landslide hazard and risk assessment. Landslides of high magnitude in non-populated areas may not be very significant, but even low-magnitude landslides in thickly populated areas may cause sizable losses. According to Varnes (1984), 'the landslide risk refers to the expected number of lives lost, persons injured, property damage, or disruption of economic activity due to landslide occurrence'. He explained landslide risk as the product of hazard and vulnerability ($R = H \times V$), where hazard (H) is a probability of landslide occurrence and vulnerability (V) is the degree of loss to a given element at risk resulting from landslides.

Rezig et al. (1996a) defined risk as the product of three components: viz., hazard, vulnerability, and cost. They further proposed that the risk is the mathematical expression of potential loss caused by a landslide of a given magnitude in an area.

The United Nations International Strategy for Disaster Reduction (2004) defines risk as the probability of harmful consequences or expected losses resulting from an interaction between natural and human hazards and vulnerable conditions. The degree of loss due to landslides is determined by the magnitude of the event and elements at risk (e.g., population, vehicles, traffic, vegetation cover, agricultural land, man-made structures). The intensity of losses varies from place to place depending upon the geometry of the landslide and the location of the elements at risk.

1.2.7 Landslide Susceptibility Zonation

Landslide susceptibility zonation is an important step in landslide investigation and landslide risk management. Varnes (1984) defines the term 'zonation' as the process of division of land surface into areas and ranking these areas according to the degree of actual or potential hazard from landslides or other mass movements. A large number of studies have been undertaken for landslide hazard zonation in various parts of the world. A brief review of a few approaches to landslide hazard zonation at various spatial scales is given as under.

Courture (2011) explained the concept of landslide hazard zonation as the 'division of land into somewhat homogeneous areas or domain and their ranking according to the degrees of actual or potential landslide susceptibility, hazard or risk or applicability of certain landslide-related regulations'.

There has been significant growth both in landslide events, particularly those induced by human activities, and in several landslide investigations in different parts of the world (Gutierrez et al., 2010). Gokceoglu and Sezer (2009) carried out a statistical assessment of international landslide literature. They argued that the publication of landslide-related articles in international journals has experienced exponential growth. They also pointed out that landslide susceptibility assessment is an important part of landslide investigation and has received more attention, with the highest number of publications in international journals. Pardeshi et al. (2013) carried out a comparative assessment of landslide hazard zonation methods, consideration of landslide causative factors, and remote sensing (RS) and geographical information system (GIS) techniques for different parts of the world. They argued that multivariate statistical methods and multi-criteria decision-making approaches are more reliable methods to assess landslide hazards than any other heuristic and bivariate statistical method.

In recent years, several attempts have been made to apply different methods of LSZ and to compare results to find the best-suited model. The advanced multivariate techniques are proven to be effective in the spatial prediction of landslides with a high degree of accuracy. Physical process-based models also perform well in LSZ mapping, even in areas with poor databases. The multi-criteria decision-making approach in LSZ mapping also plays a significant role in determining the relative importance of landslide causative factors in the slope instability process.

Over the last three decades, LSZ mapping has been carried out in different parts of the world. Several approaches have been developed for LSZ mappings, such as inventory-based mapping, heuristic approach, probabilistic assessment, deterministic approach, statistical analysis, and multi-criteria decision-making approach. However, no one method is accepted universally for the effective assessment of landslide hazards.

1.3 LANDSLIDE HAZARD ASSESSMENT

Landslide hazard assessment (LHA) refers to the evaluation of various conditions and factors responsible for slope instability and the application of different techniques to predict landslide occurrence and delineate landslide-susceptible zones. LHA involves landslide inventory, landslide susceptibility zonation, landslide vulnerability, risk mapping, and landslide mitigation measures. A brief discussion of various approaches to LHA is given in subsequent sections.

1.3.1 Distribution (Inventory) Approach

The distribution or inventory approach is the simplest approach to landslide hazard assessment. This method aims at discussing spatial patterns of landslide distribution in a given area based on historical landslide records. Inventory approach to landslides is one of the simplest qualitative approaches to LSZ mapping. It is also known as 'landslide inventory'. In this analysis, landslide inventory maps are produced, which portray spatial and temporal patterns of landslide distribution, type of movement, rate of movement, type of displaced material (soil, debris, or rock), etc. Landslide data are obtained through field survey mapping, historical records, satellite images, and aerial photo interpretation. Landslide distribution and density maps provide the basis for landslide susceptibility analysis and mitigation measures.

Different scholars have advocated the distribution approach to landslide hazard assessment (Cruden1991; Guzzetti et al., 2003; Galli et al. 2008; Guzzetti et al. 2003; Colombo et al. 2005; Varnes's 1984). The detailed description of existing literature on landslide inventory approach is reviewed in section 2.4.1.

1.3.2 Landslide Susceptibility Zonation

In the last few years, the approach towards LSZ has changed from a heuristic (knowledge based) approach to a data-driven approach (statistical) to minimise subjectivity in weightage assignment procedures and produce more objective and reproducible results (Kanungo et al., 2009). Methods based on a statistical analysis of geo-environmental factors related to landslide occurrence are preferred. The statistical methods for LSZ can be grouped into two: viz., bivariate statistical analysis and multivariate statistical analysis.

1.3.2.1 Bivariate Statistical Analysis

The bivariate statistical analysis for landslide hazard zonation compares each data layer of causative factor to the existing landslide distribution (Kanungo et al., 2009). Weights to the landslide causative factors are assigned based on landslide density. The frequency analysis approach, IVM, weights of evidence (WoE) model, weighted overlay model, etc. are important bivariate statistical methods used in LSZ mapping.

1.3.2.2 Heuristic Approach (BIS-Based LHEF Method)

The Bureau of Indian Standards (1998) has given guidelines for macro-level landslide hazard zonation in India. The BIS-based landslide hazard evaluation factors (LHEF) rating scheme for landslide susceptibility zonation is a heuristic approach (based on expert opinion) to LHA. According to the Bureau of Indian Standards (1998), landslide hazard zonation procedure can be performed using the LHEF rating for different landslide causative factors. The BIS identified six landslide causative factors for hazard zonation: viz., lithology, structure, slope morphometry, relative relief, land use and land cover, and hydrological conditions. In this method, the area under investigation is divided into small mapping units, to which numerical weights are assigned for each thematic data layer, and finally, the total estimated hazard (TEHD) is obtained by adding the weights of all the variables for each mapping unit, and a landslide hazard map is produced.

Naithani (2007) applied univariate statistical techniques for LSZ mapping in the Garhwal Himalayas. The guidelines of the Bureau of Indian Standards (BIS 14496, Part 1998) were followed to calculate the maximum LHEF ratings. The study reveals that geological characteristics are important to determine landslide susceptibility followed by hydrogeological conditions.

Anbalagan et al. (2008) worked out landslide hazard zonation mapping for urban areas of Nainital in the Kumaun Himalayas, India. The landslide susceptibility zonation was carried out using the LHEF rating scheme, which is based primarily on a heuristic approach. The mesoscale landslide susceptibility zonation for Nainital city was done to identify safe and unsafe areas with landslide hazards. Singh et al. (2011) carried out work on LSZ mapping along NH 39 in Manipur using a BIS guideline-based LHEF rating scheme. Ghosh et al. (2009) evaluated the effectiveness of the existing BIS method in the Darjeeling Himalayas by adopting the WoE model.

The BIS-based LHEF rating scheme is a very simple and cost-effective method of landslide hazard mapping. However, subjectivity in the weight assignment procedure exists in this method, which can affect the level of accuracy of the LSZ map. Moreover, this method does not consider landslide distribution, and therefore, it's very difficult to test its validity. Though BIS has proven its importance, it is mainly applied to slope facets of the area.

1.3.2.3 Multivariate Statistical Analysis

Multivariate statistical analysis for landslide hazard zonation considers the relative contribution of each thematic data layer to the total landslide susceptibility (Kanungo et al., 2009). These methods calculate the percentage of landslide area for each pixel, and the landslide absence-presence data layer is produced, followed by the application of the multivariate statistical method for reclassification of hazards for the given area. Logistic regression model, discriminant analysis, multiple regression model, conditional analysis, and ANN are the commonly used multivariate methods for LSZ mapping.

1.3.2.4 Multi-criteria Decision-Making Approach

Landslide hazard assessment is associated with consideration of several landslide preparatory and triggering factors. The identification of a relative contribution of given geoenvironmental parameters in landslide prediction modelling is a very difficult task. Hence, the application of the multi-criteria decision-making approach is consider to be the most reliable and important method for delineating landslide-susceptible zones. The analytical hierarchy process (AHP) is a popular multi-criteria decision-making process of measurement through pairwise comparisons and relies on the judgements of experts to derive priority scales (Saaty,

2008). AHP operates at five levels: viz., defining the problem, determination of goals and alternatives, constructing of pairwise comparison matrix, determining weights, and obtaining overall priority. AHP-based landslide susceptibility modelling considers different landslide causative factors are considered as alternatives. Scales of absolute numbers (from 1 to 9) are used to assign score for each landslide-related parameter based on its relative importance, and comparison matrices are constructed to compute the consistency ratio (CR) and consistency index (CI). Akgun (2011) carried out comparison of landslide susceptibility maps produced from different landslide susceptibility models using the AUC (area under curvature) method. Landslide hazard maps produced by logistic regression (LR), multi-criteria decision-making approach (MCDA), and likelihood ratio method (LRM) for Azmir, Turkey, using LR and MCDA showed similar results. Ayalew and Yamagishi (2005) compared LSZ maps using LR and AHP models to assess landslide hazards. The study revealed that when LSZ maps were compared with landslide activity maps, the AHP-based map performed better than the LR model.

1.3.2.5 Physical Process-Based Landslide Susceptibility Models

Physical processes leading to the landslide event – on simple mechanical laws – provide the basis for landslide hazard assessment through physical process-based models. These models account for the momentary groundwater response of slope to rainfall (Kuriakose, 2010). Process-based models do not need long-term landslide data and therefore can also apply to areas with incomplete landslide database (Kuriakose, 2010).

There has been a significant focus on use of physically process-based models. Transient rainfall infiltration and grid-based slope stability (TRIGRS) model (Salciarini et al. 2006), real tome susceptibility modelling (Montrasio et al. 2011).

Kuriakose (2010) compared four physical process-based models for delineating landslide-susceptible areas in the Western Ghats of Kerala, India.

Zizioli et al. (2013) compared four physical process-based models in a study on shallow landslides in northern Italy. These are deterministic models which have a high predictive capability and assess the influence of individual parameters related to landslides (Mantovani et al., 1996). These models perform well for landslides where slope behaviour can easily be predicted by simple mechanical laws, but they lack consideration of the temporal aspects of landslides (Guzzetti et al., 2003). The detailed review of existing literature on physically process-based models for landslide hazard assessment is discussed in 2.4.9.

1.3.2.6 Other Approaches

Besides heuristic, bivariate, and multivariate statistical models; multi-criteria decision-making models; and physical process-based models for landslide prediction, there has been a wide range of methods of LHA developed in recent times. The angle of reach (AOR) is one of the simplest qualitative approaches to landslide susceptibility assessment. AOR is an angle connecting the landslide crown with the margins of runout material (Dahl et al., 2010). This is an empirical method of landslide susceptibility, in which landslide geometry (shape of the scar, depth of the landslide deposits, runout distance) is used to infer the landslide volume and other parameters.

Chang and Chiang (2009) used logit model for predicting rainfall-induced landslides. The detailed discussion on their findings is described in 2.4.10.2.

Decision tree modelling is a classification technique and can be used to map landslide susceptibility in data-scares environment. Yeon et al. (2010) adopted GIS-based decision tree model for mapping landslide susceptibility in Injae, Korea. The detailed results derived from the study have been discussed in 2.4.10.3.

LSZ mapping involves several causative factors and triggering mechanisms. One of the most important challenges faced by scientists engaged in LSZ mapping is determining the appropriate method of assigning weights to landslide explanatory factors based on their relative importance. Several qualitative and quantitative methods of LSZ (discussed earlier) provide a methodological framework for assigning weights to a given parameter based on certain criteria such as an expert's knowledge (heuristic approach), the relationship of landslide causative factors with landslide occurrence (bivariate statistical approach), the relative importance of individual landslide causative factors in slope failure (multivariate statistical approach), and pairwise comparison of landslide explanatory variables. However, still there is a lack of agreement on one appropriate LSZ method which could universally be applicable in all kinds of geo-environmental conditions. This is probably due to the complexity of the landslide phenomenon in different parts of the world. The application of the multi-criteria decision-making approach of LSZ mapping can effectively be used to maintain objectivity in the assignment of weights and to determine the relative importance of an individual parameter based on interrelationships among other variables determining the degree of landslide hazards.

Another major issue related to LSZ mapping, especially in the Indian context, is the lack of availability of sufficient and reliable data for accurate landslide assessment. Therefore, the application of data-driven methods of LSZ in data-scarce environments is a major limitation in LHA. However, the need is to be cautious in the selection of appropriate methods in unique geo-environmental conditions.

There are various approaches adopted by researchers worldwide for landslide hazard zonation mapping at varying spatial scales. Generally, these approaches can be grouped into heuristic (knowledge based/expert opinion), semi-quantitative, and quantitative. Heuristic approaches give more emphasis to the rating of landslide causative factors, but actual landslide distribution is not taken into consideration. The weightage is given based on the expertise and experience of the investigator, which may give rise to subjectivity. Semi-quantitative landslide hazard zonation mapping is done by assigning weightage based on the correlation between landslide causative factors and landslide distribution. Quantitative landslide hazard zonation mapping is mainly associated with field measurements and the application of multivariate statistical techniques along with spatial and temporal probability analysis, which helps in the prediction of future landslide events with a high degree of confidence.

1.3.3 LANDSLIDE RISK ASSESSMENT

Landslide risk assessment aims at delineating and quantifying actual and potential loss caused by landslide events. Assessment of landslide is important in prioritising mitigation measures and preparedness planning. There are two approaches of landslide risk assessment: qualitative and quantitative landslide risk assessment. Qualitative landslide risk assessment deals with delineating potential landslide risk zones considering different elements at risk. Quantitative landslide risk assessment on the other hand attempts to quantify potential losses. It is essential for accurate landslide risk estimation. There are several approaches to landslide risk assessment.

Jaiswal et al. (2010) applied Gumbel distribution model to estimate landslide spatial probability for different return periods.

Rezig et al. (1996b) emphasised probabilistic landslide risk evaluation. Van Westen (2010) proposed a GIS-based landslide risk assessment method. Saaty (2008) proposed an analytic hierarchy process, an important multi-criteria decision-making approach. This mathematical approach have been extensively used for landslide hazard and risk assessment by many researchers.

Pardeshi et al. (2009) used landslide geometry to estimate volume of displaced material due to landslides. Anbalagan et al. (2008) elaborated the need for micro-level landslide hazard zonation to prepare a detailed land-use planning model. Existing literature on landslide risk assessment is discussed in 2.8.

1.4 APPLICATION OF REMOTE SENSING AND GIS IN LANDSLIDE INVESTIGATIONS

Application of remotely sensed data has been proved to be powerful in assessment of landslide hazards. In fact, remote sensing and GIS technology have been used extensively for landslide hazard assessment since past two decades. Since landslide is a localised phenomenon, it occurs in relatively smaller geographical areas. Therefore, high-resolution aerial photographs and satellite images are essential for effective delineation of landslides and further mapping.

At the advent of RS and GIS technologies in the field of spatial studies, they are now being widely used in hazard assessment. Their role in the field of landslide investigation cannot be overemphasised. The extraction of relevant spatial information related to landslide occurrence is an integral part of hazard assessment. RS data and GIS are proven to be effective tools for generating and processing spatial information. The advancement in 'Earth observation' (EO) techniques facilitates effective landslide detection, mapping, monitoring, and hazard analysis (Tofani et al., 2013).

The review of a few studies on LHA using RS data indicates that aerial photographs are widely used in landslide detection and mapping. Good-quality aerial photographs help in accurate landslide detection and mapping. However, aerial photographs may not be used in continuous landslide monitoring, since they do not provide repetitive coverage of the same area.

The recent developments in the application of satellite RS data in landslide studies in Europe have been discussed by Tofani et al. (2013). The study showed that over 70% of the total applications of RS data for landslide studies are associated with landslide detection, mapping, and monitoring. High-resolution satellite data are being effectively used for landslide detection, mapping, monitoring, and other applications.

The use of DEM is of immense importance in landslide hazard assessment. Several thematic data layers, such as slope angle, slope aspect, curvature, lineaments, drainage, and ridges can be extracted from DEM with good resolution. Landslide hazard zonation studies in recent times have used DEM with high resolution to generate spatial information data layers related to landslide hazards.

GIS is widely used in landslide hazard assessment, especially for the generation of thematic data layers, computation of different indices, assignment of weights, data integration, and generation of LSZ maps. Several LSZ methods, such as ANN, decision tree model, weighted overlay, AHP, MCDA, IVM, and physical process-based landslide hazard models, are GIS-based models to predict landslide probability.

1.5 DISASTER MANAGEMENT POLICY IN INDIA IN THE CONTEXT OF LANDSLIDE HAZARDS

According to the Geological Survey of India, about 0.42 million km^2 of land area in India (12.6% of the total land area of the country) are identified as landslide-prone areas. The Himalayan region, Western Ghats, and Eastern Ghats are important concentrations of landslide-prone areas in India. More than 22 states and two union territories are affected by landslide events in different parts of the country. Almost all landslides in India are triggered by intense precipitation, particularly during the monsoon season. In the past few decades, slope failure events along major roads and railway communication routes and those in proximity to human settlements often destroy lives and property. The formulation of an appropriate strategy for the management of risk caused by landslides in different parts of India is thus necessary.

As per the Disaster Management Act of 2005, the NDMA has been mandated to lay down policies, plans, and guidelines for disaster management. Considering the importance of landslide risk management, the NDMA has constituted a task force for the preparation of national- and local-level landslide management strategies. Recently, in September 2019, the NDMA, under the umbrella of the Union Ministry of Home Affairs, formulated and published the National Landslide Risk Management Strategy (National Disaster Management Authority, 2019).

The task force for landslide reduction in India was divided into six subgroups dedicated to landslide hazard mapping, landslide early warning system, awareness programme, capacity building and training of stakeholders, preparation of mountain zone regulation for slope stabilisation, and mitigation of landslides. The implementation strategy for landslide management is divided into two: short-term implementation strategy and long-term implementation strategy. The short-term implementation strategy includes the formation of a landslide hazard zonation monitoring committee, the cataloguing of landslide susceptibility maps, taking up pilot projects, and the formation of a group of expert agencies for mesoscale landslide mapping. Long-term implementation strategy includes landslide susceptibility mapping at a 1:10,000 scale using medium- to high-resolution satellite data, use of web-based tools for mapping landslides, etc. The landslide risk reduction strategy is proven to be a significant step for regulating and strengthening landslide disaster risk reduction to minimise losses. However, it is important to implement these strategies at the local level to ensure effective landslide mitigation measures.

1.6 SUMMARY

Landslides are important geological hazards. They cause significant loss in terms of property and life if they occur in proximity to human habitation. Based on the rate of movement and material characteristics, landslides are classified into various types. Classification of mass movement was given by Sharpe, Varnes, and Hutchinson. However, the classification of mass movement given by Varnes has been adopted universally as compared to other classifications. LHA involves the preparation of landslide inventory, landslide susceptibility zonation, landslide vulnerability assessment, and landslide risk zonation. It is carried out by using different models. However, the predictability of a landslide-prone zone depends largely upon landslide preparatory and triggering factors. RS and GIS are instrumental in landslide hazard assessment. India's landslide hazard mitigation strategy has been developed by the NDMA in 2019 to plan and implement landslide mitigation measures to minimise the losses caused by it.

REFERENCES

Akgun, A. 2011. "A Comparison of Landslide Susceptibility Maps Produced by Logistic Regression, Multi-Criteria Decision, and Likelihood Ratio Methods: A Case Study at Lzmir, Turkey." *Landslides, Springer-Verlag.* https://doi.org/10.1007/s10346-011-0283-7.

Alexander, D. 1993. *Natural Disasters.* UCL Press Ltd.

Anbalagan, R., D. Chakraborty, and A. Kohali. 2008. "Landslide Hazard Zonation (LHZ) Mapping on Meso-Scale for Systematic Town Planning in Mountainous Terrain." *Journal of Scientific & Industrial Research* 67: 486–497.

Ayalew, L., and H. Yamagishi. 2005. "The Application of GIS-Based Logistic Regression for Landslide Susceptibility Mapping in the Kakuda-Yahiko Mountains, Central Japan." *Geomorphology* 65: 15–31.

Brusden, D. 1984. "Mudslides." In D. Brusden, D. Prior (Eds.), *Slope Instability* (pp. 363–418). Wiley.

Bureau of Indian Standards. 1998. *Preparation of Landslide Hazard Zonation Maps in Mountainous Terrain – Guidelines (Part2-Macrozonation).* IS 14496-2, Hill Area Development Engineering (CED 56), New Delhi, pp. 1–20.

Chang, K., and S. Chiang. 2009. "An Integrated Model for Predicting Rainfall Induced Landslides." *Geomorphology* 105: 366–373.

Chen, X. L., H. L. Ran, and W. T. Yang. 2012. "Evaluation of Factors Controlling Large Earthquake-Induced Landslides by the Wenchuan Earthquake." *Natural Hazards and Earth System Science* 12 (12): 3645–3657. https://doi.org/10.5194/nhess-12-3645-2012.

Colombo, A., L. Lanteri, M. Ramasco, and C. Troisi. 2005. "Systematic GIS Based Landslide Inventory as the First Step for Effective Landslide Hazard Management." *Landslides* 2: 291–301.

Courture, R. 2011. *Landslide Terminology – National Technical Guidelines and Best Practices on Landslides.* Geological Survey of Canada. Open Files 6824.

Crozier, M. 1986. *Landslides – Causes, Consequences and Environment.* Croom Helm Ltd.

Cruden, D. M. 1991. "A Simple Definition of a Landslide." *Bulletin of the International Association of Engineering Geology – Bulletin de l'Association Internationale de Géologie de l'Ingénieur* 43 (1): 27–29. https://doi.org/10.1007/BF02590167.

Dahl, M.-P. J., L. E. Mortensen, A. Veihe, and N. H. Jensen. 2010. "A Simple Qualitative Approach for Mapping Regional Landslide Susceptibility in the Faroe Islands." *Natural Hazards and Earth System Sciences* 10: 159–170.

Das, I., A. Stein, N. Kerle, and V. Dadhwal. 2011. "Probabilistic Landslide Hazard Assessment Using Homogeneous Susceptible Units (HSU) along a National Highway Corridor in the Northern Himalayas, India." *Landslides* 8: 293–308.

Eeckhaut, M., P. Reichenbach, F. Guzzetti, M. Rossi, and J. Poesen. 2009. "Combined Landslide Inventory and Susceptibility Assessment Based on Different Mapping Units: An Example from the Flemish Ardennes, Belgium." *Natural Hazards and Earth System Sciences* 9: 507–521.

Galli, M., F. Ardizzone, M. Cardinali, F. Guzzetti, and P. Reichenbach. 2008. "Comparing Landslide Inventory Maps." *Geomorphology* 94: 268–289.

Ghosh, S., C. van Westen, E. Carranza, T. Ghoshal, N. Sarkar, and M. Surendranath. 2009. "A Quantitative Approach for Improving BIS (Indian) Method of Medium-Scale Landslide Susceptibility." *Journal Geological Society of India* 74: 625–638.

Gokceoglu, C., and E. Sezer. 2009. "A Statistical Assessment on International Landslide Literature (1945–2008)." *Landslides* 6: 345–351.

Gutierrez, F., M. Soldati, F. Audemard, and D. Balteanu. 2010. "Recent Advances in Landslide Investigation: Issues and Perspectives." *Geomorphology* 124: 95–101.

Guzzetti, F. 2000. "Landslide Fatalities and Evaluation of Landslide Risk in Italy." *Engineering Geology* 58: 89–107.

Guzzetti, F., P. Aleotti, and B. T. Malamud. 2003. "Comparison of three Landslide Event Inventories in Central and Northern Italy." In *4th EGS Plinius Conference* (pp. 5–9). Spain: Universitat de Les Ilies Balears.

Guzzetti, F., A. Carrara, M. Cardinali, and P. Reichenbach. 1999. "Landslide Hazard Evaluation: A Review of Current Techniques and Their Application in a Multi-Study, Central Italy." *Geomorphology* 31: 181–216.

Hansen, A. 1984. *Slope Instability* (pp. 523–602). D. Brusden, D. Prior (Eds.). John Wiley and Sons.

Highland, L. M. 2008. "Introduction the Landslide Handbook-a Guide to Understanding Landslides." *The Landslide Handbook – a Guide to Understanding Landslides*, 4–42. http://landslides.usgs.gov/.

Highland, L. M., and P. Bobrowsky. 2008. "The Landslide Handbook – a Guide to Understanding Landslides." *US Geological Survey Circular* 1325: 1–147.

Hutchinson, J. 1988. "Mass movement." In *The Encyclopaedia of Geomorphology* (pp. 688–695). R.W. Fairbridge, Reinold.

Jaiswal, P., C. van Westen, and V. Jetten. 2010. "Quantitative Assessment of Direct and Indirect Landslide Risk Along Transportation Lines in Southern India." *Natural Hazards and Earth System Sciences* 10: 1253–1267.

Kanungo, D., M. Arora, R. Gupta, and S. Sarkar. 2009. "Landslide Risk Assessment Using Concept of Danger Pixel and Fuzzy Set Theory in Darjeeling Himalayas." *Landslides* 5: 407–416.

Kuriakose, S. 2010. *Physically-Based Dynamic Modelling of the Effect of Land Use Changes on Shallow Landslide Initiation in the Western Ghats of Kerala, India.* University of Twente, International Institute for Geo-Information Science and Earth Observation.

Mantovani, F., R. Soeters, and C. vanWesten. 1996. "Remote Sensing Techniques for Landslide Studies and Hazard Zonationn Europe." *Geomorphology* 15: 213–225.

Monkhouse, F, J. 1970. *Dictionary of Geography (Monkhouse)* (2nd ed.). Edward Arnold Ltd.

Montrasio, L., R. Valentino, and G. Losi. 2011. "Towards a Real-Time Susceptibility Assessment of Rainfall-Induced Shallow Landslides on a Regional Scale." *Natural Hazards and Earth System Sciences* 11: 1927–1947.

Naithani, A. 2007. "Macro Landslide Hazard Zonation Mapping Using Univariate Statistical Analysis in Parts of Garhwal Himalaya." *Journal Geological Society of India* 70: 353–368.

National Disaster Management Authority. 2019. *National Landslide Risk Management Strategy,* Government of India, September 2019, New Delhi, pp. 1–68.

National Disaster Management Authority of India. 2009. *National Disaster Management Guidelines: Management of Landslides and Snow Avalanches.* National Disaster Management Authority, Government of India.

Panikkar, S., and V. Subramaniyan. 1997. "Landslide Hazard Analysis of the Area Around Dehra Dun and Mussoorie, Uttar Pradesh." *Current Science* 73: 1117–1123.

Pardeshi, S., S. Pardeshi, and S. Autade. 2013. "Landslide Hazard Assessment: Recent Trends and Techniques." *SpringerPlus* 2 (523). https://doi.org/10.1186/2193-1801-2-253.

Pardeshi, S., S. Pardeshi, and V. Nagare. 2009. "A Study of Effect of Landslides on Human Environment in Thane District, Maharshtra State." *The Deccan Geographer* 47 (1): 45–56.

Pelty, D. 2010. "Landslide Hazard." In A. Ayala, I. Goudie (Eds.), *Geomorphological Hazards and Disaster Prevention* (pp. 63–73). Cambridge University Press.

Pereira, S., J. Zezere, and C. Bateira. 2012. "Assessing Predictive Capacity and Conditional Independence of Landslide Predisposing Factors for Shallow Landslide Susceptibility Models." *Natural Hazards and Earth System Sciences* 12: 979–988.

Rezig, S., J. Favre, and E. Leroi. 1996a. *Landslides* (pp. 351–355). I. Sennset (Ed.). Balkema.

Rezig, S., J. Favre, and E. Leroi. 1996b. "The Probabilistic Evaluation of Landslide Risk." In I. Sennset (Ed.), *Landslides* (pp. 351–355).

Rowbotham, D., and D. Dudycha. 1998. "GIS Modelling of Slope Stability in Phewa Watershed, Nepal." *Geomorphology* 26: 151–170.

Saaty, T. 2008. "Decision Making with the Analytic Hierarchy Process." *International Journal of Services Sciences* 1 (1): 83–98.

Salciarini, D., J. Godt, W. Savage, P. Conversini, R. Baum, and J. Michel. 2006. "Modeling Regional Initiation of Rainfall-Induced Shallow Landslide in the Eastern Umbria Region of Central Italy." *Landslides* 3: 181–194.

Selby, M. J. 1982. *Hillslope Materials and Processes*. Oxford University Press.

Sharpe, C. 1938. *Landslides and Related Phenomenon*. Columbia University Press.

Singh, C., K. Behra, and W. Rocky. 2011. "Landslide Susceptibility Along NH-39 between Karong and Mao, Senapati District, Manipur." *Journal Geological Society of India* 78: 559–570.

Tofani, V., S. Segoni, A. Agostini, F. Catani, and N. Casagli. 2013. "Use of Remote Sensing for Landslide Studies in Europe." *Natural Hazards and Earth System Sciences* 13: 299–309.

United Nations International Strategy for Disaster Reduction. 2004. *Terminology of Disaster Risk Reduction*. www.unisdr.org/eng/library/lib-terminology-eng home.htm.

van Westen, C. 2010. "GIS for the Assessment of Risk from Geomorphological Hazards." In I. Alcántara-Ayala, A. S. Goudie (Eds.), *Geomorphological Hazards and Disaster Prevention* (pp. 205–219). Cambridge University Press.

Varnes, D. I. 1984. *Landslide Hazard Zonation: A Review of Principles and Practice*. Paris: United Nations Scientific and Cultural Organization.

Wu, W., and R. C. Sidle. 1995. "A Distributed Slope Stability Model for Steep Forested Basins." *Water Resources* 31: 2097–2110. https://doi.org/10.1029/95WR01136.

Yeon, Y., J. Han, and K. Ryu. 2010. "Landslide Susceptibility Mapping in Injae, Korea Using a Decision Tree." *Engineering Geology, Elsevier* 166: 274–283.

Zizioli, D., C. Meisina, R. Valentino, and L. Montrasio. 2013. "Comparison between Different Approaches to Modelling Shallow Landslide Susceptibility: A Case History in Oltrepo Pavese, Northern Italy." *Natural Hazards and Earth System Sciences* 13: 559–573.

2 Recent Advances in Methods and Techniques of Landslide Hazard Assessment

2.1 INTRODUCTION

Mitigation of any natural disaster requires having a complete database with the highest possible spatial and temporal accuracy. Landslide hazard assessment involves the preparation of a landslide database at the global and regional scale, assessment of the probability of occurrence of landslides, and evaluation of the risks associated with landslide occurrence. There have been a large number of studies carried out over the past few decades. Different approaches to landslide hazard assessment have been developed for producing accurate landslide prediction models. This chapter deals with the review of recent trends in landslide hazard assessment, concerning landslide inventory preparation, landslide susceptibility zonation, and landslide risk assessment.

2.2 SPATIAL PATTERNS OF LANDSLIDE HAZARDS IN THE WORLD AND INDIA (BASED ON THE GLOBAL LANDSLIDE DATABASE)

A complete and reliable landslide database is essential for landslide hazard prediction and landslide risk assessment. There are various global landslide databases available that provide the global occurrence of landslides, along with the number of fatalities and monetary losses caused by landslides. EM-DAT is one of the popular disaster databases introduced in 1988, which has records of natural and technological disasters from AD 1900 till date. EM-DAT has records of more than 25,600 disasters available at the country and sub-country levels. However, it needs to be mentioned that the EM-DAT database records only those disasters in the global disaster database which satisfy the minimum criteria of losses in terms of the number of fatalities and the percentage human population affected by the given disaster.

DOI: 10.1201/9781003323488-2

The Office for Foreign Disaster Assistance (OFDA)/Centre for Research on Epidemiology of Disaster (CRED) is another popular global disaster database that records different natural disasters occurring globally. 'Our World in Data' is also an important international disaster database, which is a project of UK-based NGO named Global Change Data Lab.

Existing global disaster databases are not only capable of demonstrating actual and potential landslide-prone zones but also useful in the validation of landslide prediction models. The global patterns of landslide distribution and landslides in India are briefly described in subsequent paragraphs of this section.

2.2.1 LANDSLIDES IN THE WORLD

According to the EM-DAT global disaster database, the highest number of landslide occurrences are attributed to China, followed by the Indian Himalayas, the Western Ghats of India, Peru (South America), Cambodia (South America), Indonesia, Japan, Italy, and many other countries. China is the country with the highest number of fatal landslide occurrences (74), followed by Indonesia (60), India (46), Cambodia (43), and the Philippines (31). The maximum number of fatal landslides is observed in Asia. Peru has the highest number of reported fatalities (9,987) caused by landslides globally, followed by China (5,655), India (4,437), Cambodia (3,046), etc. Brazil (4.004 million people) and India (3.99 million people) are among the top countries in the number of affected people by landslide occurrence. Peru (4.6 billion USD) and China (2.7 billion USD) are the top two countries in terms of monetary losses caused by landslides.

Landslide distribution in the world is attributed to the Indian Himalayas, China, the Western Ghats, the Central and Southern Alps, the Rockies, and the Andes mountains (Peru and Cambodia) in North and South America.

2.2.2 LANDSLIDES IN INDIA

About 12.6% (0.4 million km^2) of the total geographical area of India is prone to landslide hazards. The Himalayan mountains and the Western Ghats are important landslide-prone areas in India. Landslides in the Eastern Himalayan terrain belong to earthquake-prone zones (zones IV–V), whereas those occurring in the Western Ghats are triggered by intense monsoonal rainfall.

Landslides in India have claimed 517 lives during five years from 2007 to 2011 and caused numerous economic losses. Economic losses caused by landslides in India are often underestimated because landslides are considered to be a secondary natural disaster.

Besides rainfall and seismicity, landslides in India are also caused by different preparatory parameters, such as slope gradient, slope aspect, lithology, structural discontinuities, vegetation cover, and land use. However, the morphometric characteristics of landslides in the Himalayan region and Western Ghats differ from each other.

2.3 LANDSLIDE INVENTORIES

Preparation of a complete landslide inventory is one of the most important preliminary exercises in the process of landslide hazard assessment. The concept of landslide inventory is defined by Cruden (1991) as 'the simplest form of landslide information which records the location and where known, the date of occurrence of a type of landslide that has left traces of the event in an area'. The landslide inventory map also shows slope failure by a single event, or it may show the cumulative effects of several events (Guzzetti, 2000).

2.3.1 NEED FOR LANDSLIDE INVENTORIES

A complete and reliable landslide inventory is required for effective landslide hazard assessment, particularly in developing data-driven models for landslide prediction and for validating the accuracy of existing landslide prediction models.

2.3.2 BASIC DETAILS OF LANDSLIDE INVENTORIES

Landslide inventories include different kinds of information about landslide occurrence. Most landslide inventories commonly record details about the date and time of landslide occurrence, geotechnical information about the landslides, landslide geometry parameters, details of losses caused by landslides, lithological and structural details of slopes, etc. Landslide inventories also intend to classify slope failures based on their morphological characteristics.

2.3.3 DATA SOURCES FOR LANDSLIDE INVENTORIES

There are various sources of information for generating landslide inventory. One of the most used data sources for landslide inventory is newspaper archives. Newspaper archives give detailed information about the date, time, and details of the occurrence of landslides along with brief information about the losses caused by landslide occurrence. However, one of the most important limitations of this source of information is that only those landslides are reported in newspapers that have caused significant losses and those in proximity to human-inhabited areas. Most landslides occurring in remote areas often remain unreported. The filed investigation is another important source of landslide inventory. A field survey is conducted physically at the landslide location using a field sheet. This is a proven data source that provides detailed and accurate information about landslide occurrence along with the observations of geotechnical parameters. The unmanned aerial vehicle (UAV) is an important tool to generate detailed landslide inventory for landslides located in remote areas. Need-based UAV surveys have been one of the recent developments in landslide data sources.

2.3.4 Types of Landslide Inventories

Landslide inventories are classified into various types. Some of them are briefly described as follows:

- **Event Inventory:** This type of inventory is restricted to a particular landslide event which provides detailed information about landslide occurrence.
- **Temporal Inventories:** This type of landslide inventory records past landslide events that occurred during a given period. It helps in obtaining annual landslide frequency, which is used as input data for quantitative landslide risk assessment.
- **Multi-disaster Inventories:** This type of landslide inventory is produced along with other natural calamities. This type of landslide inventory is prepared in case a landslide is a secondary hazard to other major disasters like earthquakes, floods, etc.
- **Real-Time Inventories:** Real-time landslide inventory is the most difficult kind of inventory to obtain. Real-time landslide inventory is restricted to real-time videos and photographs. Morphological details of landslides are difficult to obtain due to the inaccessibility of landslide locations.
- **Geotechnical Inventories:** This type of landslide inventory includes geotechnical details such as landslide morphology, geotechnical properties of the slope, etc.
- **Regional Inventories:** Regional landslide inventories are intended to produce spatial and temporal landslide databases at a regional scale. News archives, government records, railway slip registers, and records of regional disaster management are some examples.
- **Secondary Database:** Secondary landslide inventory includes landslide data obtained from various data sources such as global landslide databases, satellite image interpretation, and landslide distribution maps produced by government departments.

2.4 LANDSLIDE HAZARD ZONATION: REVIEW OF METHODS AND TECHNIQUES

Delineation of potential landslide hazard zones is one of the most important steps in landslide hazard assessment. There have been significant developments in methodologies of landslide hazard zonation. One of the earliest attempts of providing guidelines for landslide hazard zonation is made by Varnes (1984), who defined the term 'zonation' as the process of division of land surface into areas and ranking these areas according to the degree of actual or potential hazard from landslides or other mass movements. A large number of studies have been undertaken for landslide hazard zonation in various parts of the world. A brief review of a few approaches to landslide hazard zonation at various spatial scales is given is discussed in 1.2.7.

Gokceoglu and Sezer (2009) have attempted to do statistical assessment of literature on landslide hazard assessment and methodological developments over the decades. There has been a significant growth in the publications of literature on landslides particularly on different case studies on landslide hazard zonation. Pardeshi et al. (2013) carried out a comparative assessment of landslide hazard zonation methods, consideration of landslide causative factors, RS, and GIS techniques for different parts of the world. They have reviewed various methodologies of landslide susceptibility zonation and elaborated the significance of muli-cirteria decision-making approach over heuristic and statistical approaches to delineating landslide hazard prone areas.

In recent times, significant developments in methodologies used for landslide hazard zonation have taken place. Particularly, at the advent of developments in geo-spatial technologies combined with data science, the predictability of landslide hazard maps have been increased significantly. The review of existing literature on various methods of landslide hazard assessment is described in subsequent sections.

2.4.1 DISTRIBUTION APPROACH

Distribution analysis is one of the simplest qualitative approaches to Landslide Hazard Zonation (LHZ) mapping. It is also known as 'landslide inventory'. In this analysis, landslide inventory maps are produced, which portray spatial and temporal patterns of landslide distribution, type of movement, rate of movement, type of displaced material (soil, debris, or rock), etc. Landslide data are obtained through field survey mapping, historical records, satellite images, and aerial photo interpretation. Landslide distribution and density maps provide the basis for landslide susceptibility analysis and mitigation measures.

Cruden (1991) defined landslide inventory as 'the simplest form of landslide information which records the location and where known, the date of occurrence, type of landslides that have left identifiable traces in the area'. A landslide inventory map also shows a slope failure by a single event, or it may show the cumulative effects of many events (Guzzetti et al., 2003).

Landslide inventory plays a significant role in landslide hazard assessment. The quality and completeness of the landslide inventory influence the reliability of the landslide investigation. Galli et al. (2008) compared landslide inventory maps prepared for different parts of Italy. Landslide distribution inventory, geomorphological maps, and multi-temporal landslide inventories were compiled, and relationships among them were established. The results of the study revealed that a complete landslide inventory map provides high predictive power for landslide susceptibility assessment.

Guzzetti et al. (2003) discussed three landslide event inventories and compared them using universal frequency-area statistics. They discussed the significance of completeness and resolution of landslide inventory maps in landslide investigations. The results of the study portray that the number of landslide events rapidly increased with increasing landslide area up to a maximum value and decreased as power-law function.

Colombo et al. (2005) prepared a landslide inventory by systematic surveys using aerial photo interpretation and a GIS database to process the data using ARPA (Agenzia Regionale per la Protezione Ambientale – Regional Agency for Environmental Protection) archives. They classified landslides based on Varnes's (1984) scheme of classification of mass movement.

2.4.2 Heuristic Approach

This is a knowledge-driven approach to LHZ, in which weightage is assigned to each landslide causative factor based on prior knowledge and the experience of the researcher. The most commonly used heuristic method for landslide hazard zonation is that developed by the Bureau of Indian Standards (BIS) in 1998. The BIS-based landslide hazard evaluation factors (LHEF) rating scheme has been applied for LHZ in the Garhwal Himalayas, North-East India, the Kumaun Himalayas, and also in the hilly tracks of the Nilgiri range in Tamil Nadu (Naithani, 2007; Anbalagan et al., 2008; Singh et al., 2011). However, the LHEF rating scheme may not be applied universally in all parts of India due to differences in geo-environmental conditions. Considering this limitation, the Geological Survey of India modified the LHEF rating scheme in 2005. Heuristic methods for LHZ have been used to map landslide susceptibility in Himalayan regions and in parts of the Nilgiri Hills in Tamil Nadu (Lallianthanga and Lalbiakmawia, 2013).

One of the most important limitations of the heuristic approach to LHZ is that it does not consider the actual spatial distribution of slope failures, and hence, it is often difficult to validate the results. Moreover, a high level of subjectivity does exist, as the accuracy of the LHZ map produced from these methods largely depends upon the experience and knowledge of the investigator. The statistical approach to LHZ came out of the need for quantitative landslide hazard assessment and to bring objectivity to the assignment of ratings.

2.4.3 Bivariate Statistical Methods (IVM, Maximum Likelihood, WoE, Discriminant Analysis, Frequency Ratio Method, etc.)

In the last few years, the approach towards LHZ has changed from a heuristic (knowledge based) approach to a data-driven approach (statistical approach) to minimise subjectivity in weightage assignment and produce more objective and reproducible results (Kanungo et al., 2009). Methods based on a statistical analysis of geo-environmental factors related to landslide occurrence are preferred. The statistical methods for LHZ can be grouped into two: viz., bivariate statistical analysis and multivariate statistical analysis.

2.4.3.1 Bivariate Statistical Analysis

The bivariate statistical analysis for landslide hazard zonation compares each data layer of causative factor to the existing landslide distribution (Kanungo et al., 2009). Weights to the landslide causative factors are assigned based on landslide density. The frequency analysis approach, information value model, weights of evidence model, weighted overlay model, etc., are the important bivariate statistical methods used in LHZ mapping.

2.4.3.1.1 Weights of Evidence Model

Weights of evidence (WoE) is a log-linear form of the Bayesian probability model for landslide susceptibility assessment, which uses landslide occurrence as training points to derive prediction outputs. It calculates both the unconditional and conditional probability of landslide hazards. This method is based on the calculation of positive and negative weights to define the degree of spatial association between landslide occurrence and each explanatory variable class. The WoE model has been used for landslide susceptibility since the 1990s (Blahut et al., 2010). It uses different combinations of landslide causative factors to describe their interrelation with landslide distribution.

Blahut et al. (2010) applied the WoE model to LSZ mapping in the Valtellina valley of the central Italian Alps. The model was applied for different combinations of factor maps. Four landslide susceptibility maps were prepared and compared using success rate curves. The best-performing model was then selected with an AUC (area under curvature) value of 88%. Sterlacchini et al. (2011) carried out LHZ mapping in the alpine environment of the Italian Alps using the WoE model. The model was validated by using success rate curves and prediction curves, which gave a success rate of up to 88%. Anthropogenic factors (land use and road network) are considered for modelling landslide susceptibility using the WoE method by Piacentini et al. (2012). Martha et al. (2013) applied this method to assess spatial landslide probability in the Rudraprayag district of the Garhwal Himalayas, India, using semiautomatically created landslide inventories.

WoE, a statistical method for landslide susceptibility modelling, has proved to be a useful spatial data prediction model in many research works published in the recent past (Piacentini et al., 2012; Schicker and Moon, 2012; Martha et al., 2013; Ghosh et al., 2009).

2.4.3.1.2 Weighted Overlay Method

The weighted overlay is a simple bivariate statistical method, wherein weights are assigned based on the relationship of landslide causative factors with the landslide frequency. Sarkar et al. (1995) developed a methodology of LHZ for the Rudraprayag district in the Garhwal Himalayas, India. Numerical weights are assigned to causative factors based on the relationships to the landslide frequency. Finally, the weights are used to overlay data layers for the production of the LHZ map.

Panikkar and Subramaniyan (1997a) carried out a landslide hazard assessment using a GIS-based weighted overlay method in the area around Dehradun and Mussoorie in Uttarakhand, India. The study revealed that rapid deforestation and urbanisation have triggered landslides in the study area. This method is used to determine the relative importance of landslide causative factors in landslide occurrence (Parise, 2002; Preuth et al., 2010; Cardinali et al., 2002).

Landslide hazard susceptibility mapping along the Mumbai-Goa Highway was carried out by Nagarajan et al. (2000) based on terrain characteristics, weathering, and drainage density. The study revealed that steep escarpments are prone to rockfall with very high susceptibility, followed by debris slide zones on the slopes between 40° and 60°.

To rank and assign weightage for landslide explanatory variables, the bivariate discriminant function can be used for producing a landslide susceptibility map (Nagarajan et al., 2000).

2.4.3.1.3 Frequency Ratio Approach

Frequency ratio is one of the bivariate statistical approaches of landslide susceptibility assessment, which is based on observed relationships between landslide distribution and each causative factor related to landslides. This method can be used to establish a spatial correlation between landslide location and landslide explanatory factors (Lee, 2005). The frequency ratio for each causative factor is calculated based on the relationship of landslide causative factors with landslide occurrence. The landslide susceptibility index (LSI) is computed by summing the frequency ratio values of each factor.

Lee (2005) applied this model to landslide susceptibility in the Penang region of Malaysia. He compared landslide susceptibility maps produced by the frequency ratio model and the logistic regression model. Goswami et al. (2011) used frequency-area statistics to assess the spatial distribution of landslides in southwest Calabria, Italy. Lee and Pradhan (2006) applied a frequency ratio model to map landslide susceptibility in the Penang region, Malaysia. The verification results showed 80.03% accuracy and found that the incorporation of precipitation data in LHZ mapping improves the prediction accuracy of the landslide susceptibility map.

Dai and Lee (2001) attempted to establish a relationship between landslide magnitude and frequency to predict landslides in Hong Kong. The results indicate that landslide volume greater than 4 m^3 is represented by the power-law function, depending on the volume of slope failure. They also established a relationship between 12 hours of incident rainfall with landslide volume.

2.4.3.1.4 Information Value Model (IVM)

The information value model (IVM) is a bivariate statistical method for the spatial prediction of landslides based on relationships between landslide occurrence and related parameters (Sarkar et al., 2006). The information values are determined for each subclass of landslide-related parameters, based on the presence of a landslide in a given mapping unit. Several studies have applied this method for LHZ mapping.

Zêzere (2002) carried out a landslide susceptibility assessment considering landslide typology in North Lisbon, Portugal. He found that information values for roads and fluvial channels are found in the high landslide susceptibility class. The study revealed that anthropogenic activities play a significant role in landslide occurrence, and the magnitude of landslides depends largely upon the typology of landslides. Wang and Sassa (2005) compared landslide susceptibility maps for the Minamata area of Japan produced by logistic regression and IVM in a GIS environment. Sarkar et al. (2006) presented a GIS-based spatial data analysis for landslide hazard mapping in the Sikkim Himalayas. They used an IVM to integrate thematic data layers, and subsequently, numerical weights were assigned. Sharma et al. (2009) carried out GIS-based landslide susceptibility zonation for the Sikkim Himalayas using IVM. The accuracy assessment of the landslide susceptibility map was confirmed with the help of the model at the highest degree of accuracy for the high susceptibility class.

Pereira et al. (2012) used IVM to evaluate the role of different combinations of landslide causative factors in the occurrence of shallow landslides in parts of northern Portugal. IVM-based 120 landslide susceptibility maps were produced and compared to determine the 'best-fit model' for landslide susceptibility in the study area. Recently, Balsubaramani and Kumaraswamy (2013) applied this method for landslide hazard zonation mapping in the Giri valley of Himachal Pradesh using high-resolution satellite data. Sarkar et al. (2006) carried out landslide susceptibility mapping in the Sikkim Himalayas using IVM in a GIS environment. The study showed that quantitative weightage assigned to each layer of causative factor gives more reliable results.

In recent times, the use of remotely sensed data and GIS are proven to be effective in applying IVM for landslide susceptibility mapping (Akbar and Ha, 2011; Kanungo et al., 2009; Champatiray et al., 2007; Bălteanu et al., 2010).

2.4.4 Multivariate Statistical Methods (Decision Tree, Logistic Regression, Fuzzy Logic, Threshold Models, etc.)

According to Kanungo et al. (2009), the application of multivariate statistical analysis takes into consideration, the relative contribution of each preparatory and triggering factors to map landslide susceptibility. These methods calculate the percentage of landslide area for each pixel, and a landslide absence-presence data layer is produced, followed by the application of the multivariate statistical method for reclassification of hazards for the given area. Logistic regression model, discriminant analysis, multiple regression model, conditional analysis, and artificial neural networks are the commonly used multivariate methods for LHZ mapping.

2.4.4.1 Logistic Regression (LR) Analysis

Logistic regression (LR) is useful for predicting the presence or absence of a characteristic or outcome based on the values of a set of predictor variables. This model is suited when the dependent variable (e.g., landslide event) is dichotomous (Wang and Sassa, 2005). It can be of two types: viz., binary logistic (when the dependent variable is dichotomous and the independent variable is of any type) and multinomial logistic regression (dependent variable with more than two classes). In the case of landslide susceptibility mapping, the LR model finds the best-fitting model to describe the relationship between the presence and absence of landslides and the set of independent variables such as slope angle, slope aspect, lithology, and land use (Ayalew and Yamagishi, 2005). It generates the model statistics and coefficient of formulae useful in defining susceptibility. If the coefficient is positive, the landslide event is likely to occur. LR is a statistical model of slope instability built on the assumption that factors that caused slope failure in a region are the same as those which will generate landslides in the future (Guzzetti et al., 1999).

Rowbotham and Dudycha (1998) applied the LR model to landslide suscep-tibility zonation for Hong Kong. Guzzetti et al. (1999) applied this method to model landslide susceptibility for the Umbria region in central Italy. Tolga et al. (2005) carried out a landslide susceptibility assessment in the Black Sea region of Turkey using the LR model. They used the unique condition unit as a mapping unit for susceptibility classification. Jaiswal et al. (2010) applied the LR model to study landslide hazards assessment along the transportation corridors of the Nilgiri Hills in Southern India.

Recently, several studies used a GIS-based LR model for landslide susceptibil-ity analyses with the pixel as a mapping unit. Several studies have applied the LR model for LHZ mapping with comparatively high success rates (Chau et al., 2004; Wang and Sassa, 2005; Ayalew and Yamagishi, 2005; García-Rodríguez et al., 2008; Ghosh, 2011; Ohlmacher and Davis, 2003; Nefeslioglu et al., 2008; Chung and Fabbri, 2008; Das et al., 2011; Akgun, 2011; Mancini et al., 2010; Erener et al., 2010; Guzzetti et al., 1999; Atkinson and Massari, 2011; Greco et al., 2007; Chang and Chiang, 2009; S. Lee, 2005; Schicker and Moon, 2012; Das et al., 2012; Meusburger and Alewell, 2009; Lee et al., 2010; Dai and Lee, 2002.

2.4.4.1.1 *Discriminant Analysis Method*

Discriminant analysis (DA) is one of the frequently used statistical models for LHZ. It allows us to determine the maximum difference for each indepen-dent variable (e.g., landslide causative factor) between the landslide group and non-landslide group and to determine the weights for these factors (Lee et al., 2008). Slope units are classified into landslide-affected and landslide-free classes, and then the relative importance of each variable is expressed by computing the standardised discriminant function coefficient (SDFC). SDFC shows the relative importance of each variable in the discriminant function as a predictor of slope instability. A variable with a high coefficient is strongly associated with the pres-ence or absence of landslide.

Several investigations for landslide susceptibility using DA have been carried out in different parts of the world. Guzzetti et al. (2005) applied DA for landslide susceptibility zonation using 46 thematic variables in a GIS environment. Calvello et al. (2013) carried out landslide susceptibility zonation for the Tammaro catch-ment of Southern Italy using DA.

Lee et al. (2008) used DA for landslide hazard zonation mapping of central western Taiwan. Ohlmacher and Davis (2003) prepared an LHZ map using LR and DA in a GIS environment for the Kanas basin, USA. Eeckhaut et al. (2009) applied DA landslide susceptibility assessment based on different mapping units.

2.4.4.1.2 *Artificial Neural Network Method*

Landslides are governed by several preparatory (causative) and triggering complexly interrelated factors. The interrelationships between these factors and landslides are nonlinear (Ercanoglu, 2005). To get an accurate landslide susceptibility assessment, more accurate methods are needed. An artificial neural network (ANN) is a system based on the capability to learn a particular phenomenon similar to a human being. ANN has over three layers of neurons connected by weights. This model uses a 'backpropagation learning algorithm', which defines rules for the assignment of weights. The weight of each variable is then adjusted to minimise errors. ANN is a nonlinear model and proven to be more effective in landslide hazard assessment (Catani et al., 2005; Ercanoglu, 2005; Pradhan and Lee, 2009, 2010; Tien Bui et al., 2012).

The ANN model was applied by Pradhan and Lee (2010) for landslide susceptibility zonation in Malaysia. Similarly, Ercanoglu (2005) produced a landslide susceptibility map using the 'backpropagation ANN model' in the NeuralNet module of Idrisi Kilimanjaro for the West Black Sea region, Turkey. The outcome of the model after validation indicated 82.5% correct results. Catani et al. (2005) applied the ANN model to landslide susceptibility zonation in the Arno River basin of central Italy. Landslide preparatory factor layers were overlaid to define unique condition units (UCU). The final susceptibility map showed over 85% correctly recognised areas susceptible to landslides.

Chang and Liu (2004) performed an ANN model for landslide susceptibility zonation in central Taiwan using high-resolution satellite data. Few recent studies applied a GIS-based ANN model for landslide susceptibility in different parts of Malaysia (Pradhan and Lee, 2009; Tien Bui et al., 2012). Hence, the ANN model can effectively be implemented in landslide hazard assessment in a GIS environment to improve landslide prediction capability. Arora et al. (2004) proposed the ANN black box approach for landslide hazard zonation mapping for Bhagirathi (Ganga) valley, India.

2.4.5 OTHER MULTIVARIATE TECHNIQUES

Rotingliano et al. (2011) discussed the role of diagnostic areas in LSZ mapping for a Sicilian chain of Italy. The causative factors of landslides were combined to identify UCUs, and then diagnostic areas were selected based on landslide types.

Clerici et al. (2002) applied a conditional analysis method for landslide hazard zonation mapping of the Parma River basin, Italy, using GRASS (Geographical Research Analysis Support System) commands in a GIS environment.

Multivariate statistical techniques for landslide susceptibility were applied by Rotingliano et al. (2011) in the Guddemmi River basin of Sicily. A multivariate approach to landslide susceptibility zonation has widely been used for the past few years and proven to be a more objective method for assessing landslide hazards in complex geo-environmental settings Conoscenti et al., 2008; Eeckhaut et al., 2009; Ercanoglu et al., 2004; Ayalew and Yamagishi, 2005).

The application of multivariate statistical methods in LHZ mapping gives more accurate results, but it includes complex calculations. These methods allow for assessing the comparative contribution of each causative factor in landslide occurrence. Therefore, these methods are more objective in the assignment of weightage in the LHZ mapping procedure.

2.4.6 PROBABILISTIC APPROACHES

Probabilistic landslide hazard assessment helps to determine the spatial, temporal, and size probability of landslides (Guzzetti et al., 2005). Probabilistic methods of LHZ mapping bring objectivity in assigning weights. In the probabilistic approach of landslide susceptibility zonation, the spatial distribution of landslides is compared with various explanatory (causative) variables within the probabilistic framework (Kanungo et al., 2009). It includes Bayesian probability, certainty factor, favourability function, etc. The degree of relationship between each thematic data layer with landslide distribution is transformed to a value on the basic probability distribution function. This approach is quantitative, but a certain degree of subjectivity exists in the weight assignment procedure (Kanungo et al., 2009).

Guzzetti et al. (2005) used a probabilistic model and proposed a landslide hazard assessment in the Stafford River basin of Italy. They applied statistical methods to obtain the spatial and temporal probability of landslides. They computed the probability of landslide size and temporal and spatial probability of landslides using the frequency-area distribution function. The Poisson probability model was applied to determine the exceedance probability of landslides in each mapping unit.

Jaiswal et al. (2010) also carried out a quantitative landslide hazard assessment along a transport route in the Nilgiri Hills, India. They performed frequency-volume statistics to obtain the probability of landslide magnitude for different return periods. The results of the study indicated that the probability of landslide frequency and volume can be estimated using rainfall magnitude data.

The landslide hazard zonation model was proposed by Jaiswal (2011). The model was designed for LSZ mapping along the railroad in the areas of the Nilgiri range in south India. It was based on historical records of the railway department, technical reports, and fieldwork to determine spatial and temporal landslide probability. The relationship between several landslide events and the return period has been established by using the Gumbel distribution model. The authors suggested that landslide magnitude needs to be quantified based on absolute values of landslide velocity, intensity, and volume of displaced material. However, it is very difficult to obtain reliable information about such parameters, as most of these are site specific and influenced by local conditions.

Das et al. (2011) assessed HSU (homogeneous susceptibility unit) based landslide hazards using spatial, temporal, and landslide size probabilities in the

Bhagirathi River basin of the north Himalayas, India. A high-resolution satellite dataset was used to define HSU for LHZ mapping. Recently, Jaiswal and van Westen (2013) attempted to assess landslide susceptibility in the Nilgiri Hills, India, using spatial probability to produce hazard and risk information for planning risk reduction measures. Further, they also developed a landslide early warning system based on a rainfall database.

Since the last decade, several studies have attempted to apply a probabilistic approach to quantitative landslide hazard zonation (Polemio and Sdao, 1999; Guzzetti et al., 2006; Chleborad et al., 2006; Floris and Bozzano, 2008).

2.4.7 Multi-Criteria Decision-Making Approach

Landslide hazard assessment involves the consideration of several landslide explanatory variables. It is a critical task to determine the relative contribution of an individual parameter in landslide occurrence. Therefore, the application of the multi-criteria decision-making approach is of utmost importance in LHZ mapping. The analytical hierarchy process (AHP) is a multi-criteria decision-making process of measurement through pairwise comparisons and relies on the judgements of experts to derive priority scales (Saaty, 2008). AHP operates at five levels: viz., defining the problem, determination of goals and alternatives, constructing of pairwise comparison matrix, determining weights, and obtaining overall priority. In LHZ, different landslide causative factors are considered as alternatives. Absolute numbers (from 1 to 9) are assigned to each landslide-related parameter based on its relative importance, and comparison matrices are constructed to compute the consistency ratio (CR) and consistency index (CI). Akgun (2011) compared landslide hazard maps produced by logistic regression (LR), multi-criteria decision-making approach (MCDA), and likelihood ratio method (LRM) for Azmir, Turkey, using AUC method. The correlation coefficients (r) were found to be 0.86, 0.62, and 0.58 for LR*LRM, LR*MCDA, and LRM*MCDA, respectively. The LRM and MCDA showed similar results. Ayalew et al. (2005) compared LSZ maps using LR and AHP models to assess landslide hazards. The study revealed that if there is an increase in the number of susceptibility classes, the LR model gives more details than AHP. However, when these maps were compared with the landslide activity map, the AHP-based map performed better than the LR model. In recent times, several attempts have been made to apply GIS-based AHP to map landslide susceptibility in various parts of the world (Mondal and Maiti, 2012; Ma et al., 2013; Kavzoglu et al., 2014; Pardeshi et al., 2013).

2.4.8 Rainfall Threshold Model

The rainfall threshold for landslides refers to the minimum intensity or duration of rainfall necessary to cause landslides (Varnes, 1984). Cumulative rainfall, antecedent rainfall, rainfall intensity, and rainfall duration are the most commonly used parameters to design rainfall thresholds. The critical rainfall threshold model (Qcr) is based on soil properties, slope angle, upslope drainage, wet soil bulk density, and density of water. Several studies on landslide susceptibility assessment have used the rainfall threshold model to predict landslides. The rainfall threshold decreases with increasing seasonal accumulation and becomes constant at 11 mm/day (Gabet et al., 2004).

Chleborad et al. (2006) applied cumulative rainfall threshold (CT) for the prediction of landslides in Seattle, Washington, USA. The model was compared with historical records of rainfall and landslide events. The results indicated that CT captured over 90% of the historical landslide events.

Floris and Bozzano (2008) proposed a modification in the conventional rainfall threshold model for landslide hazard assessment. Based on historical records of landslide events and rainfall, rainfall exceedance thresholds were estimated for two complex landslides in south Apennines, Italy. Chang and Chiang (2009) proposed an integrated model for landslide susceptibility, combining deterministic, statistical, and rainfall threshold models for typhoon-induced landslides in Taiwan. Gabet et al. (2004) applied rainfall threshold for landslides in the Nepal Himalayas, considering daily and seasonal rainfall thresholds for modelling.

Dahal and Hasegawa (2008) studied over 670 landslides that occurred from 1951 to 2006 in the Nepal Himalayas to analyse the rainfall threshold. Coe et al. (2004) and (Polemio and Sdao, 1999) have also applied rainfall thresholds for landslide susceptibility assessment.

2.4.9 Physical Process-Based Landslide Susceptibility Models

Physical process-based models for LHA describe physical processes leading to the landslide event and are based on simple mechanical laws. These models account for the transient groundwater response of slope to rainfall. They do not need long-term landslide data and therefore can also be applied to areas with incomplete landslide inventories.

Salciarini et al. (2006) applied the transient rainfall infiltration and grid-based slope stability (TRIGRS) model for modelling rainfall-induced shallow landslides in the central Umbria region of central Italy. They have chosen known rainfall events and past landslide records to calibrate the model, and simulations were performed, and it was argued that high-resolution digital elevation models and information about the spatial distribution of physical properties of the surface are needed for better simulation in the TRIGRS model.

The real-time susceptibility to shallow landslides has been assessed by Montrasio et al. (2011) for Emilion Apennine in north Italy. They compared SLIP (shallow landslide instability prediction) and TRIGRS models of landslide susceptibility analysis in a GIS environment. The results of the study showed that both models have similar predictive capabilities.

Kuriakose (2010) carried out a detailed study to compare four physical process-based models: viz., SHALSTAB (shallow landsliding stability), SINMAP (stability index mapping), TRIGRIS, and STARWAR + PROBSTAB (storage and redistribution of water on agricultural and revegetated slope + probability of stability) models in the Western Ghats of Kerala, India. The study revealed that the STARWAR + PROBSTAB model is the most suitable model for the assessment of spatiotemporal probabilities of shallow landslides.

Listo and Carvalho Vieira (2012) applied the SHALSTAB model for susceptibility mapping of shallow landslides in São Paulo, Brazil. Recently, the high-resolution slope stability simulator (HIRESSS) model was used to predict landslides based on hydrological parameters (Mercogliano et al., 2013).

In a recent study on shallow landslides in northern Italy, Zizioli et al. (2013) compared four physical process-based models: viz., SHALSTAB, SINMAP, TRIGRIS, and SLIP. The results indicate that the SLIP model analyses multi-temporal shallow landslides with very low computations, and SINMAP gives more unrealistic results as compared to other models. These are deterministic models which have a high predictive capability and assess the influence of individual parameters related to landslides (Mantovani et al., 1996; Jelínek and Wagner, 2007). These models perform well for landslides where slope behaviour can easily be predicted by simple mechanical laws, but they lack consideration of the temporal aspects of landslides (Guzzetti, 2003).

2.4.10 OTHER APPROACHES

2.4.10.1 AOR Approach

The angle of reach (AOR) is a simple qualitative approach to landslide susceptibility assessment. AOR is an angle connecting the landslide crown with the margins of runout material (Dahl et al., 2010). This is an empirical method of landslide susceptibility, in which landslide geometry (shape of the scar, depth of the landslide deposits, runout distance) is used to infer the landslide volume and other parameters.

Dahl et al. (2010) applied a simple qualitative method to prepare a landslide susceptibility map for the Faeroe Islands. The AOR approach was followed by these authors for landslide characterisation. More emphasis was given to estimating landslide volume based on geometric characteristics of slope failure areas. The correlation between AOR and landslide geometry has been established to determine the magnitude of landslide-prone areas.

Dahl et al. (2010) attempted to map landslide susceptibility for the Faeroe Islands using the AOR method. The field data were collected from 67 landslides, and landslide volume was estimated using landslide geometry. The data about landslide geometry is then statistically analysed to identify and map landslide initiation and runout areas. The landslide susceptibility map was finally compared with the landslide distribution map, and it was found that 69% of actual landslide initiation areas were mapped correctly. Though the AOR method helps in the qualitative assessment of landslide susceptibility, it may not provide many accurate results as compared to the statistical methods. Moreover, the use of this method is time consuming and may not be viable for mapping landslide susceptibility at a regional scale.

2.4.10.2 Logit Model

Chang and Chiang (2009) proposed a logit model for predicting rainfall-induced landslides. This method combines process-based models and statistical models along with consideration of rainfall data for landslide hazard modelling. Chang and Chiang (2009) developed this model using rainfall intensity difference (RID) and rainfall duration. This model was applied to map landslide hazards in the Baichi watershed of north Taiwan. Chang and Chiang (2009) claimed that the logit model is much better than traditional models for landslide susceptibility zonation.

2.4.10.3 Decision Tree Approach

The decision tree is a popular classification technique and can be used to map landslide susceptibility with limited data input. The GIS-based decision tree model has been applied for landslide susceptibility zonation in the Injae region of Korea by Yeon et al. (2010). The algorithm was prepared using causative data layers as input in a GIS environment, and the prediction model was generated using the decision tree classification scheme. Causative data layers were compared with the landslide distribution to quantify the degree of association of each causative factor to landslide occurrence of a defined magnitude. The study showed that the leaf node ranking method can effectively be used to identify landslide-prone areas and to predict landslide events.

LHZ mapping involves several causative factors and triggering mechanisms. One of the most important challenges faced by scientists engaged in LHZ mapping is determining the appropriate method of assigning weights to landslide explanatory factors based on their relative importance. Several qualitative and quantitative methods of LHZ (discussed earlier) provide a methodological framework for assigning weights to a given parameter based on certain criteria such as an expert's knowledge (heuristic approach), the relationship of landslide causative factors with landslide occurrence (bivariate statistical approach), the relative importance of individual landslide causative factor in slope failure (multivariate statistical approach), and pairwise comparison of landslide explanatory variables. However, still there is a lack of agreement on one appropriate LHZ method which could universally be applicable in all kinds of geo-environmental conditions. This is probably due to the complexity of the landslide phenomenon in different parts of the world. The application of the multi-criteria decision-making approach of LHZ mapping can effectively be used to maintain objectivity in the assignment of weights and to determine the relative importance of an individual parameter based on interrelationships among other variables determining the degree of landslide hazards.

Another major issue related to LHZ mapping, especially in the Indian context, is the lack of availability of sufficient and reliable data for accurate landslide assessment. Therefore, the application of data-driven methods of LHZ in data-scarce environments is a major limitation in landslide hazard assessment. However, the need is to be cautious in the selection of appropriate methods in unique geo-environmental conditions.

There are various approaches adopted by researchers worldwide for landslide hazard zonation mapping at varying spatial scales. Generally, these approaches can be grouped into heuristic (knowledge based/expert opinion), semi-quantitative, and quantitative landslide hazard assessment. Heuristic approaches give more emphasis on the rating of landslide causative factors, but actual landslide distribution is not taken into consideration. The weightage is given based on the expertise and experience of the investigator, which may give rise to subjectivity. Semi-quantitative landslide hazard zonation mapping is done by assigning weightage based on the correlation between landslide causative factors and landslide distribution. Quantitative landslide hazard zonation mapping is mainly associated with field measurements and the application of multivariate statistical techniques along with spatial and temporal probability analysis, which helps in the prediction of future landslide events with a high degree of confidence.

2.5 TRIGGERING AND PREDISPOSING FACTORS FOR LHZ

Landslide hazard zonation is an integral part of landslide hazard assessment. Delineation of the land surface based on its susceptibility to landslides of different magnitudes is based on various causative factors and triggering mechanisms.

Conditions and processes that are responsible for slope failure can be used to estimate their relative contribution to slope failures (Varnes, 1984).

According to Wu and Sidle (1995), preparatory variables and triggering variables are important types of factors responsible for slope instability. Chen et al. (2012) discussed the role of geological settings on geometry of earthquake-induced landslides. A detailed description of the causative factors considered for landslide hazard zonation by a few researchers is given in subsequent paragraphs.

Naithani (2007) argued that major landslides in the Garhwal Himalayas are associated with the North Almora Thrust and foliation joints dipping out of the slope. However, he did not consider a landslide distribution map for validation of the BIS-based LHEF (landslide hazard evaluation factors) methodology of landslide hazard zonation. Singh et al. (2011) considered soil erosivity as an additional parameter for hazard assessment in parts of Manipur, India. The findings of the study reveal that very high landslide hazard areas are mostly associated with road cutting, whereas landslide hazard areas with low magnitude are situated away from roads.

A review of the literature published in recent years indicates that several preparatory variables are considered for LHZ mapping, but unfortunately, not much effort is made to develop scientific methods for the selection of variables for hazard zonation. Pereira et al. (2012) assessed the predictive capacity of seven landslide predisposing factors with landslide inventory (slope angle, inverse wetness index, land use, aspect, curvature, geomorphology, and lithology) for susceptibility assessment using 120 combinations of predisposing factors. They found that out of seven landslide predisposing factors only three variables – viz., slope angle, inverse wetness index, and land use – give the best suitable model for landslide inventory.

2.6 THE MAPPING UNIT

The mapping unit refers to a portion of land containing a set of ground conditions which differ from the adjacent unit across the defined geographical area (Hansen, 1984). Mapping units are grouped into five: viz., grid-cell (pixel), terrain unit, unique condition unit, slope units, and topographic units (Guzzetti et al., 1999).

Terrain units are used by Rowbotham and Dudycha (1998) and Panikkar and Subramaniyan (1997a).

Unique Condition Units are used by Guzzetti et al. (1999), Tolga et al. (2005), Conoscenti et al. (2008), and Sterlacchini et al. (2011).

Slope units (sub-basin units, main slope units, and slope facets) are commonly used as mapping units for delineating landslide-susceptible zones. Many studies have used slope units for landslide susceptibility mapping (Guzzetti et al., 1999; Dahl et al., 2010; Parise, 2002; Naithani, 2007; Kannan et al., 2011).

Eeckhaut et al. (2009) compared landslide susceptibility maps produced by using the multivariate approach for mapping units.

Das et al. (2011) and Das (2011) used HSUs (homogenous susceptible units) for landslide susceptibility mapping.

Eeckhaut et al. (2009) compared landslide susceptibility maps based on different mapping units and found that TMU-based landslide susceptibility maps show larger susceptible areas than grid-based LHZ maps.

So far, various mapping units have been used for susceptibility zonation in different parts of the world, but still, there is a lack of agreement to determine the best-suited mapping unit universally for LHZ mapping.

2.7 UAV-BASED LANDSLIDE HAZARD ASSESSMENT

Methodologies for preparing landslide inventories, landslide susceptibility zonation, and landslide risk assessment. This section aims at reviewing recent trends in landslide hazard assessment across the globe.

Reliable and complete landslide database is a key component in adopting a landslide inventory approach. High-resolution satellite data have been used widely in recent times. Yang et al. (2022) conducted event-specific landslide investigations in South Taiwan using high-resolution data of LiDAR and SPOT. Dewitte et al. (2021) used historical landslide data for NTK Rift, Burundi, Africa, using multi-temporal DIDINAR and SRTM images.

They have also used UAVs for preparing landslide inventory. Wang et al. (2022) used penetration test and fault zone and Doppler array test for the characterisation of the Chengtian landslide, in China. Franceschini et al. (2022) have used automated web data mining and semantic engine to classify geotagging news of landslide events. They also carried out content analysis for the preparation of landslide inventory in Italy. Ghorbanzadeh et al. (2022) used object-based image analysis for landslide inventory in Taiwan. Tanyaş et al. (2021) used Sentinel-2 data and TANDEM-X to prepare an inventory of earthquake-triggered landslides in New Zealand. Hughes et al. (2021) attempted to use bathymetric and sub-bottom data (multi-beam echosounder) to delineate underwater paleo-landslides in British Columbia, Canada.

Li et al. (2022a) used an object detection algorithm and image segmentation for detecting loess landslides in North China. Li et al. (2022b) attempted to study landslide geometry parameters using automatic landslide profile analysis. Marino et al. (2021) did a geotechnical investigation of landslides in South Italy using hydrological monitoring with an automatic weather system. Meena et al. (2022) attempted landslide detection using a machine learning algorithm in the Nepal Himalayas using high-resolution satellite data of RapidEye, ALOS, and PALSAR. Paulin et al. (2022) used LiDAR, and WS-DNR for the estimation of landslide volume in Whatcom County, Washington.

In recent times, UAV surveys are extensively used for the identification of landslides and near-real-time monitoring of processes associated with slope instability. The method uses aerial imagery collected by UAVs for preliminary investigations of landslides. This is a more appropriate, faster, and more accurate method of landslide mapping and monitoring of landslide movement. UAV surveys also help in producing two-dimensional and three-dimensional products like orthophotos and digital surface models (DSMs) and probably can be used for landslide detection in quasi-real time. Several studies on landslide detection and mapping have been conducted using UAVs in the past few years (Ilinca et al., 2022; Wang et al., 2022; Liang et al., 2022; Xu et al., 2021b).

Tempa et al. (2021) used DJI Photon 4 Pro + obsidian for a UAV survey for analysing landslide hazards in the Rinchending Goenpa area of Bhutan. They also compared the applicability of the weighted overlay method and knowledge-driven method of landslide susceptibility. Niethammer et al. (2010) used radio-controlled UAVs for landslide monitoring in the Southern French Alps. They have also produced a Mikrokopter system to detect and analyse landslides in the study area.

This brief review of recent studies on landslide hazard assessment reveals that there has been a significant development in methodologies used for landslide inventory, susceptibility zonation, and risk assessment. A range of methods from knowledge-driven approaches to data-driven models is applied for landslide investigations. Further, attempts are also being made to compare the accuracy of landslide assessment models. Near-real-time monitoring of landslides using sophisticated instruments has been an important development in recent times. High-resolution satellite data, combined with field investigations, are proven databases for accurate mapping of landslides.

2.8 LANDSLIDE RISK ASSESSMENT

Several approaches have been proposed for the assessment of risk related to landslide hazards. A few of them are discussed as under.

Jaiswal et al. (2010) applied quantitative techniques for landslide risk assessment along transportation routes in the Nilgiri Hills, India. The authors used the Gumbel distribution model to estimate landslide spatial probability for different return periods. The probability of landslides was estimated using volume-frequency distribution. Direct and indirect risks were calculated by quantifying each element at risk separately (i.e., specific risk). By combining specific risks for all individual elements at risk, the final total risk assessment map was prepared. However, the authors admit that there may be uncertainty in actual and estimated landslide risk on account of insufficient and unreliable data.

Rezig et al. (1996a) emphasised probabilistic landslide risk evaluation. Van Westen (2010) proposed a GIS-based landslide risk assessment method. He gave a detailed account of spatial data requirements for risk assessment and element at-risk data sources. He explained how multi-criteria decision-making processes – for example, analytic hierarchy process of Saaty (2008) – are useful for risk assessment. He emphasises the significance of remotely sensed data in the risk evaluation process.

Pardeshi et al. (2009) attempted to estimate the volume of slope-forming material using cone geometry and sign rule for landslides along the Wada-Khodala Road in the Thane district of Maharashtra. The authors quantified elements at risk along the road and assess risk due to landslides along the road in terms of monetary losses and indirect losses as traffic delay and disruption in economic activities. The study carried out by Anbalagan et al. (2008) in the Kumaun Himalayas revealed that slopes falling in high and very high landslide hazard classes should be prohibited from all kinds of construction. Further, they explained the need for micro-level landslide hazard zonation to prepare a detailed land-use planning model.

2.9 RS AND GIS APPLICATION IN LANDSLIDE HAZARD ASSESSMENT

Remote sensing and geographical information systems (GIS) are powerful tools to assess landslide hazards and are being used extensively in landslide research since the last decade. Aerial photographs and high-resolution satellite data are useful in the detection, mapping, and monitoring of landslide processes. GIS-based LHZ models help not only to map and monitor landslides but also to predict future slope failures. The advancement in geospatial technologies has opened the doors for detailed and accurate assessment of landslide hazards.

With the advent of RS and GIS technologies in the field of spatial studies, they are now being widely used in hazard assessment. Their role in the field of landslide investigation cannot be overemphasised. The extraction of relevant spatial information related to landslide occurrence is an integral part of hazard assessment. RS data and GIS are proven to be effective tools for generating and processing spatial information. The advancement in 'Earth observation' (EO) techniques facilitates effective landslide detection, mapping, monitoring, and hazard analysis (Tofani et al., 2013).

The review of a few studies on landslide hazard assessment using RS data indicates that aerial photographs are widely used in landslide detection and mapping

(Panikkar and Subramaniyan, 1997a; Rowbotham and Dudycha, 1998; Clerici et al., 2002; Guzzetti et al., 2003; Chau et al., 2004; Ayalew and Yamagishi, 2005; Guzzetti et al., 2005; Miller and Burnett, 2007; Galli et al., 2008; Pradhan and Lee, 2009; Yeon et al., 2010; Rotingliano et al., 2011). Good-quality aerial photographs help in accurate landslide detection and mapping. However, aerial photographs may not be used in continuous landslide monitoring, since they do not provide repetitive coverage of the same area.

The recent developments in the application of satellite RS data in landslide studies in Europe have been discussed by Tofani et al. (2013). The study showed that over 70% of the total applications of RS data for landslide studies are associated with landslide detection, mapping, and monitoring. High-resolution satellite data are being effectively used for landslide detection, mapping, monitoring, and other applications (Nagarajan et al., 2000; Naithani, 2007; Chand, 2008; Akbar and Ha, 2011; Mondal and Maiti, 2012; Xu et al., 2012; Ma et al., 2013; Balsubaramani and Kumaraswamy, 2013).

The use of the digital elevation model is of immense importance in landslide hazard assessment. Several thematic data layers, such as slope angle, slope aspect, curvature, lineaments, drainage, and ridges can be extracted from DEM with good resolution. Landslide hazard zonation studies in recent times have used DEM with high resolution to generate spatial information data layers related to landslide hazards (Rezig et al., 1996a; Rowbotham and Dudycha, 1998; Nagarajan et al., 2000; Clerici et al., 2002; Coe et al., 2004; Guzzetti et al., 2005; Ayalew and Yamagishi, 2005; Tolga et al., 2005; Jelínek and Wagner, 2007; Naithani, 2007; Miller and Burnett, 2007; Chand, 2008; Ghosh et al., 2009; Jaiswal et al., 2010; Dahl et al., 2010; Yeon et al., 2010; Erner et al., 2010; Akbar and Ha, 2011; Rotingliano et al., 2011; Ma et al., 2013; Balsubaramani and Kumaraswamy, 2013; Calvello et al., 2013). Few studies also used radar techniques (e.g., DInSAR, PSInSAR) for landslide hazard assessment (Catani et al., 2005).

GIS is widely used in landslide hazard assessment, especially for the generation of thematic data layers, computation of different indices, assignment of weights, data integration, and generation of LSZ maps. Several LSZ methods, such as ANN, decision tree model, weighted overlay, AHP, MCDA, IVM, and physical process-based landslide hazard models, are GIS-based models to predict landslide probability (Chang and Liu, 2004; Ayalew et al., 2005; Pradhan and Lee, 2009; Yeon et al., 2010; Akgun, 2011; Mondal and Maiti, 2012; Ma et al., 2013; Kavzoglu et al., 2014).

2.10 RECENT DEVELOPMENTS IN LANDSLIDE INVESTIGATIONS

There has been a significant development in landslide investigations in terms of methodologies for landslide inventories, landslide susceptibility zonation, landslide risk assessment, and the resolution of spatial data used for analysing factors controlling slope instability processes. This section of the chapter is dedicated to discussing and reviewing recent trends in landslide hazard investigations. For this, the latest research papers published in reputed international journals are reviewed.

2.10.1 LANDSLIDE INVENTORY

Seeking accurate and precise landslide information is one of the key issues in adopting a landslide inventory approach. There has been a wide range of tools used for landslide detection and inventory. However, high-resolution satellite data such as LiDAR, SPOT, DINSAR, DTMs, DEMs, ALOS DEMs, Sentinel-1, Sentinel-2, TANDEM, and Pleiades 1A have been used widely in recent times. Yang et al. (2022) conducted event-specific landslide investigations in South Taiwan using high-resolution data from LiDAR and SPOT. Dewitte et al. (2021) used historical landslide data for NTK Rift, Burundi, Africa, using multi-temporal DINSAR and SRTM images. They have also used UAVs for preparing landslide inventory. Wang et al. (2022) used penetration test and fault zone and Doppler array test for the characterisation of the Chengtian landslide, in China. Franceschini et al. (2022) have used automated web data mining and semantic engine to classify geotagging news of landslide events. They also carried out content analysis for the preparation of landslide inventory in Italy. Ghorbanzadeh et al. (2022) used object-based image analysis for landslide inventory in Taiwan. Tanyaş et al. (2021) used Sentinel-2 data and TANDEM-X to prepare an inventory of earthquake-triggered landslides in New Zealand. Hughes et al. (2021) attempted to use bathymetric and sub-bottom data (multi-beam echosounder) to delineate underwater paleo-landslides in British Columbia, Canada.

Li et al. (2022a) used an object detection algorithm and image segmentation for detecting loess landslides in North China. Li et al. (2022b) attempted to study landslide geometry parameters using automatic landslide profile analysis. Marino et al. (2021) did a geotechnical investigation of landslides in South Italy using hydrological monitoring with an automatic weather system. Meena et al. (2022) attempted landslide detection using a machine learning algorithm in the Nepal Himalayas using high-resolution satellite data of RapidEye, ALOS, and PALSAR. Paulin et al. (2022) used LiDAR and WS-DNR for the estimation of landslide volume in Whatcom County, Washington.

The use of high-resolution satellite data, combined with field investigations, has proven to be instrumental in a few landslide studies in the Indian Himalayas (Roy et al., 2022; Sharma et al., 2022).

2.10.2 LANDSLIDE SUSCEPTIBILITY ZONATION

The latest research papers published in reputed international journals like *Landslides* and *Geomorphology* have been reviewed to understand recent developments in landslide hazard assessment. The review of the latest published papers on landslides reveals that there has been a wide range of methodological advancements concerning landslide susceptibility zonation in recent times. A brief discussion of recent trends in landslide hazard assessments is given in subsequent paragraphs.

Dynamic hazard mapping has proven to be effective in using the support vector machine (SVM) method for accurate prediction of landslides (Zhou et al., 2022; Meena et al., 2022).

Chen et al. (2022) developed a mass movement warning system using soil water index (SWI) to predict landslides in Taiwan. The rainfall threshold model has also been widely used to predict landslides. This model has proved to be the most accurate and precise in determining potential landslide-prone zones. Giuseppe et al. (2021) have used a rainfall threshold model to predict landslides in Emilia-Romagna, the Apennines, Italy. They have also compared validation results of landslide susceptibility models such as frequency ratio (FR), weight of evidence (WoE), and logistic regression (LR) models.

Jain et al. (2021) have used a rainfall data model and debris flow model for semiautomatic landslide prediction in Kerala, India. They have used high-resolution satellite data (Carto DEM, Pleiades 1A, SPOT, Sentinel-1) to generate thematic data layers of geo-environmental parameters affecting landslide occurrence. Peres and Cancelliere (2021) have used the rainfall threshold model and intensity duration model using TRIGRS and MATLAB to develop a landslide prediction map for Sicily, Italy. Rosi et al. (2021) used rainfall and landslide data, combined with DEM, to generate a 3D rainfall threshold model for Emilia-Romagna, North Italy.

Lee et al. (2021) used tropical rainfall measuring mission (TRMM) data for temporal landslide prediction modelling, employing the extreme value distribution (rainfall threshold) model and machine learning to predict rainfall-induced landslides in Korea. They have also used the topographical wetness index for the prediction of rainfall-induced landslides in the study area. Lee et al. (2021) have recently used a distinct element method for soil creep modelling in the south-east Tibetan plateau of China. They have used Google Earth images and DEM for the extraction of information about geo-environmental parameters. Lima et al. (2021) compared logistic regression and mixed effect modelling to predict landslides in Austria. Besides lithological and geomorphic factors, they have also used the topographic wetness index as one of the landslide causative factors. Liu et al. (2022) used time series analysis and wavelet analysis to identify potential landslide-prone areas in the Tibetan plateau of China. They have used high-resolution DEMs, Sentinel-1 and -2 data, and InSAR images to perform landslide prediction modelling. Ng et al. (2021) used a machine learning approach for the spatiotemporal prediction of landslides in Hong Kong. Nguyen and Kim (2021) used the ensemble method and convolutional neural network (CNN) model for landslide prediction in Mount Umyeon, Korea. They have used digital aerial photos and LiDAR images to perform analysis.

Xu et al. (2021a) adopted the deep learning method, multi-scale feature fusion with an encoder-decoder network (MFFENet) and adversarial domain adaptation network (ADANet) for the prediction of earthquake-induced landslides. Rodrigues et al. (2021) applied natural language processing (NLP) and machine learning (SVM) for landslide susceptibility zonation in Recife, North Brazil. Song et al. (2022) have recently used multiple surge load models for modelling debris flow impact.

The review of recent trends in landslide susceptibility zonation modelling reveals that data-driven methods, threshold modelling, and machine learning have been widely used to predict landslides with the highest possible accuracy. It has also been observed that high-resolution satellite and airborne spatial data have been extensively used for landslide susceptibility modelling. The use of complex landslide models has not only improved the accuracy of landslide prediction but the application of data-driven methods has also helped in generating a complete and reliable landslide database.

2.10.3 LANDSLIDE RISK ASSESSMENT

Landslide risk assessment is an important step in landslide hazard assessment. In recent times, there have been several methodological advancements in landslide vulnerability and the delineation of landslide risk–prone areas. A brief review of recently published literature is discussed in the following paragraphs.

Ferlisi et al. (2021) carried out a quantitative landslide risk assessment in Campania, South Italy. They have also suggested engineering solutions to mitigate the negative impact of landslide occurrence, such as sealing cracks in road pavements, use of asphalt concrete, spraying bitumen, and foundations of stabilised granular mixture to stabilise the slope. Caleca et al. (2022) have recently carried out a quantitative risk assessment in the Arno River basin of central Italy using different databases, such as the digital terrain model, IFFI (inventory of Italian landslides), ISTAT census section, OMI (real estate market observatory) database, VAM (agriculture values) database, and OSM (open street map) database. He et al. (2022) studied landslide-prone areas in Wenchuan, China, and suggested landslide mitigation measures such as the construction of a cascade check dam, branch channel, and divergence channel.

Delgado Garcia et al. (2022) used the public database to assess the impact of landslides on the economy and public health and to determine other socio-economic implications in Columbia. They have used the Bayesian information criterion (BIC) model to evaluate landslide risk and impact. Perrone et al. (2021) used MATLAB, electrical resistivity tomography (ERT), and seismic refraction tomography (SRT) for the evaluation of landslide risk in the Basilicata region of South Italy. Strouth and McDougall (2021) carried out a social landslide risk assessment using FN criterion in Western Canada. Capobianco et al. (2022) used web-based tools for landslide mitigation in Bode, England. They have used the LaRiMiT risk mitigation tool to suggest landslide mitigation in the study area.

The brief review of recent studies on landslide risk assessment indicates that there have been developments in tools and techniques for quantitative landslide risk assessment. Web-based tools, socio-economic and public databases of elements at risk due to landslides have improved the accuracy of landslide risk assessment.

2.11 SUMMARY

Landslide hazard assessment requires significant developments in methods, techniques, and models for landslide prediction. The accuracy and predictivity of landslide hazard models depend largely upon the completeness and reliability of landslide databases. There have been significant advancements in landslide hazard modelling and landslide information data sources. The use of UAVs for landslide databases has proven significant in recent times. High-resolution satellite data have now been put into use for landslide hazard assessment extensively.

REFERENCES

Akbar, Tahir Ali, and Sung Ryong Ha. 2011. "Landslide Hazard Zoning along Himalayan Kaghan Valley of Pakistan-by Integration of GPS, GIS, and Remote Sensing Technology." *Landslides* 8 (4): 527–540. https://doi.org/10.1007/s10346-011-0260-1.

Akgun, A. 2011. "A Comparison of Landslide Susceptibility Maps Produced by Logistic Regression, Multi-Criteria Decision, and Likelihood Ratio Methods: A Case Study at Lzmir, Turkey." *Landslides, Springer-Verlag.* https://doi.org/10.1007/s10346-011-0283-7.

Anbalagan, R., D. Chakraborty, and A. Kohali. 2008. "Landslide Hazard Zonation (LHZ) Maping on Meso-Scale for Systematic Town Planning in Mountainous Terrain." *Journal of Scientific & Industrial Research* 67: 486–497.

Arora, M., A. Dasgupta, and R. Gupta. 2004. "An Artificial Neural Network Approach for Landslide Hazard Zonation in the Bhagirathi (Ganga) Valley, Himalaya." *International Journal of Remote Sensing* 25: 559–572.

Atkinson, P. M., and R. Massari. 2011. "Autologistic Modelling of Susceptibility to Landsliding in the Central Apennines, Italy." *Geomorphology* 130 (1–2): 55–64. https://doi.org/10.1016/j.geomorph.2011.02.001.

Ayalew, L., and H. Yamagishi. 2005. "The Application of GIS-Based Logistic Regression for Landslide Susceptibility Mapping in the Kakuda-Yahiko Mountains, Central Japan." *Geomorphology* 65: 15–31.

Ayalew, L., H. Yamagishi, H. Marui, and T. Kanno. 2005. "Landslides in Sado Island of Japan: Part II. GIS-Based Susceptibility Mapping with Comparisons of Results from Two Methods and Verifications." *Engineering Geology* 81 (4): 432–445. https://doi.org/10.1016/j.enggeo.2005.08.004.

Balsubaramani, K., and K. Kumaraswamy. 2013. "Application of Geospatial Technology and Information Value Technique in Landslide Hazard Zonation Mapping: A Case Study of Giri Valley, Himachal Pradesh." *Disaster Advances* 6: 38–47.

Bălteanu, Dan, Viorel Chendeş, Mihaela Sima, and Petru Enciu. 2010. "A Country-Wide Spatial Assessment of Landslide Susceptibility in Romania." *Geomorphology* 124 (3–4): 102–112. https://doi.org/10.1016/j.geomorph.2010.03.005.

Blahut, Jan, Cees J. van Westen, and Simone Sterlacchini. 2010. "Analysis of Landslide Inventories for Accurate Prediction of Debris-Flow Source Areas." *Geomorphology* 119 (1–2): 36–51. https://doi.org/10.1016/j.geomorph.2010.02.017.

Caleca, Francesco, Veronica Tofani, Samuele Segoni, Federico Raspini, Ascanio Rosi, Marco Natali, Filippo Catani, and Nicola Casagli. 2022. "A Methodological Approach of QRA for Slow-Moving Landslides at a Regional Scale." *Landslides* (August 2021): 1539–1561. https://doi.org/10.1007/s10346-022-01875-x.

Calvello, Michele, Leonardo Cascini, and Sabrina Mastroianni. 2013. "Landslide Zoning over Large Areas from a Sample Inventory by Means of Scale-Dependent Terrain Units." *Geomorphology* 182: 33–48. https://doi.org/10.1016/j.geomorph.2012.10.026.

Capobianco, Vittoria, Marco Uzielli, Bjørn Kalsnes, Jung Chan Choi, James Michael Strout, Loretta von der Tann, Ingar Haug Steinholt, Anders Solheim, Farrokh Nadim, and Suzanne Lacasse. 2022. "Recent Innovations in the LaRiMiT Risk Mitigation Tool: Implementing a Novel Methodology for Expert Scoring and Extending the Database to Include Nature-Based Solutions." *Landslides* 19 (7): 1563–1583. https://doi.org/10.1007/s10346-022-01855-1.

Cardinali, M., P. Reichenbach, F. Guzzetti, F. Ardizzone, G. Antonini, M. Galli, M. Cacciano, M. Castellani, and P. Salvati. 2002. "A Geomorphological Approach to the Estimation of Landslide Hazards and Risks in Umbria, Central Italy." *Natural Hazards and Earth System Sciences* 2 (1–2): 57–72. https://doi.org/10.5194/nhess-2-57-2002.

Catani, F., N. Casagli, L. Ermini, G. Righini, and G. Menduni. 2005. "Landslide Hazard and Risk Mapping at Catchment Scale in the Arno River Basin." *Landslides* 2 (4): 329–342. https://doi.org/10.1007/s10346-005-0021-0.

Champatiray, P., S. Dimri, R. Lakhera, and S. Sati. 2007. "Fuzzy Based Methods for Landslide Hazard Assessment in Active Seismic Zone of Himalaya." *Landslides, Springer-Verlag* 4: 101–110.

Chand, Dipender Singh. 2008. "Landslide Monitoring in Space and Time Using Optical Satellite Imagery and Dem Derived Parameters: Case Study from Garhwal Himalaya, Uttarakhand." In *International Institute for Geoinformation Science and Earth Observation, Enschede, Netherlands and Indian Institute of Remote Sensing, National Remote Sensing Agency, Department of Space, Dehradun, India* p. 111.

Chang, K. T., and S. Chiang. 2009. "An Integrated Model for Predicting Rainfall Induced Landslides." *Geomorphology* 105: 366–373.

Chang, K. T., and J. K. Liu. 2004. "Landslide Features Interpreted by Neural Network Method Using a High-Resolution Satellite Image and Digital Topographic Data." *Proceedings of ISPRS XX Congress, Commission VII TS WG VII/5, Istanbul, Turkey* 1: 574–579.

Chau, K. T., Y. L. Sze, M. K. Fung, W. Y. Wong, E. L. Fong, and L. C. P. Chan. 2004. "Landslide Hazard Analysis for Hong Kong Using Landslide Inventory and GIS." *Computers and Geosciences* 30 (4): 429–443. https://doi.org/10.1016/j.cageo.2003.08.013.

Chen, Chi-Wen, Ching Hung, Guan-Wei Lin, Jun-Jih Liou, Shih-Yao Lin, Hsin-Chi Li, Yung-Ming Chen, and Hongery Chen. 2022. "Preliminary Establishment of a Mass Movement Warning System for Taiwan Using the Soil Water Index." *Landslides, Springer-Verlag* 19: 1779–1789.

Chen, X. L., H. L. Ran, and W. T. Yang. 2012. "Evaluation of Factors Controlling Large Earthquake-Induced Landslides by the Wenchuan Earthquake." *Natural Hazards and Earth System Science* 12 (12): 3645–3657. https://doi.org/10.5194/nhess-12-3645-2012.

Chleborad, Alan F., Rex L. Baum, Jonathan W. Godt, and US Geological Survey. 2006. *Rainfall Thresholds for Forecasting Landslides in the Seattle, Washington, Area-Exceedance and Probability.* http://scholar.google.com/scholar?q=related:94WNYvf6TfEJ:scholar.google.com/&hl=en&num=30&as_sdt=0,5.

Chung, Chang Jo, and Andrea G. Fabbri. 2008. "Predicting Landslides for Risk Analysis – Spatial Models Tested by a Cross-Validation Technique." *Geomorphology* 94 (3–4): 438–452. https://doi.org/10.1016/j.geomorph.2006.12.036.

Clerici, Aldo, Susanna Perego, Claudio Tellini, and Paolo Vescovi. 2002. "A Procedure for Landslide Susceptibility Zonation by the Conditional Analysis Method." *Geomorphology* 48 (4): 349–364. https://doi.org/10.1016/S0169-555X(02)00079-X.

Coe, J, J. Godt, R. Baum, R. Bucknam, and J. Michael. 2004. "Landslide Susceptibility from Topography in Guatemala." *Landslides: Evaluation and Stabilization/Glissement de Terrain: Evaluation et Stabilisation* 2: 69–78. https://doi.org/10.1201/b16816-8.

Colombo, A., L. Lanteri, M. Ramasco, and C. Troisi. 2005. "Systematic GIS Based Landslide Inventory as the First Step for Effective Landslide Hazard Management." *Landslides* 2: 291–301.

Conoscenti, Christian, Cipriano Di Maggio, and Edoardo Rotigliano. 2008. "GIS Analysis to Assess Landslide Susceptibility in a Fluvial Basin of NW Sicily (Italy)." *Geomorphology* 94 (3–4): 325–339. https://doi.org/10.1016/j.geomorph.2006.10.039.

Courture, R. 2011. *Landslide Terminology – National Technical Guidelines and Best Practices on Landslides. Open Files 6824, Geological Survey of Canada.* Canada.

Cruden, D. M. 1991. "A Simple Definition of a Landslide." *Bulletin of the International Association of Engineering Geology – Bulletin de l'Association Internationale de Géologie de l'Ingénieur* 43 (1): 27–29. https://doi.org/10.1007/BF02590167.

Dahal, Ranjan Kumar, and Shuichi Hasegawa. 2008. "Representative Rainfall Thresholds for Landslides in the Nepal Himalaya." *Geomorphology* 100 (3–4): 429–443. https://doi.org/10.1016/j.geomorph.2008.01.014.

Dahl, M. P. J., L. E. Mortensen, A. Veihe, and N. H. Jensen. 2010. "A Simple Qualitative Approach for Mapping Regional Landslide Susceptibility in the Faroe Islands." *Natural Hazards and Earth System Sciences* 10 (2): 159–170. https://doi.org/10.5194/nhess-10-159-2010.

Dai, F. C., and C. F. Lee. 2001. "Frequency-Volume Relation and Prediction of Rainfall-Induced Landslides." *Engineering Geology* 59 (3–4): 253–266. https://doi.org/10.1016/S0013-7952(00)00077-6.

Dai, F. C., and C. F. Lee. 2002. "Landslide Characteristics and Slope Instability Modeling Using GIS, Lantau Island, Hong Kong." *Geomorphology* 42 (3–4): 213–228. https://doi.org/10.1016/S0169-555X(01)00087-3.

Das, I. 2011. "Spatial Statistical Modelling for Assessing Landslide Hazard and Vulnerability." In *ITC Dissertation Number 192.* University of Twenty, International Institute for Geo-Information Science and Earth Observation, Enschede, Netherlands.

Das, I., Alfred Stein, Norman Kerle, and V. K. Dadhwal. 2011. "Probabilistic Landslide Hazard Assessment Using Homogeneous Susceptible Units (HSU) along a National Highway Corridor in the Northern Himalayas, India." *Landslides* 8 (3): 293–308. https://doi.org/10.1007/s10346-011-0257-9.

Das, I., Alfred Stein, Norman Kerle, and Vinay K. Dadhwal. 2012. "Landslide Susceptibility Mapping along Road Corridors in the Indian Himalayas Using Bayesian Logistic Regression Models." *Geomorphology* 179: 116–125. https://doi.org/10.1016/j.geomorph.2012.08.004.

Delgado Garcia, Helbert, Petly David, Bermudez Mouricio, and Sepulveda Sergio. 2022. "Fatal Landslides in Colombia (from Historical Times to 2020) and Their Socio-economic Impacts." *Landslides, Springer-Verlag* 19: 1689–1716.

Dewitte, Olivier, Antoine Dille, Arthur Depicker, Désiré Kubwimana, Jean Claude Maki Mateso, Toussaint Mugaruka Bibentyo, Judith Uwihirwe, and Elise Monsieurs. 2021. "Constraining Landslide Timing in a Data-Scarce Context: From Recent to Very Old Processes in the Tropical Environment of the North Tanganyika-Kivu Rift Region." *Landslides* 18 (1): 161–177. https://doi.org/10.1007/s10346-020-01452-0.

Eeckhaut, M., P. Reichenbach, F. Guzzetti, M. Rossi, and J. Poesen. 2009. "Combined Landslide Inventory and Susceptibility Assessment Based on Different Mapping Units: An Example from the Flemish Ardennes, Belgium." *Natural Hazards and Earth System Sciences* 9: 507–521.

Ercanoglu, Murate. 2005. "Landslide Susceptibility Assessment of SE Bartin (West Black Sea Region, Turkey) by Artificial Neural Networks." *Natural Hazards and Earth System Science* 5 (6): 979–992. https://doi.org/10.5194/nhess-5-979-2005.

Ercanoglu, M., C. Gokceoglu, and T. W. J. Van Asch. 2004. "Landslide Susceptibility Zoning North of Yenice (NW Turkey) by Multivariate Statistical Techniques." *Natural Hazards* 32 (1): 1–23. https://doi.org/10.1023/B:NHAZ.0000026786.85589.4a.

Erener, Arzu, H. Sebnem, and B. Düzgün. 2010. "Improvement of Statistical Landslide Susceptibility Mapping by Using Spatial and Global Regression Methods in the Case of More and Romsdal (Norway)." *Landslides* 7 (1): 55–68. https://doi.org/10.1007/s10346-009-0188-x.

Ferlisi, Settimio, Antonio Marchese, and Dario Peduto. 2021. "Quantitative Analysis of the Risk to Road Networks Exposed to Slow-Moving Landslides: A Case Study in the Campania Region (Southern Italy)." *Landslides* 18 (1): 303–319. https://doi.org/10.1007/s10346-020-01482-8.

Floris, Mario, and Francesca Bozzano. 2008. "Evaluation of Landslide Reactivation: A Modified Rainfall Threshold Model Based on Historical Records of Rainfall and Landslides." *Geomorphology* 94 (1–2): 40–57. https://doi.org/10.1016/j.geomorph.2007.04.009.

Franceschini, Rachele, Ascanio Rosi, Filippo Catani, and Nicola Casagli. 2022. "Exploring a Landslide Inventory Created by Automated Web Data Mining: The Case of Italy." *Landslides* 19 (4): 841–853. https://doi.org/10.1007/s10346-021-01799-y.

Gabet, Emmanuel J., Douglas W. Burbank, Jaakko K. Putkonen, Beth A. Pratt-Sitaula, and Tank Ojha. 2004. "Rainfall Thresholds for Landsliding in the Himalayas of Nepal." *Geomorphology* 63 (3–4): 131–143. https://doi.org/10.1016/j.geomorph.2004.03.011.

Galli, Mirco, Francesca Ardizzone, Mauro Cardinali, Fausto Guzzetti, and Paola Reichenbach. 2008. "Comparing Landslide Inventory Maps." *Geomorphology* 94 (3–4): 268–289. https://doi.org/10.1016/j.geomorph.2006.09.023.

García-Rodríguez, M. J., J. A. Malpica, B. Benito, and M. Díaz. 2008. "Susceptibility Assessment of Earthquake-Triggered Landslides in El Salvador Using Logistic Regression." *Geomorphology* 95 (3–4): 172–191. https://doi.org/10.1016/j.geomorph.2007.06.001.

Ghorbanzadeh, Omid, Hejar Shahabi, Alessandro Crivellari, Saeid Homayouni, Thomas Blaschke, and Pedram Ghamisi. 2022. "Landslide Detection Using Deep Learning and Object-Based Image Analysis." *Landslides* 19 (4): 929–939. https://doi.org/10.1007/s10346-021-01843-x.

Ghosh, Saibal. 2011. "Knowledge Guided Empirical Prediction of Landslide Hazard." In *ITC Dissertation Number 190.* University of Twenty, International Institute for Geo-Information Science and Earth Observation, Enschede, Netherlands.

Ghosh, Saibal, C. J. van Westen, E. J. M. Carranza, T. B. Ghoshal, N. K. Sarkar, and M. Surendranath. 2009. "A Quantitative Approach for Improving the BIS (Indian) Method of Medium-Scale Landslide Susceptibility." *Journal of the Geological Society of India* 74 (5): 625–638. https://doi.org/10.1007/s12594-009-0167-9.

Giuseppe, Ciccarese, Mulas Marco, and Corsini Alessandro. 2021. "Combining Spatial Modelling and Regionalization of Rainfall Thresholds for Debris Flows Hazard Mapping in the Emilia-Romagna Apennines (Italy)." *Landslides* 18 (11): 3513–3529. https://doi.org/10.1007/s10346-021-01739-w.

Gokceoglu, C., and E. Sezer. 2009. "A Statistical Assessment on International Landslide Literature (1945–2008)." *Landslides* 6: 345–351.

Goswami, Rajasmita, Neil C. Mitchell, and Simon H. Brocklehurst. 2011. "Distribution and Causes of Landslides in the Eastern Peloritani of NE Sicily and Western Aspromonte of SW Calabria, Italy." *Geomorphology* 132 (3–4): 111–122. https://doi.org/10.1016/j.geomorph.2011.04.036.

Greco, R., M. Sorriso-Valvo, and E. Catalano. 2007. "Logistic Regression Analysis in the Evaluation of Mass Movements Susceptibility: The Aspromonte Case Study, Calabria, Italy." *Engineering Geology* 89 (1–2): 47–66. https://doi.org/10.1016/j.enggeo.2006.09.006.

Gutiérrez, Francisco, Mauro Soldati, Franck Audemard, and Dan Bălteanu. 2010. "Recent Advances in Landslide Investigation: Issues and Perspectives." *Geomorphology* 124 (3–4): 95–101. https://doi.org/10.1016/j.geomorph.2010.10.020.

Guzzetti, F. 2000. "Landslide Fatalities and Evaluation of Landslide Risk in Italy." *Engineering Geology* 58: 89–107.

Guzzetti, Fausto. 2003. "Landslide Hazard Assessment and Risk Evaluation: Limits and Prospectives." *Proceedings of the 4th EGS Plinius Conference* (October 2002): 1–4.

Guzzetti, F., P. Aleotti, B. Malamud, and D. L. Turcotte. 2003. "Comparison of Three Landslide Event Inventories in Central and Northern Italy." In *4th EGS Plinius Conference* (pp. 5–9). Mallorca, Spain: Universitat de les Ilies Balears.

Guzzetti, F., A. Carrara, M. Cardinali, and P. Reichenbach. 1999. "Landslide Hazard Evaluation: A Review of Current Techniques and Their Application in a Multi-Study, Central Italy." *Geomorphology* 31: 181–216.

Guzzetti, Fausto, M. Galli, P. Reichenbach, F. Ardizzone, and M. Cardinali. 2006. "Landslide Hazard Assessment in the Collazzone Area, Umbria, Central Italy." *Natural Hazards and Earth System Science* 6 (1): 115–131. https://doi.org/10.5194/nhess-6-115-2006.

Guzzetti, Fausto, Paola Reichenbach, Mauro Cardinali, Mirco Galli, and Francesca Ardizzone. 2005. "Probabilistic Landslide Hazard Assessment at the Basin Scale." *Geomorphology* 72 (1–4): 272–299. https://doi.org/10.1016/j.geomorph.2005.06.002.

Hansen, A. 1984. *Slope Instability* (pp. 523–602). D. Brusden, D. Prior (Eds.). New York: John Wiley and Sons.

He, Jian, Limin Zhang, Ruilin Fan, Shengyang Zhou, Hongyu Luo, and Dalei Peng. 2022. "Evaluating Effectiveness of Mitigation Measures for Large Debris Flows in Wenchuan, China." *Landslides* 19 (4): 913–928. https://doi.org/10.1007/s10346-021-01809-z.

Hughes, K. E., M. Geertsema, E. Kwoll, M. N. Koppes, N. J. Roberts, J. J. Clague, and S. Rohland. 2021. "Previously Undiscovered Landslide Deposits in Harrison Lake, British Columbia, Canada." *Landslides* 18 (2): 529–538. https://doi.org/10.1007/s10346-020-01514-3.

Ilinca, Viorel, Ionuț Șandric, Zenaida Chițu, Radu Irimia, and Ion Gheuca. 2022. "UAV Applications to Assess Short-Term Dynamics of Slow-Moving Landslides under Dense Forest Cover." *Landslides* 19 (7): 1717–1734. https://doi.org/10.1007/s10346-022-01877-9.

Jain, Nirmala, Tapas R. Martha, Kirti Khanna, Priyom Roy, and K. Vinod Kumar. 2021. "Major Landslides in Kerala, India, during 2018–2020 Period: An Analysis Using Rainfall Data and Debris Flow Model." *Landslides* 18 (11): 3629–3645. https://doi.org/10.1007/s10346-021-01746-x.

Jaiswal, P. 2011. "Landslide Risk Quantification along Transportation Corridors Based on Historical Information." In *ITC Dissertation Number 191*. University of Twenty, International Institute for Geo-Information Science and Earth Observation, Enschede, Netherlands.

Jaiswal, Pankaj, and Cees J. van Westen. 2013. "Use of Quantitative Landslide Hazard and Risk Information for Local Disaster Risk Reduction along a Transportation Corridor: A Case Study from Nilgiri District, India." *Natural Hazards* 65 (1): 887–913. https://doi.org/10.1007/s11069-012-0404-1.

Jaiswal, P., C. J. van Westen, and V. Jetten. 2010. "Quantitative Assessment of Direct and Indirect Landslide Risk along Transportation Lines in Southern India." *Natural Hazards and Earth System Science* 10 (6): 1253–1267. https://doi.org/10.5194/nhess-10-1253-2010.

Jelínek, Róbert, and Peter Wagner. 2007. "Landslide Hazard Zonation by Deterministic Analysis (Veľká Čausa Landslide Area, Slovakia)." *Landslides* 4 (4): 339–350. https://doi.org/10.1007/s10346-007-0089-9.

Kannan, M., E. Saranathan, and R. Anbalagan. 2011. "Macro Landslide Hazard Zonation Mapping – Case Study from Bodi – Bodimettu Ghats Section, Theni District, Tamil Nadu – India." *Journal of the Indian Society of Remote Sensing* 39 (4): 485–496. https://doi.org/10.1007/s12524-011-0112-4.

Kanungo, D., M. Arora, R. Gupta, and S. Sarkar. 2009. "Landslide Risk Assessment Using Concept of Danger Pixel and Fuzzy Set Theory in Darjeeling Himalayas." *Landslides* 5: 407–416.

Kavzoglu, Taskin, Emrehan Kutlug Sahin, and Ismail Colkesen. 2014. "Landslide Susceptibility Mapping Using GIS-Based Multi-Criteria Decision Analysis, Support Vector Machines, and Logistic Regression." *Landslides* 11 (3): 425–439. https://doi.org/10.1007/s10346-013-0391-7.

Kuriakose, S. 2010. *Physically-Based Dynamic Modelling of the Effect of Land Use Changes on Shallow Landslide Initiation in the Western Ghats of Kerala, India.* University of Twente, International Institute for Geo-Information Science and Earth Observation.

Lallianthanga, R. K., and F. Lalbiakmawia. 2013. "Micro-Level Landslide Hazard Zonation of Saitual Town, Mizoram, India Using Remote Sensing and GIS Techniques *1,2." *International Journal of Engineering Sciences & Research Technology* 2 (9): 184–194.

Lee, C. T., C. C. Huang, J. F. Lee, K. L. Pan, M. L. Lin, and J. J. Dong. 2008. "Statistical Approach to Storm Event-Induced Landslides Susceptibility." *Natural Hazards and Earth System Science* 8 (4): 941–960. https://doi.org/10.5194/nhess-8-941-2008.

Lee, Jung Hyun, Hanbeen Kim, Hyuck Jin Park, and Jun Haeng Heo. 2021. "Temporal Prediction Modeling for Rainfall-Induced Shallow Landslide Hazards Using Extreme Value Distribution." *Landslides* 18 (1): 321–338. https://doi.org/10.1007/s10346-020-01502-7.

Lee, S. T. 2005. "Cross-Verification of Spatial Logistic Regression for Landslide Susceptibility Analysis: A Case Study of Korea." *Proceedings, 31st International Symposium on Remote Sensing of Environment, ISRSE 2005: Global Monitoring for Sustainability and Security.* St. Petersburg, Russia, 20–24 June 2005. http://www.scopus.com/inward/record.url?eid=2-s2.0-84879728712&partnerID=tZOtx3y1.

Lee, S. T., and Biswajeet Pradhan. 2006. "Probabilistic Landslide Hazards and Risk Mapping on Penang Island, Malaysia." *Journal of Earth System Science* 115 (6): 661–672. https://doi.org/10.1007/s12040-006-0004-0.

Lee, S. T., T. T. Yu, W. F. Peng, and C. L. Wang. 2010. "Incorporating the Effects of Topographic Amplification in the Analysis of Earthquake-Induced Landslide Hazards Using Logistic Regression." *Natural Hazards and Earth System Science* 10 (12): 2475–2488. https://doi.org/10.5194/nhess-10-2475-2010.

Li, Huajin, Yusen He, Qiang Xu, Jiahao Deng, Weile Li, and Yong Wei. 2022a. "Detection and Segmentation of Loess Landslides via Satellite Images: A Two-Phase Framework." *Landslides* 19 (3): 673–686. https://doi.org/10.1007/s10346-021-01789-0.

Li, Langping, Hengxing Lan, Alexander Strom, and Renato Macciotta. 2022b. "Landslide Longitudinal Shape: A New Concept for Complementing Landslide Aspect Ratio." *Landslides* 19 (5): 1143–1163. https://doi.org/10.1007/s10346-021-01828-w.

Liang, Xin, Samuele Segoni, Kunlong Yin, Juan Du, Bo Chai, Veronica Tofani, and Nicola Casagli. 2022. "Characteristics of Landslides and Debris Flows Triggered by Extreme Rainfall in Daoshi Town during the 2019 Typhoon Lekima, Zhejiang Province, China." *Landslides* 19 (7): 1735–1749. https://doi.org/10.1007/s10346-022-01889-5.

Lima, Pedro, Stefan Steger, and Thomas Glade. 2021. "Counteracting Flawed Landslide Data in Statistically Based Landslide Susceptibility Modelling for Very Large Areas: A National-Scale Assessment for Austria." *Landslides* 18 (11): 3531–3546. https://doi.org/10.1007/s10346-021-01693-7.

Listo, Fabrizio de Luiz Rosito, and Bianca Carvalho Vieira. 2012. "Mapping of Risk and Susceptibility of Shallow-Landslide in the City of São Paulo, Brazil." *Geomorphology* 169–170: 30–44. https://doi.org/10.1016/j.geomorph.2012.01.010.

Liu, Ya, Haijun Qiu, Dongdong Yang, Zijing Liu, Shuyue Ma, Yanqian Pei, Juanjuan Zhang, and Bingzhe Tang. 2022. "Deformation Responses of Landslides to Seasonal Rainfall Based on InSAR and Wavelet Analysis." *Landslides* 19 (1): 199–210. https://doi.org/10.1007/s10346-021-01785-4.

Ma, Fengshan, Jie Wang, Renmao Yuan, Haijun Zhao, and Jie Guo. 2013. "Application of Analytical Hierarchy Process and Least-Squares Method for Landslide Susceptibility Assessment along the Zhong-Wu Natural Gas Pipeline, China." *Landslides* 10 (4): 481–492. https://doi.org/10.1007/s10346-013-0402-8.

Mancini, F., C. Ceppi, and G. Ritrovato. 2010. "GIS and Statistical Analysis for Landslide Susceptibility Mapping in the Daunia Area, Italy." *Natural Hazards and Earth System Science* 10 (9): 1851–1864. https://doi.org/10.5194/nhess-10-1851-2010.

Mantovani, F., R. Soeters, and C. van Westen. 1996. "Remote Sensing Techniques for Landslide Studies and Hazard Zonationn Europe." *Geomorphology* 15: 213–225.

Marino, Pasquale, Giovanni Francesco Santonastaso, Xuanmei Fan, and Roberto Greco. 2021. "Prediction of Shallow Landslides in Pyroclastic-Covered Slopes by Coupled Modeling of Unsaturated and Saturated Groundwater Flow." *Landslides* 18 (1): 31–41. https://doi.org/10.1007/s10346-020-01484-6.

Martha, Tapas R., Cees J. van Westen, Norman Kerle, Victor Jetten, and K. Vinod Kumar. 2013. "Landslide Hazard and Risk Assessment Using Semi-Automatically Created Landslide Inventories." *Geomorphology* 184: 139–150. https://doi.org/10.1016/j.geomorph.2012.12.001.

Meena, Sansar Raj, Lucas Pedrosa Soares, Carlos H. Grohmann, Cees van Westen, Kushanav Bhuyan, Ramesh P. Singh, Mario Floris, and Filippo Catani. 2022. "Landslide Detection in the Himalayas Using Machine Learning Algorithms and U-Net." *Landslides* 19 (5): 1209–1229. https://doi.org/10.1007/s10346-022-01861-3.

Mercogliano, P., S. Segoni, G. Rossi, B. Sikorsky, V. Tofani, P. Schiano, F. Catani, and N. Casagli. 2013. "Brief Communication A Prototype Forecasting Chain for Rainfall Induced Shallow Landslides." *Natural Hazards and Earth System Science* 13 (3): 771–777. https://doi.org/10.5194/nhess-13-771-2013.

Meusburger, K., and C. Alewell. 2009. "On the Influence of Temporal Change on the Validity of Landslide Susceptibility Maps." *Natural Hazards and Earth System Science* 9 (4): 1495–1507. https://doi.org/10.5194/nhess-9-1495-2009.

Miller, Daniel J., and Kelly M. Burnett. 2007. "Effects of Forest Cover, Topography, and Sampling Extent on the Measured Density of Shallow, Translational Landslides." *Water Resources Research* 43 (3): 1–23. https://doi.org/10.1029/2005WR004807.

Mondal, Sujit, and Ramkrishna Maiti. 2012. "Landslide Susceptibility Analysis of Shiv-Khola Watershed, Darjiling: A Remote Sensing & GIS Based Analytical Hierarchy Process (AHP)." *Journal of the Indian Society of Remote Sensing* 40 (3): 483–496. https://doi.org/10.1007/s12524-011-0160-9.

Montrasio, L., R. Valentino, and G. Losi. 2011. "Towards a Real-Time Susceptibility Assessment of Rainfall-Induced Shallow Landslides on a Regional Scale." *Natural Hazards and Earth System Sciences* 11: 1927–1947.

Nagarajan, R., A. Roy, R. Vinodkumar, and M. Khire. 2000. "Landslide Hazard Susceptibility Mapping Based on Terrain and Climatic Factors for Tropical Monsoon Region." *Engineering Geology* 58: 275–287.

Naithani, A. 2007. "Macro Landslide Hazard Zonation Mapping Using Univariate Statistical Analysis in Parts of Garhwal Himalaya." *Journal Geological Society of India* 70: 353–368.

Nefeslioglu, Hakan A., Tamer Y. Duman, and Serap Durmaz. 2008. "Landslide Susceptibility Mapping for a Part of Tectonic Kelkit Valley (Eastern Black Sea Region of Turkey)." *Geomorphology* 94 (3–4): 401–418. https://doi.org/10.1016/j.geomorph.2006.10.036.

Ng, C. W. W., B. Yang, Z. Q. Liu, J. S. H. Kwan, and L. Chen. 2021. "Spatiotemporal Modelling of Rainfall-Induced Landslides Using Machine Learning." *Landslides* 18 (7): 2499–2514. https://doi.org/10.1007/s10346-021-01662-0.

Nguyen, Ba Quang Vinh, and Yun Tae Kim. 2021. "Regional-Scale Landslide Risk Assessment on Mt. Umyeon Using Risk Index Estimation." *Landslides* 18 (7): 2547–2564. https://doi.org/10.1007/s10346-021-01622-8.

Niethammer, U., S. Rothmund, M. R. James, J. Travelletti, M. Joswig, and Lancaster Environment Centre. 2010. "Uav-Based Remote Sensing of Landslides." In *International Archives of Photogrammetry, Remote Sensing and Spatial Information Sciences, Vol. XXXVIII, Part 5 Commission V Symposium*. Newcastle upon Tyne, UK.

Ohlmacher, Gregory C., and John C. Davis. 2003. "Using Multiple Logistic Regression and GIS Technology to Predict Landslide Hazard in Northeast Kansas, USA." *Engineering Geology* 69 (3–4): 331–343. https://doi.org/10.1016/S0013-7952(03)00069-3.

Panikkar, S., and V. Subramaniyan. 1997a. "4 SV Pannikar DDun(1997).Pdf." *Current Science* 72 (12): 1117–1121.

Panikkar, S., and V. Subramaniyan. 1997b. "Landslide Hazard Analysis of the Area around Dehra Dun and Mussoorie, Uttar Pradesh." *Current Science* 73: 1117–1123.

Pardeshi, S., S. Pardeshi, and S. Autade. 2013. "Landslide Hazard Assessment: Recent Trends and Techniques." *SpringerPlus* 2 (523). https://doi.org/10.1186/2193-1801-2-253.

Pardeshi, S., S. Pardeshi, and V. Nagare. 2009. "A Study of Effect of Landslides on Human Environment in Thane District, Maharshtra State." *The Deccan Geographer* 47 (1): 45–56.

Parise, M. 2002. "Landslide Hazard Zonation of Slopes Susceptible to Rock Falls and Topples." *Natural Hazards and Earth System Sciences* 2 (1–2): 37–49. https://doi.org/10.5194/nhess-2-37-2002.

Paulin, Gabriel Legorreta, Katherine A. Mickelson, Trevor A. Contreras, William Gallin, Kara E. Jacobacci, and Marcus Bursik. 2022. "Assessing Landslide Volume Using Two Generic Models: Application to Landslides in Whatcom County, Washington, USA." *Landslides* 19 (4): 901–912. https://doi.org/10.1007/s10346-021-01825-z.

Pereira, S., J. Zezere, and C. Bateira. 2012. "Assessing Predictive Capacity and Conditional Independence of Landslide Predisposing Factors for Shallow Landslide Susceptibility Models." *Natural Hazards and Earth System Sciences* 12: 979–988.

Peres, David J., and Antonino Cancelliere. 2021. "Comparing Methods for Determining Landslide Early Warning Thresholds: Potential Use of Non-Triggering Rainfall for Locations with Scarce Landslide Data Availability." *Landslides* 18 (9): 3135–3147. https://doi.org/10.1007/s10346-021-01704-7.

Perrone, Angela, Filomena Canora, Giuseppe Calamita, Jessica Bellanova, Vincenzo Serlenga, Serena Panebianco, Nicola Tragni, et al. 2021. "A Multidisciplinary Approach for Landslide Residual Risk Assessment: The Pomarico Landslide (Basilicata Region, Southern Italy) Case Study." *Landslides* 18 (1): 353–365. https://doi.org/10.1007/s10346-020-01526-z.

Piacentini, Daniela, Francesco Troiani, Mauro Soldati, Claudia Notarnicola, Daniele Savelli, Stefan Schneiderbauer, and Claudia Strada. 2012. "Statistical Analysis for Assessing Shallow-Landslide Susceptibility in South Tyrol (South-Eastern Alps, Italy)." *Geomorphology* 151–152: 196–206. https://doi.org/10.1016/j.geomorph.2012.02.003.

Polemio, Maurizio, and Francesco Sdao. 1999. "The Role of Rainfall in the Landslide Hazard: The Case of the Avigliano Urban Area (Southern Apennines, Italy)." *Engineering Geology* 53 (3–4): 297–309. https://doi.org/10.1016/S0013-7952(98)00083-0.

Pradhan, Biswajeet, and Saro Lee. 2009. "Landslide Risk Analysis Using Artificial Neural Network Model Focussing on Different Training Sites." *International Journal of Physical Sciences* 4 (1): 01–015.

Pradhan, Biswajeet, and Saro Lee. 2010. "Regional Landslide Susceptibility Analysis Using Back-Propagation Neural Network Model at Cameron Highland, Malaysia." *Landslides* 7 (1): 13–30. https://doi.org/10.1007/s10346-009-0183-2.

Preuth, Thomas, Thomas Glade, and Alain Demoulin. 2010. "Stability Analysis of a Human-Influenced Landslide in Eastern Belgium." *Geomorphology* 120 (1–2): 38–47. https://doi.org/10.1016/j.geomorph.2009.09.013.

Rezig, S., J. Favre, and E. Leroi. 1996. "The Probabilistic Evaluation of Landslide Risk." In I. Sennset (Ed.), *Landslides* (pp. 351–355). Rotterdam: Balkema.

Rodrigues, Saulo Guilherme, Maisa Mendonça Silva, and Marcelo Hazin Alencar. 2021. "A Proposal for an Approach to Mapping Susceptibility to Landslides Using Natural Language Processing and Machine Learning." *Landslides* 18 (7): 2515–2529. https://doi.org/10.1007/s10346-021-01643-3.

Rosi, Ascanio, Samuele Segoni, Vanessa Canavesi, Antonio Monni, Angela Gallucci, and Nicola Casagli. 2021. "Definition of 3D Rainfall Thresholds to Increase Operative Landslide Early Warning System Performances." *Landslides* 18 (3): 1045–1057. https://doi.org/10.1007/s10346-020-01523-2.

Rotingliano, E., V. Agnesi, C. Cappadonia, and C. Conoscenti. 2011. "The Role of Diagnostic Areas in the Assessment of Landslide Susceptibility Models: A Test in the Sicilian Chain." *Natural Hazards* 58 (NA): 981–999.

Rowbotham, D., and D. Dudycha. 1998. "GIS Modelling of Slope Stability in Phewa Watershed, Nepal." *Geomorphology* 26: 151–170.

Roy, Priyom, Nirmala Jain, Tapas R. Martha, and K. Vinod Kumar. 2022. "Reactivating Balia Nala Landslide, Nainital, India – a Disaster in Waiting." *Landslides* 19 (6): 1531–1535. https://doi.org/10.1007/s10346-022-01881-z.

Saaty, T. 2008. "Decision Making with the Analytic Hierarchy Process." *International Journal of Services Sciences* 1 (1): 83–98.

Salciarini, Diana, Jonathan W. Godt, William Z. Savage, Pietro Conversini, Rex L. Baum, and John A. Michael. 2006. "Modeling Regional Initiation of Rainfall-Induced Shallow Landslides in the Eastern Umbria Region of Central Italy." *Landslides* 3 (3): 181–194. https://doi.org/10.1007/s10346-006-0037-0.

Sarkar, S., D. Kanungo, and G. Mehrotra. 1995. "Landslide Hazard Zonation: A Case Study of Garhwal Himalaya, India." *Mountain Research and Development* 15: 301–309.

Sarkar, S., D. Kanungo, A. Patra, and P. Kumar. 2006. "GIS Based Landslide Susceptibility Mapping- A Case Study in Indian Himalaya." In *Disaster Mitigation of Debris Flows, Slope Failures and Landslides* (pp. 617–624). Tokyo, Japan: Universal Academy Press.

Schicker, Renée, and Vicki Moon. 2012. "Comparison of Bivariate and Multivariate Statistical Approaches in Landslide Susceptibility Mapping at a Regional Scale." *Geomorphology* 161–162: 40–57. https://doi.org/10.1016/j.geomorph.2012.03.036.

Sharma, Amol, Chander Prakash, and Anuj Nautiyal. 2022. "Recent Landslide News of Himachal Pradesh, India." *Landslides* 19 (1): 221–224. https://doi.org/10.1007/s10346-021-01802-6.

Sharma, L., N. Patel, M. Ghosh, and P. Debnath. 2009. "Geographical Information System Based Landslide Probabilistic Model with Trivariate Approach – A Case Study in Sikkim Himalaya." In *Eighteenth United Nations Regional Cartographic Conference for Asia and the Pacific*. Bangkok: Economic and Social Council, United Nations.

Singh, C., K. Behra, and W. Rocky. 2011. "Landslide Susceptibility along NH-39 between Karong and Mao, Senapati District, Manipur." *Journal Geological Society of India* 78: 559–570.

Song, Dongri, Yitong Bai, Xiao Qing Chen, Gordon G. D. Zhou, Clarence E. Choi, Alessandro Pasuto, and Peng Peng. 2022. "Assessment of Debris Flow Multiple-Surge Load Model Based on the Physical Process of Debris-Barrier Interaction." *Landslides* 19 (5): 1165–1177. https://doi.org/10.1007/s10346-021-01778-3.

Sterlacchini, S., C. Ballabio, J. Blahut, M. Masetti, and A. Sorichetta. 2011. "Spatial Agreement of Predicted Patterns in Landslide Susceptibility Maps." *Geomorphology* 125 (1): 51–61. https://doi.org/10.1016/j.geomorph.2010.09.004.

Strouth, Alex, and Scott McDougall. 2021. "Societal Risk Evaluation for Landslides: Historical Synthesis and Proposed Tools." *Landslides* 18 (3): 1071–1085. https://doi.org/10.1007/s10346-020-01547-8.

Tanyaş, Hakan, Dalia Kirschbaum, Tolga Görüm, Cees J. van Westen, Chenxiao Tang, and Luigi Lombardo. 2021. "A Closer Look at Factors Governing Landslide Recovery Time in Post-Seismic Periods." *Geomorphology* 391. https://doi.org/10.1016/j.geomorph.2021.107912.

Tempa, Karma, Kinley Peljor, Sangay Wangdi, Rupesh Ghalley, Kelzang Jamtsho, Samir Ghalley, and Pratima Pradhan. 2021. "UAV Technique to Localize Landslide Susceptibility and Mitigation Proposal: A Case of Rinchending Goenpa Landslide in Bhutan." *Natural Hazards Research* 1 (4): 171–186. https://doi.org/10.1016/j.nhres.2021.09.001.

Tien Bui, Dieu, Biswajeet Pradhan, Owe Lofman, Inge Revhaug, and Oystein B. Dick. 2012. "Landslide Susceptibility Assessment in the Hoa Binh Province of Vietnam: A Comparison of the Levenberg-Marquardt and Bayesian Regularized Neural Networks." *Geomorphology* 171–172: 12–29. https://doi.org/10.1016/j.geomorph.2012.04.023.

Tofani, V., S. Segoni, A. Agostini, F. Catani, and N. Casagli. 2013. "Technical Note: Use of Remote Sensing for Landslide Studies in Europe." *Natural Hazards and Earth System Science* 13 (2): 299–309. https://doi.org/10.5194/nhess-13-299-2013.

Tolga, Can, Hakan A. Nefeslioglu, Candan Gokceoglu, Harun Sonmez, and Tamer Y. Duman. 2005. "Susceptibility Assessments of Shallow Earthflows Triggered by Heavy Rainfall at Three Catchments by Logistic Regression Analyses." *Geomorphology* 72 (1–4): 250–271. https://doi.org/10.1016/j.geomorph.2005.05.011.

van Westen, C. 2010. "GIS for the Assessment of Risk from Geomorphological Hazards." In I. A. Ayala, A. S. Goudie (Eds.), *Geomorphological Hazards and Disaster Prevention* (pp. 205–219). Cambridge University Press.

Varnes, D. I. 1984. *Landslide Hazard Zonation: A Review of Principles and Practice*. Paris: United Nations Scientific and Cultural Organization.

Wang, Fawu, Zixin Zhao, Ye Chen, Guolong Zhu, Kounghoon Nam, and Zhenhua Ye. 2022. "Guocun Landslide in a Slate Slope with Dip Structure on 27 March 2021 in Tonglu County, Zhejiang Province, China." *Landslides* 19 (6): 1435–1447. https://doi.org/10.1007/s10346-022-01846-2.

Wang, H. B., and K. Sassa. 2005. "Comparative Evaluation of Landslide Susceptibility in Minamata Area, Japan." *Environmental Geology* 47 (7): 956–966. https://doi.org/10.1007/s00254-005-1225-2.

Wu, W., and R. Sidle. 1995. "A Distributed Slope Stability Model for Steep Forested Basins." *Water Resource Research* 31 (8): 2097–2110.

Xu, Chong, Fuchu Dai, Xiwei Xu, and Yuan Hsi Lee. 2012. "GIS-Based Support Vector Machine Modeling of Earthquake-Triggered Landslide Susceptibility in the Jianjiang River Watershed, China." *Geomorphology* 145–146: 70–80. https://doi.org/10.1016/j.geomorph.2011.12.040.

Xu, Ling, Dongdong Yan, and Tengyuan Zhao. 2021a. "Probabilistic Evaluation of Loess Landslide Impact Using Multivariate Model." *Landslides* 18 (3): 1011–1023. https://doi.org/10.1007/s10346-020-01521-4.

Xu, Yuankun, David L. George, Jinwoo Kim, Zhong Lu, Mark Riley, Todd Griffin, and Juan de la Fuente. 2021b. "Landslide Monitoring and Runout Hazard Assessment by Integrating Multi-Source Remote Sensing and Numerical Models: An Application to the Gold Basin Landslide Complex, Northern Washington." *Landslides* 18 (3): 1131–1141. https://doi.org/10.1007/s10346-020-01533-0.

Yang, Che-Ming, Wei-An Chao, Meng-Chia Weng, Yu-Yao Fu, Jui-Ming Chang, and Wei-Kai Huang. 2022. "Che-Ming Yang 2022.Pdf." *Landslides, Springer-Verlag* 19: 1807–1811.

Yeon, Y., J. Han, and K. Ryu. 2010. "Landslide Susceptibility Mapping in Injae, Korea Using a Decision Tree." *Engineering Geology, Elsevier* 166: 274–283.

Zêzere, J. L. 2002. "Landslide Susceptibility Assessment Considering Landslide Typology. A Case Study in the Area North of Lisbon (Portugal)." *Natural Hazards and Earth System Sciences* 2 (1–2): 73–82. https://doi.org/10.5194/nhess-2-73-2002.

Zhou, Chao, Ying Cao, Xie Hu, Kunlong Yin, Yue Wang, and Filippo Catani. 2022. "Enhanced Dynamic Landslide Hazard Mapping Using MT-InSAR Method in the Three Gorges Reservoir Area." *Landslides* 19 (7): 1585–1597. https://doi.org/10.1007/s10346-021-01796-1.

Zizioli, D., C. Meisina, R. Valentino, and L. Montrasio. 2013. "Comparison between Different Approaches to Modeling Shallow Landslide Susceptibility: A Case History in Oltrepo Pavese, Northern Italy." *Natural Hazards and Earth System Sciences* 13 (3): 559–573. https://doi.org/10.5194/nhess-13-559-2013.

3 Review of Landslide Investigations in the World

Recent Trends

3.1 INTRODUCTION

Landslides are common geological hazards in rugged topographies across the globe. Slope failures occurring in proximity to human habitation often cause losses in terms of damage to man-made structures, human lives, vegetation, crops, soil loss, disruption to transportation routes, time, and effort. Human and socioeconomic losses occur when the human population and man-made structures are exposed to landslides. According to Varnes (1984), during 1971–1975, 14% of the total casualties from natural hazards globally are associated with slope failure events.

High-quality landslide databases, combined with complete and reliable landslide inventories, are essential for the delineation of actual and potential landslide-prone zones, validation of landslide susceptibility models, estimation of potential losses, and application of data-driven methods of landslide hazard assessment.

This chapter deals with global patterns of landslide distribution and their characteristics, identification of the major concentration of actual landslide-affected regions in the world, and review of recent trends (concerning landslide inventory, landslide susceptibility zonation, landslide risk assessment, etc.) in landslide hazard assessment. The chapter also highlights how geospatial technology has been used in landslide hazard assessment. The chapter also attempts to throw light on the methods and techniques of landslide hazard assessment developed over the years.

3.2 GLOBAL PATTERNS OF LANDSLIDE DISTRIBUTION AND CHARACTERISTICS

Global disaster databases are proven data sources on occurrences and losses caused by various natural disasters. Global landslide databases are available with EM-DAT International Database, NASA Global Landslide Catalog, and Global Fatal Landslide Database (GFLD).

DOI: 10.1201/9781003323488-3

Although EM-DAT is an important disaster database from AD 1900 till present, the actual number of landslides and associated losses are often underestimated because of the minimum threshold defined by EM-DAT to record landslide events in the global disaster database. The GFLD is an extensive global fatal landslide database which has been compiled since 2004, including details such as date of occurrence, description of landslide location, the absolute location of the landslide (latitudes and longitudes, geographical region), the number of fatalities, injuries, and triggering factors.

Based on the information obtained from the global disaster database and other sources of information on landslides, we could observe the following landslide-prone regions in the world.

1 North America
2 South America
3 Central America
4 Alpine regions of Italy
5 France
6 Switzerland and Austria in Europe
7 Himalayan regions of India
8 Nepal Himalayas
9 China
10 Japan
11 Indonesia

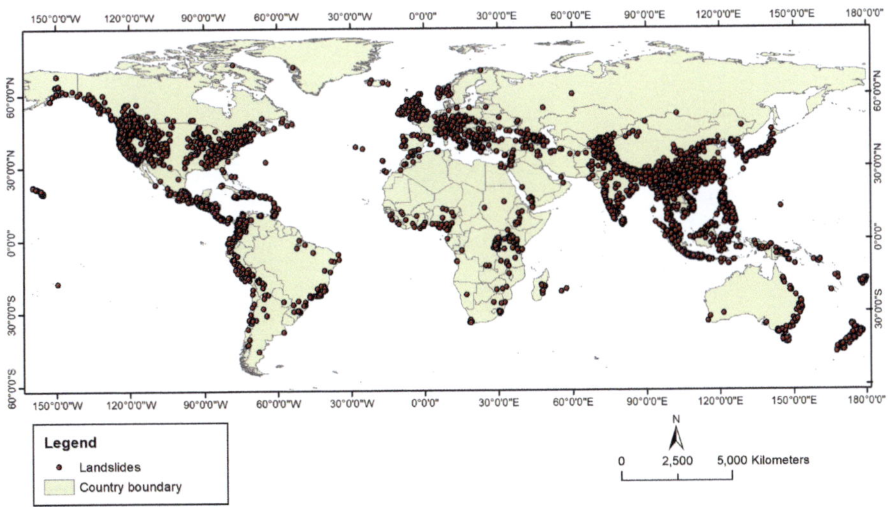

FIGURE 3.1 Global Distribution of Landslides.

3.2.1 NORTH AMERICA

Landslides in the United States are an important geological hazard is in almost every state. According to Froude and Petley (2018), 93 landslide-related fatalities occurred in the country from 2004 to 2016. According to NASA's global landslide database, occurred between 1988 and 2016, most landslides in the United States are triggered by high-intensity rainfall and a few by the freeze-and-thaw process. According to NASA, mudslides (711), landslides (573), complex slope failures (59), rockfalls (44), debris flow (44), and snow avalanches (4) are important types of slope failures. A mudslide occurred near Whitman, due to a heavy downpour, on 22 March 2014 and was one of the largest fatal landslides, causing 43 fatalities and 8 injuries.

The Appalachian Mountains, Rocky Mountains, Pacific coastal ranges, Washington, Alaska, Hawaii, and south-western hilly areas in the United States are prone to debris flow on slopes greater than 25° (Brabb, 1999). Sixty-seven per cent of the mapped landslide inventory is found in three West Coast states: Washington, Oregon, and California. Landslides in Alaska happen in the permafrost and glacial regions. A few landslides in Alaska are induced by earthquakes.

South-western parts of the Rockies (Colorado, Utah, New Mexico) are another major concentration of landslides. Steep-sloping terrain, poorly consolidated sediments, and thawing of snow are the major factors responsible for slope failures in this region. The Appalachians are one of the oldest mountains in the world. Kentucky, West Virginia, Virginia, Ohio, Georgia, and South California are

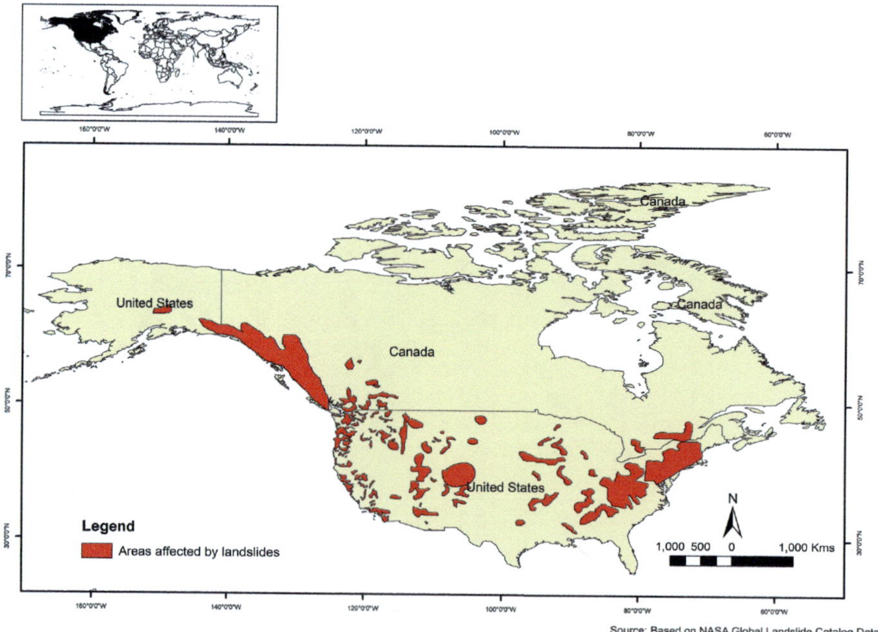

FIGURE 3.2   Landslide-Affected Areas in North America.

other important provinces affected by landslide events. The Pacific north-western region of the country also experiences numerous slope failure events along the hillsides in the coastal rugged topography. Oregon, Washington, and California are the provinces in this region.

3.2.2 South America

The terrain of South America is characterised by steep unstable slopes formed due to tectonic upliftment. According to the Enhanced Durham Fatal Landslide Database (EFLD), nearly 95% of the total landslides in South America are triggered by heavy rainfall, whereas only 15% are identified to be triggered by tropical storms. Considering the tectonic settings of the Andes Mountain range, this region is seismically very active. However, only 4% of the total fatal landslides are identified as earthquake-induced in South America. Brazil (19.48%) and Columbia (18%) are among the top countries in South America with the highest number of fatal landslides.

The highest number of fatal landslide events in Latin America is observed along the slopes of the Andes Mountains. The majority of landslide sites in South America are covered predominantly by igneous and metamorphic rocks. Although Brazil is a tectonically and seismically passive region, the concentration

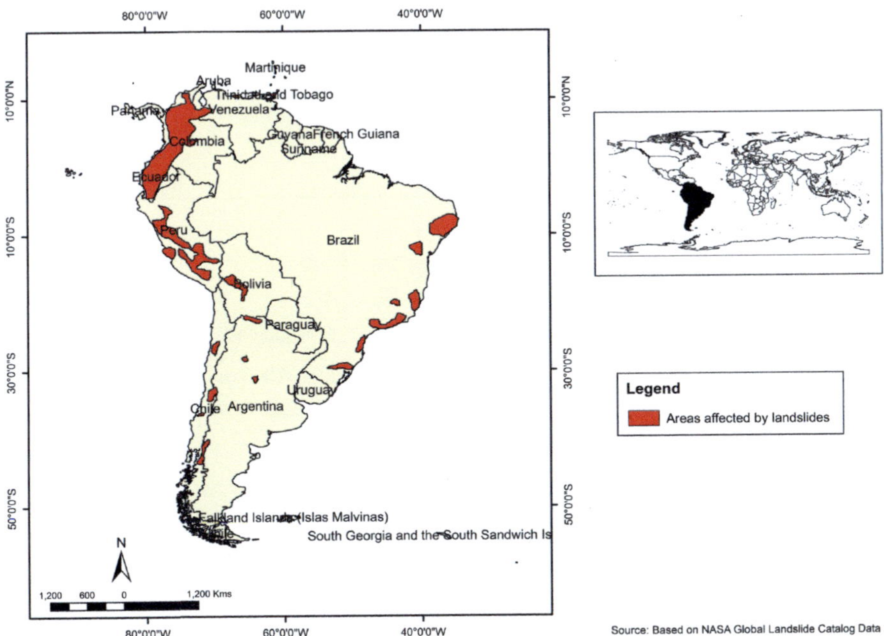

FIGURE 3.3 Landslide-Affected Areas in South America (Based on EFLD Database).

Note: Black dots represent landslide locations.

of landslide events in this region is high, probably due to the reason that slope instability in this region is triggered by high-intensity rainfall. It is probably due to the fact that landslide occurrences in the Andes Mountain range in Bolivia, northern Chile, and Argentina are characterised by low annual rainfall.

3.2.3 CENTRAL AMERICA

A hilly terrain characterises Central America, particularly the provinces of Guatemala, El Salvador, Nicaragua, Honduras, and Costa Rica. Central America is a comparatively dry region with scanty rainfall as compared to other regions in America. However, most of the landslides in this region are associated with high-intensity cyclonic rainfall events (e.g., Hurricane Eta, on 6 November 2020, caused numerous landslides in the provinces of Guatemala and Honduras).

Due to the undulating topography of Central America, almost 80% of its total geographical area is prone to landslides of different intensities. Landslides in Central America are concentrated in El Salvador, Honduras, Costa Rica, and Belize provinces. Besides rainfall, Central America also experiences seismic activities, which cause earthquake-triggered landslides in this region. Soil falls and slides are common in the steeply sloping areas of volcanic soils of Guatemala and El Salvador. However, large-size shallow translational landslides are the predominant slope failure types in the laterite terrain of Costa Rica and Panama (Bommer and Rodriguez, 2002).

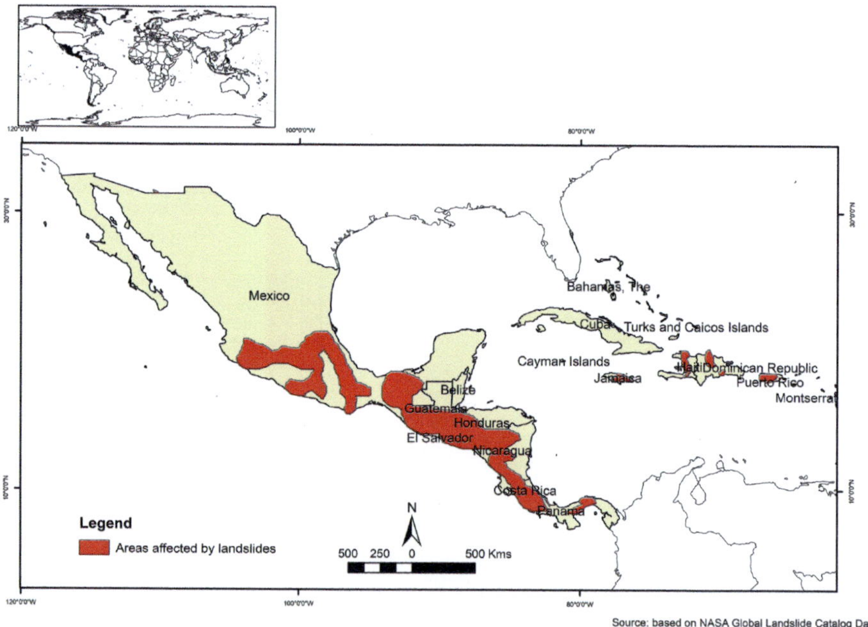

FIGURE 3.4   Landslide-Affected Areas in Central America.

3.2.4 ALPINE REGIONS OF ITALY

Italy is one of the most important concentrations of landslide hazards in the world. The Italian Landslide Inventory (IFFI) is a project carried out by ISPRA since 2005 to produce a national landslide inventory for Italy. To date, 620,808 landslide events, affecting an area of 23,700 km² (7.9% of the total geographical area of Italy) have been recorded. Emilia-Romagna, Friuli-Venezia Giulia, Liguria, Piemonte, Sicilia, Basilicata, and Lombardia are the important landslide-affected provinces in Italy.

The highest concentration of reported landslides in Italy is observed in Toscana province. Northern Italy has the highest number of landslides, particularly in Veneto, Lombardia, Emilia-Romagna, and Liguria. The alpine region of Italy is characterised by high relative relief and rocks (predominantly, granite, metamorphic, limestones, and dolomite), which experience specific mass movement types, such as rockslides, rock avalanches, and rockfalls. Extreme rainfall events and snow melting processes are the important landslide triggering factors in alpine regions of Italy.

3.2.5 FRANCE

Landslide is an important geological hazard in France, particularly in the southern parts of France. Topples, rockfall, and subsidence are the common slope failure types in France. Land subsidence is a common slope failure type in the plains, whereas rockfall, debris slides, and debris flow are commonly observed along soft or hard rock cliffs.

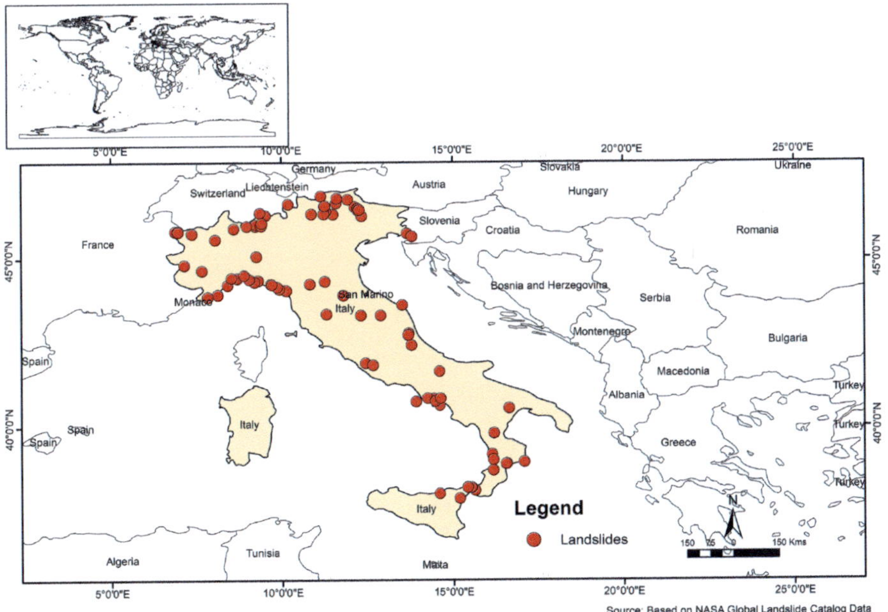

FIGURE 3.5 Landslide-Affected Areas in Italy.

The degradation of permafrost slopes of the French Alps results in the slope instability process. The increased amount of precipitation and its intensity have accelerated slope instability processes in this region.

Short-term intense precipitation in France is responsible for shallow translational landslides, whereas long-term changes in precipitation have affected and caused deep-seated landslides due to the subsurface flow of rainwater between the weathered surface material and unweathered rocks beneath.

3.2.6 SWITZERLAND AND AUSTRIA

Switzerland is a small country located in the Alps, which is landslide-prone area in the world. Debris flows, slides, rockfalls, and snow avalanches are the principal slope failure types in Switzerland's Alps. The geological setting of Switzerland resulted from the collision of African and European plates. About 57% of the total geographical area is covered by the Alps, 30% by the Swiss plateau, and 13% by the Jura.

Weathering, combined with increased water pressure in fault systems, is the primary landslide preparatory factor in this region, and rainstorm and seismicity are important landslide triggering mechanisms. Rapid permafrost degradation is an important triggering factor in Switzerland. About 29% of the total recorded landslides in Switzerland's Alps occurred during March–May, 36% in summer (June–August), 15% in autumn (September–November), and 20% in winter (December–February). Large catastrophic landslides in the Swiss Alps occur due to a reduction in the snow volume of retreating glaciers.

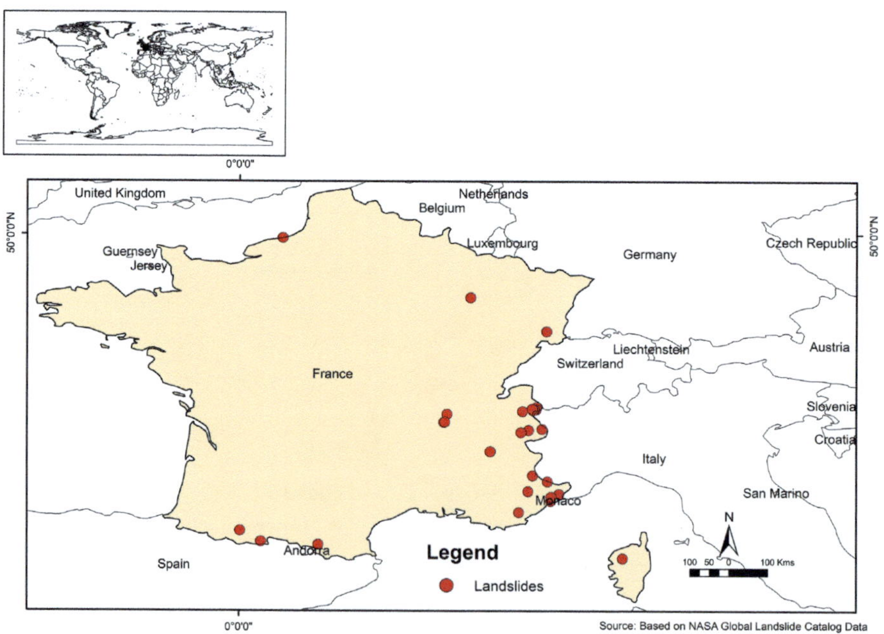

FIGURE 3.6 Landslide-Affected Areas in France.

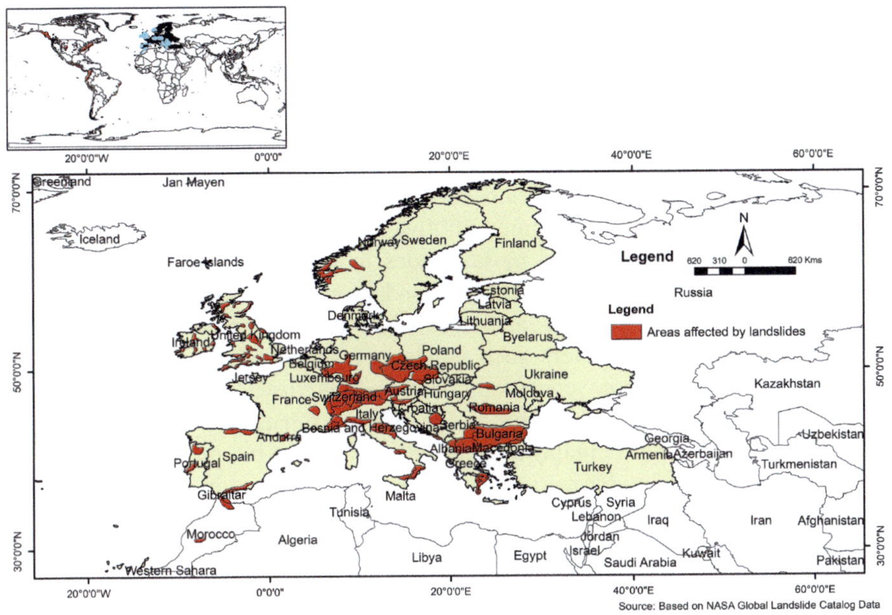

FIGURE 3.7 Landslide-Affected Areas in Western Europe.

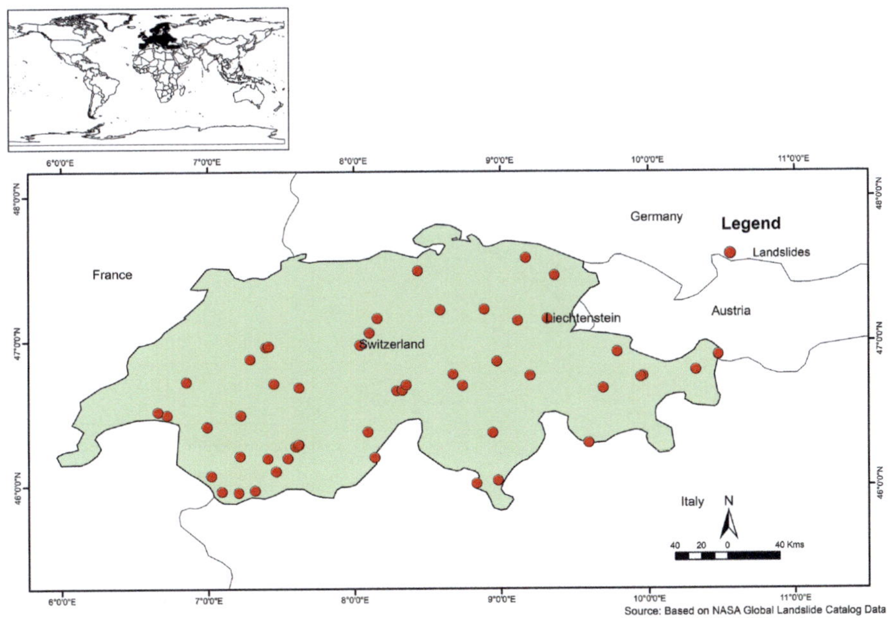

FIGURE 3.8 Landslide-Affected Areas in Switzerland.

3.2.7 HIMALAYAN REGIONS OF INDIA

About 12.6% (0.42 million km²) of the total land area in India is prone to landslide hazards. The Himalayan mountains and Western Ghats are important landslide-prone areas in India. Landslides in the Eastern Himalayan terrain are triggered by earthquakes, whereas those occurring in the Western Ghats are triggered by intense monsoonal rainfall.

Landslides in India have claimed 517 lives in just five years between 2007 and 2011. Economic losses caused by landslides in India are often underestimated because a landslide is considered as a secondary natural disaster.

Besides rainfall and earthquakes, landslides in India are also caused by different preparatory parameters such as slope gradient, slope aspect, lithology, structural discontinuities, vegetation cover, and land use. However, the morphological characteristics of landslides in the Western Ghats and Himalayan terrain differ from each other. Himalayan landslides are caused mainly due to seismicity and slope undercutting by natural drainage, whereas those in the Western Ghat escarpment are triggered by high-intensity rainfall combined with slope deformation due to human activities such as construction of roads and railways along hillsides.

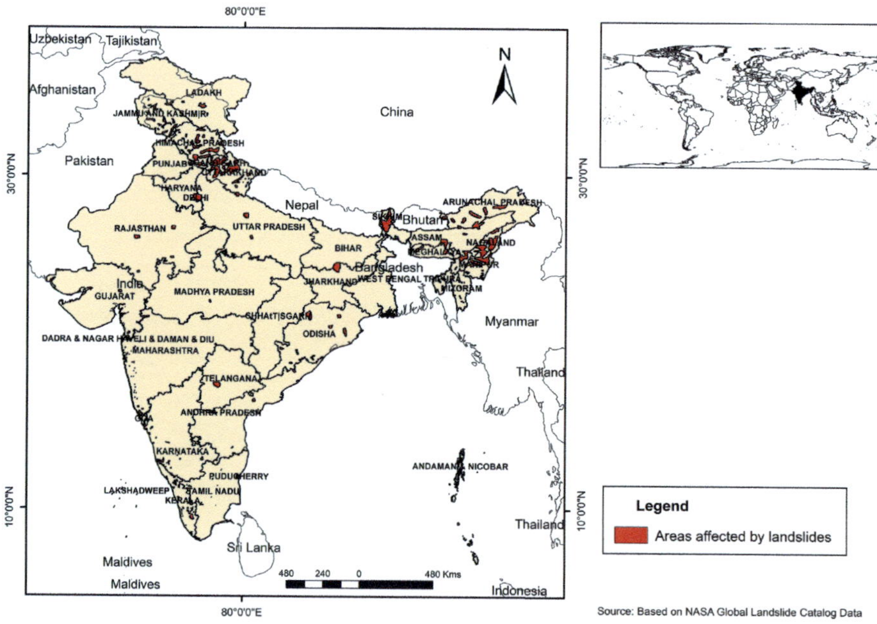

FIGURE 3.9 Landslide-Affected Areas in India.

3.2.8 Nepal Himalayas

Nepal is the country with highest relative relief on the earth.

3.2.9 China

China is facing landslide hazards in many parts, particularly in Western and Southern China. Several landslide events in China cause disruptions in society in terms of traffic interruption and clogging of river channels, mines, villages, and towns. The western region of China is characterised by extensive landslides with tremendous socioeconomic impacts. The geomorphic characteristics of China are favourable for the initiation of landslides and related mass movements. Debris flow is a common type of slope failure in China, particularly around the hilly Qinghai-Tibet plateau.

Southern areas of China and the upper reach of the Yangtze River witness the most frequent landslides, whereas the north-western and north-eastern regions of China have relatively few landslides. The majority of landsides in China are triggered by intense rainfall events.

Predominant landslide events in China are distributed in Sichuan, Yunan, Guizhou, South-East Tibet and Gansu, the middle and southern parts of Shaanxi,

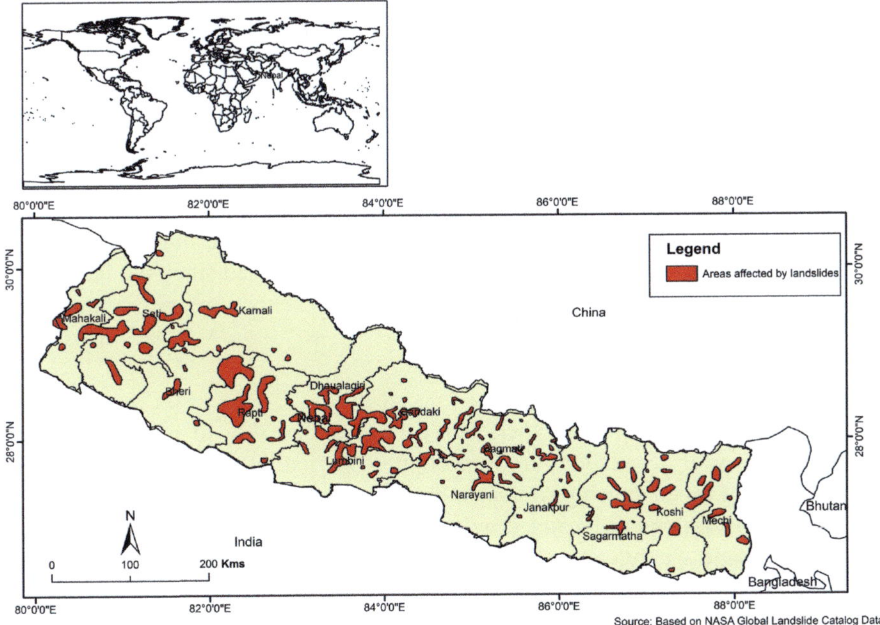

FIGURE 3.10 Landslide-Affected Areas in Nepal.

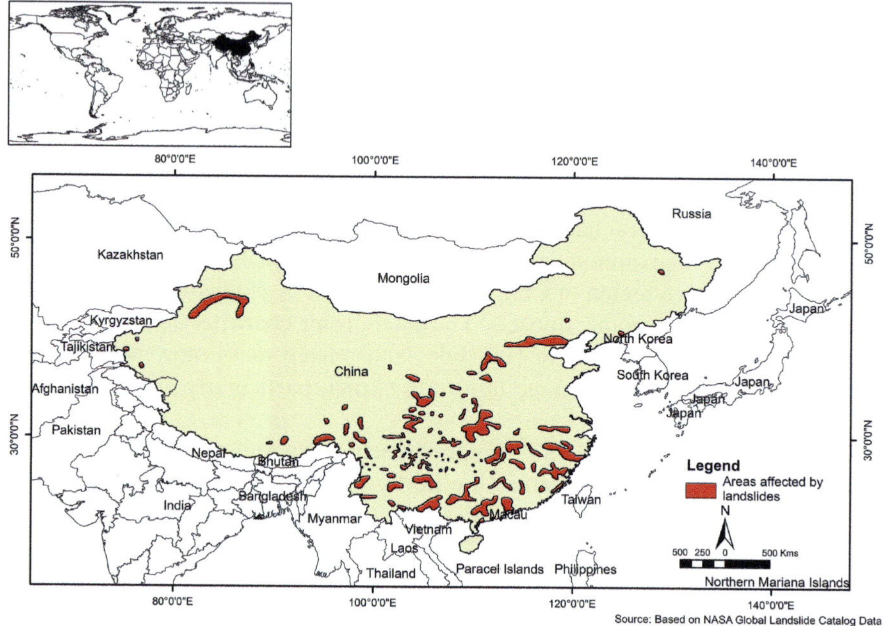

FIGURE 3.11 Landslide-Affected Areas in China.

and southern Ningxia. Landslides are sparsely distributed in southern and eastern Chinese provinces. Besides intense rainfall events, neo-tectonic uplifting of the loess plateau region of North China, combined with frequent earthquakes of high magnitude and intensity, are also proven to be major landslide triggering factors.

Physiography, geological settings, climate, tectonism, and earthquakes are important landslide preparatory and triggering factors in China. The trends of landslide occurrence from 1950 to 1998 (Li, 2022) indicated that the frequency of landslide occurrence in China has significantly increased since the late 1970s. In the 21st century, the highest number of landslide events were reported in 2008.

3.2.10 JAPAN

Japan is an Asian country experiencing huge losses due to natural disasters almost every year. Landslides are important geological hazards in the mountainous regions of Japan causing socioeconomic losses and sometimes fatalities too.

High-erosion potential, caused by mountain-building processes, frequent and high-magnitude earthquakes, and abundance of precipitation (average annual precipitation of 1,750 mm) in Japan are the major triggering factors responsible for slope instability processes in Japan. Hokkaido, Iwata, Niigata, Saka, Fukuoka, and Kumamoto are the important landslide-prone areas in Japan.

According to Yoshimatsu and Abe (2006), Japan is spread over a total geographical area of 3,70,000 km², of which 70% is mountainous. Although Japan

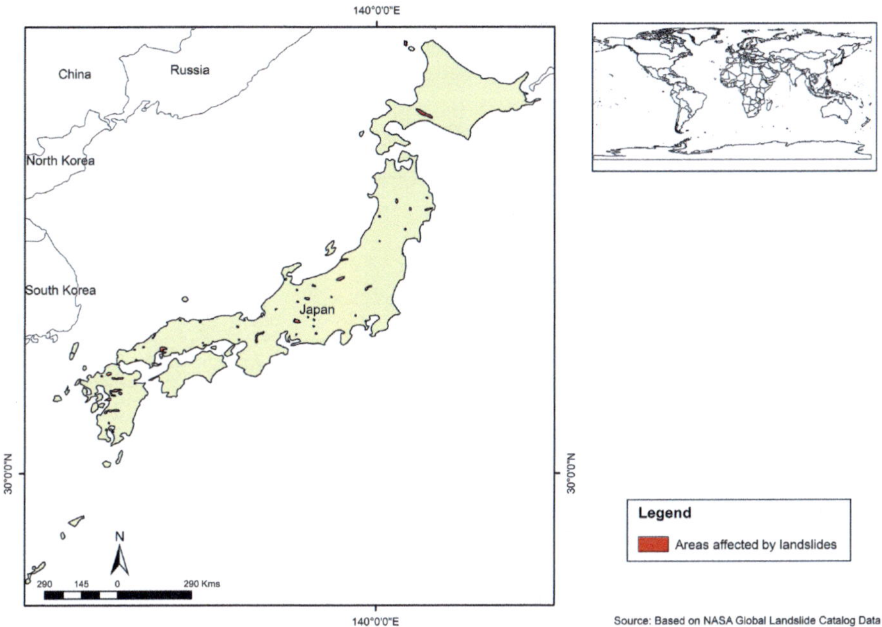

FIGURE 3.12 Landslide-Affected Areas in Japan.

is a small country (0.1% of the earth's total land area), it accounts for 10% of the world's active volcanoes and 10% of the world's earthquakes.

Lithologically, granite and metamorphic rocks of the Palaeozoic to Mesozoic era are the dominant rock types in Japan. Landslides in Japan are attributed to a slope gradient from 5° to 6°. The entire country of Japan is under the influence of tectonic forces, where landslides tend to occur in fault or fracture zones.

There are several slope failure events occurring on slopes consisting of pyroclastic material and weathered granite rocks. Such slopes have been reported to fail during typhoons and heavy rainstorms during the rainy season. Sudden collapse of slope-forming material normally causes serious damage to human settlements, whereas slow-moving landslides cause comparatively less damage.

3.2.11 INDONESIA

Indonesia is a South Asian country having diverse physiographic settings. Landslides are a common phenomenon in Indonesia, particularly during the monsoon season. Heavy rainfall and human interventions are important landslide triggering factors in Indonesia. Like many other countries in the world, landslide is an important natural hazard in Indonesia, and it has one of the highest number of landslide-related fatalities in the world.

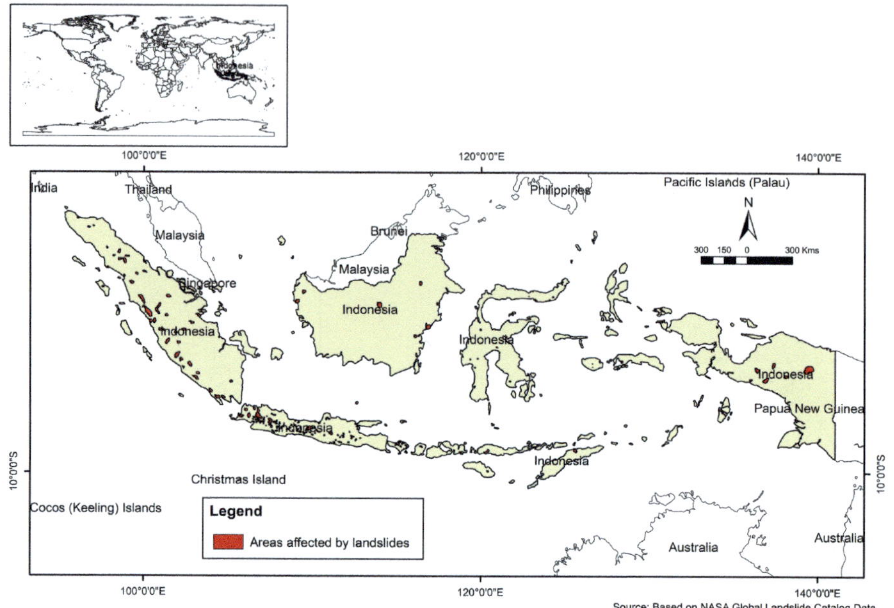

FIGURE 3.13   Landslide-Affected Areas in Indonesia.

Indonesia is the fourth-largest populated country in the world with a high population density. A significantly large population of Indonesia is forced to be inhabit areas prone to natural disasters, including landslides. It is probably the reason why landslide fatalities in Indonesia are more as compared to other countries in Asia.

Sumatra, Java, Kalimantan, Sulawesi, and Papua islands are the important landslide-prone areas in Indonesia. Indonesia receives heavy monsoonal rainfall, ranging from 1,000 to 1,500 mm annually. The annual rainfall increases progressively from the south-east parts of Indonesia to south-west regions of the country. High-intensity concentrated rainfall causes numerous landslides in the hilly terrain of Indonesia.

Mountain regions are observed in the central islands of Indonesia, including Sumatra, Java, Sulawesi, and Papua, where the slopes of volcanic cones are prone to landslides.

3.3 OVERVIEW OF LANDSLIDE HAZARD ASSESSMENT IN THE WORLD

The global patterns of landslide distribution suggest that different parts of the world are prone to slope failures of different intensities. North America, Central and South America, the European Alps, the Himalayan region of Asia, the undulated terrains of different regions, plate boundaries across the globe, and seismic zones of the Pacific are major landslide-prone areas in the world. Besides understanding the global patterns of landslide distribution, it is also important to understand the methodological developments in landslide hazard assessment in different parts of the world. For this, a detailed review of recently published literature in reputed journals, including *Landslides, Geomorphology, Natural Hazard and Earth System Sciences*, and *Engineering Geology*, has been done. The focus of the review of recent literature is to understand the recent trends in landslide hazard assessment: particularly, landslide inventory approaches, landslide susceptibility zonation, and landslide risk assessment. The brief discussion on methodological developments in landslide hazard assessment in recent times is given in subsequent sections of this chapter.

3.4 REVIEW OF RECENT STUDIES ON LANDSLIDE INVENTORIES

Building landslide inventory, including spatial and temporal aspects of landslide events, is an important task in landslide hazard assessment. Complete and reliable landslide inventories help in identifying actual and potential landslide-prone zones and in validating landslide susceptibility models. Over decades now, landslide inventories have been produced through generating a landslide database using past landslide records, news archives, filed investigations, etc.

With the developments in geospatial technologies and data-driven models for landslide hazard assessment, various methods and techniques have been developed to generate an accurate and reliable landslide database. Identification of landslides, geotechnical investigations, consideration of geometrical (morphological) parameters of landslide events, estimation of volume of displaced material, estimation of rate and type of movement, identification of pre-landslide signatures, details of losses caused by landslide events, landslide density and frequency, etc. are some of the important factors involved in landslide inventory.

The review of recent studies on landslide inventory reveals that high-resolution satellite data, such as LiDAR, SPOT, SRTM and ASTER DEM, DInSAR, Carto DEM, and Sentinel-1 and 2, are used for identification of landslides and landslide inventory. Global landslide databases are also being used as input data in many recent studies on landslide inventories. Unmanned aerial vehicles (UAVs) are proven tools for the detection and mapping of landslides almost in real-time conditions. Several recent studies have used different types of UAVs for mapping landslides in remote locations. UAVs have also made it possible to assess short-term dynamics in slow-moving landslides. Meena et al. (2022) have recently carried out a study on landslide inventory in the Nepal Himalayas, wherein automated landslide detection using a machine learning algorithm was used to detect landslide movements automatically. Object-based landslide detection algorithm using image segmentation has also been a proven method used by Li et al. (2022) to identify loess landslides in Northern China. Li et al. (2022) attempted automatic landslide profile analysis in parts of China. Automated web data mining for landslide detection and mapping, combined with geotagging of landslide news archives and content analysis, has been used recently in Italy by Franceschini et al. (2022).

Based on the review of recent studies on landslide inventory, it is evident that developments in geospatial technologies have made revolutionary changes in the detection and mapping of landslides. Besides landslide mapping, details of geo-technical and morphological characteristics of landslides are also being put to use for landslide inventories. A fully automated landslide inventory system would be the future of landslide inventory systems.

3.5 REVIEW OF RECENT STUDIES ON LANDSLIDE SUSCEPTIBILITY ZONATION

Landslide susceptibility zonation refers to the process of delineation of potential landslide-prone zones based on the information of landslide preparatory and triggering parameters (sometimes, the distribution of actual landslide events) for a given region. There have been a wide range of methodologies used for landslide susceptibility zonation across the globe during the past few decades. The subsequent paragraphs aim to discuss recent trends in landslide susceptibility zonation, particularly after 2020.

Like landslide inventory, there has been a significant development in methodological approaches to landslide susceptibility mapping in different parts of the world. The review of recent studies on landslide susceptibility zonation reveals that the accuracy of landslide susceptibility maps has been improved by using high-resolution geospatial information. Besides heuristic and bivariate statistical methods, data-driven models and sophisticated machine learning-based models are being used to predict landslides. Dynamic landslide hazard mapping, rainfall threshold models, semiautomatic debris flow models, distinct element method, convolutional neural network (CNN) model, deep learning method, natural language processing (NLP), fuzzy logic model, 3D rainfall threshold model, multiple surge model, entertainment model, general linear models, global precipitation model, finite element method, critical rainfall threshold model, etc. are few of the advanced landslide susceptibility models that have been used to delineate potential landslide-prone zones.

Methodological developments in landslide susceptibility zonation have led to significant changes in landslide hazard assessment. The use of machine learning and threshold modelling has improved the predictability of landslides much more effectively than other conventional landslide susceptibility methods. The availability of high-resolution and accurate spatial database and detailed landslide inventories has improved the predictability of landslide susceptibility zonation.

3.6 RECENT APPROACHES TO LANDSLIDE RISK ASSESSMENT

Landslide risk assessment is an important part of landslide hazard assessment. Landslide risk assessment helps in identifying potential zones where the probability of landslide occurrence is high and in determining whether these zones are in proximity to human habitation. Landslide risk may be associated with human fatalities and injuries, loss of property, disruption to transportation and communication routes, damage to agricultural crops and patterns of land use and land cover, etc. There have been a limited number of attempts made for evaluating the risk associated with landslides, particularly in the past few decades. Most of the studies were limited to qualitative landslide risk assessments. However, due to the availability of complete and accurate landslide database and socioeconomic datasets, it has become easy to quantify potential losses caused by landslides in a given region. A few of the recent studies on landslide risk assessment have been reviewed. A brief discussion on recent trends in landslide risk assessment is given in subsequent paragraphs of this section.

Ferlisi et al. (2021) have conducted quantitative landslide risk assessment in South Italy using high-resolution DInSAR data. The study came out with engineering measures to mitigate landslide hazards, such as sealing of cracks in the road pavements, asphalt concrete, spraying of bitumen on roadside slopes, and foundations of stabilised granular mixture. Delgado Garcia et al. (2022) used Bayesian information criterion (BIC) model for landslide risk and impact assessment using a public database. Strouth and McDougall (2021) conducted social risk assessment associated with landslides using FN criterion. Gorokhovich and Vustianiuk (2021) have conducted a study on the implication of slope and aspect on landslide risk caused by cyclone-triggered landslides in Puerto Rico and Ukraine.

This brief review of recent studies on landslide risk assessment reveals that more emphasis is being given to quantify risk associated with landslides. Besides fatalities and economic losses caused by landslides, attempts are also being made to evaluate the social risk associated with landslides. A multidisciplinary approach in evaluating landslide risk would give different dimensions to landslide risk assessment. Hence, there is a need to involve psychological, cultural, and other social dimensions in landslide risk assessment, and attempts can also be made to quantify such qualitative attributes for better understanding of the impacts caused by landslides. Accurate and reliable landslide risk maps are useful for planners and administrators in prioritising landslide mitigation strategies.

3.7 USE OF GEO-INFORMATION IN LANDSLIDE INVESTIGATIONS

The advancements in geospatial technologies have opened the doors for detailed and accurate assessment of landslide hazards. Particularly, the availability of high-resolution data for landslide hazard assessment has increased the predictability of landslide susceptibility models. Hence, used of geospatial technologies in landslide hazard assessment has been proved to be powerful tools for accurate and reliable landslide prediction models based on complex spatial and mathematical analyses.

Remote sensing and GIS technologies are now being widely used in hazard assessment and developments of different spatial data models are helping to increase the accuracy of landslide prediction map for appropriate mitigation measures. The advancement in 'Earth observation' (EO) techniques facilitates effective landslide detection, mapping, monitoring and hazard analysis (Tofani et al., 2013). The review of a few studies on landslide hazard assessment using RS data indicates that aerial photographs are widely used in landslide detection and mapping. Good-quality aerial photographs help in accurate landslide detection and mapping. However, aerial photographs may not be used in continuous landslide monitoring, since they do not provide repetitive coverage of the same area. In recent times, UAVs have been used to obtain need-based high-resolution spatial information, which is particularly suitable for small geographical areas.

Developments in the application of geospatial technologies in landslide studies in Europe has been discussed by Tofani et al. (2013). The study showed that over 70% of the total applications of RS data for landslide studies are associated with landslide detection, mapping, and monitoring. High-resolution satellite data are being effectively used for landslide detection, mapping, monitoring, and other applications.

The use of high-resolution digital elevation models (DEM) is of immense importance in landslide hazard assessment. Several thematic data layers, such as slope angle, slope aspect, curvature, lineaments, drainage, and ridges can be extracted from DEM with good resolution. Landslide hazard zonation studies in recent times have used DEM with high resolution to generate spatial information data layers related to landslide hazards.

GIS is widely used in landslide hazard assessment, especially for the generation of thematic data layers, computation of different indices, assignment of weights, data integration, and generation of landslide susceptibility zonation (LSZ) maps. Several LSZ methods, such as artificial neural networks (ANN), decision tree model, weighted overlay, AHP, MCDA, IVM, and physical process-based landslide hazard models are GIS-based models to predict landslide probability.

3.8 SUMMARY

Landslides are an important class of geological hazard that cause numerous losses in terms of lives and property. Almost all countries across the globe are facing the impacts of landslides, although at different intensities. The available global landslide databases are instrumental in delineating the actual distribution of landslides. Global patterns of landslide suggest that mountainous terrains; plate boundaries; hilly regions; and slopes characterised by loose, unconsolidated slope-forming materials are susceptible to different types of slope failures.

There has been a significant development in landslide hazard assessment, particularly, in terms of input parameters, high-resolution geospatial data, and data-driven methodologies. It is evident that the predictability of landslide susceptibility models has increased due to the use of data-driven approaches for landslide hazard assessment. However, there is a need to bring a multidisciplinary approach to landslides hazard assessment to expand the horizon of landslide investigations in view of its application for landslide mitigation measures.

TABLE 3.1

Reference Matrix of Landslide Causative Factors Considered for Landslide Hazard Zonation

Causative Factors (Layers) Considered for Landslide Hazard Assessment

S. No.	Reference	Slope	Rainfall	Structure	Lithology	Soil	Land Cover	Land Use	Hydrological Conditions	Drainage Network	Landslides	Relief	Seismicity	Geomorphology	Other Factors
1	(Rowbotham and Dudycha, 1998)	●		●	●	●		●		●					
2	(Nagarajan et al., 2000)	●	●	●			●			●		●			
3	Dai and Lee (2001)			●							●				
4	Sarkar et al. (2006)	●		●	●	●		●		●	●				
5	Naithani (2007)	●		●	●		●	●	●	●		●			
6	Anbalagan et al. (2008)	●		●	●	●	●	●	●	●		●			
7	Pradhan and Lee (2010)	●	●	●	●	●	●	●		●					NDVI
8	Jaiswal et al. (2010)	●	●	●				●			●				
9	Dahl et al. (2010)			●		●									Landslide Geometry
10	Yeon et al. (2010)	●				●	●				●				
11	Akbar and Ha (2011)	●			●	●	●	●		●					
12	Rotigliano et al. (2011)	●			●				●						
13	Sarkar et al. (2006)	●		●				●		●	●		●	●	Neotectonics
14	Panikkar and Subramanyan (1997)	●				●				●	●	●			

(Continued)

TABLE 3.1 (*Continued*)

Reference Matrix of Landslide Causative Factors Considered for Landslide Hazard Zonation

S. No.	Reference	Causative Factors (Layers) Considered for Landslide Hazard Assessment													
		Slope	Rainfall	Structure	Lithology	Soil	Land Cover	Land Use	Hydrological Conditions	Drainage Network	Landslides	Relief	Seismicity	Geomorphology	Other Factors
15	Chiang et al.		•								•			•	
16	Ercanoglu (2005)	•			•				•			•		•	NDVI
17	Chau et al. (2004)	•	•		•	•					•	•		•	
18	Clerici et al. (2002)	•	•		•		•	•							
19	Balsubramani and Kumaraswamy (2013)	•		•	•		•	•			•				
20	Ghosh et al. (2009)	•		•	•		•	•	•			•			
21	Wang and Sassa (2005)	•			•	•	•	•			•			•	
22	Coe et al. (2004)	•										•			Platform Curvature
23	Ayalew et al. (2005)	•			•						•	•			
24	Ercanoglu et al. (2004)	•	•		•	•	•	•	•			•			Weathering
25	Floris and Bozzano (2008)		•		•						•				
26	García Rodríguez et al. (2008)	•	•		•		•	•				•		•	
27	Calvello et al. (2013)	•		•	•					•	•			•	
28	Mantovani et al. (1996)	•	•	•	•		•	•			•	•	•	•	
29	Ghosh (2011)	•		•	•	•	•	•		•				•	
30	Ohlmacher and Davis (2003)	•			•	•					•				

	Reference															
31	Guzzetti et al. (2006)															Plan Curvature
32	Haken et al. (2008)															
33	Das et al. (2011)															
34	Jaiswal and van Westen (2013)															
35	Chang et al. (2007)															Road Buffer
36	Mondal and Maiti (2012)															
37	Sharma et al. (2009)															
38	Parise (2002)															
39	Jelínek et al. (2007)															
40	Ray (2000)															
41	Liu and Chang (2005)															
42	Das et al. (2011)															Weathering
43	Champatiray (2007)															
44	Mathew (2008)															
45	Kanungo (2006)															
46	Arrora (2004)															
47	Lee (2004)															
48	Mhaskare (2008)															
49	Sarkar et al. (2006)															
50	Catani et al. (2005)															
51	Akgun (2011)															
52	Mancini et al. (2010)															Platform Curvature
53	Guzzetti et al. (2006)															
54	Erener et al. (2010)															
55	Eeckhaut et al. (2009)															
56	Zêzere (2002)															

(Continued)

TABLE 3.1   (Continued)
Reference Matrix of Landslide Causative Factors Considered for Landslide Hazard Zonation

Causative Factors (Layers) Considered for Landslide Hazard Assessment

S. No.	Reference	Slope	Rainfall	Structure	Lithology	Soil	Land Cover	Land Use	Hydrological Conditions	Drainage Network	Landslides	Relief	Seismicity	Geomorphology	Other Factors
57	Ayalew and Yamagishi (2005)	●		●	●		●				●	●			
58	Jaboyedoff et al. (2005)										●				
59	Kannan et al. (2011)	●		●	●	●	●		●			●			
60	Guzzetti et al. (1999)	●	●		●						●	●	●		
61	García Rodríguez et al. (2008)	●	●		●		●				●	●	●	●	
62	Das (2011)	●			●	●	●			●	●				Weathering
63	Atkinson and Massari (1998)	●		●	●	●		●			●	●			
64	Pradhan and Lee (2009)	●		●	●	●				●	●	●			NDVI
65	Greco et al. (2007)	●					●					●			Rock Type
66	Chang and Chiang (2009)		●								●				
67	Polemio and Sdao (1999)		●		●						●				
68	Chleborad et al. (2006)		●								●				
69	Tolga et al. (2005)	●	●				●	●		●					
70	Gabet et al. (2004)	●	●									●			
71	Lee (2005)	●		●	●		●	●			●				NDVI
72	Mantovani et al. (1996)	●	●		●		●	●	●		●				

No.	Study	
73	Ruff and Czurda (2008)	Soil Erosion and Neotectonics
74	Bălteanu et al. (2010)	
75	Martha et al. (2013)	
76	Conoscenti et al. (2008)	WI & Stream Power Index
77	Dahal and Hasegawa (2008)	
78	Preuth et al. (2010)	
79	Blahut et al. (2010)	
80	Sterlacchini et al. (2011)	
81	Atkinson and Massari (2011)	
82	Goswami et al. (2011)	
83	Saez et al. (2012)	
84	Xu et al. (2012)	
85	Listo and Vieira (2012)	
86	Piacentini et al. (2012)	Platform Curvature
87	Schicker and Moon (2012)	
88	Bui et al. (2012)	
89	Das et al. (2012)	Weathering
90	Neuhausev et al. (2012)	
91	Mercogliano et al. (2013)	
92	Gunther and Thiel (2009)	
93	Meusburger and Alewell (2009)	Wetness Index
94	Montrasio et al. (2011)	
95	Pereira et al. (2012)	Wetness Index

(Continued)

TABLE 3.1   (*Continued*)

Reference Matrix of Landslide Causative Factors Considered for Landslide Hazard Zonation

Causative Factors (Layers) Considered for Landslide Hazard Assessment

S. No.	Reference	Slope	Rainfall	Structure	Lithology	Soil	Land Cover	Land Use	Hydrological Conditions	Drainage Network	Landslides	Relief	Seismicity	Geomorphology	Other Factors
96	Cardinali et al. (2002)										•				LS Frequency and Intensity
97	Lee et al. (2008)	•	•	•	•					•	•	•			
98	Lee et al. (2010)	•			•			•			•	•	•		
99	Salciarini et al. (2006)	•	•		•	•					•				
100	Lee and Pradhan (2006)	•	•	•	•	•	•	•		•					NDVI
101	Kuriakose (2010)	•				•	•	•			•				Friction Angle
102	Chao Zhou et al. (2022)	•		•	•					•		•			Slope Aspect, Topographical Relief Index (TRI)
103	Che Ming et al. (2022)	•			•						•	•		•	
104	Chi wen Chen (2021)		•			•					•				Soil Wetness Index (SWI)
105	Dewttee et al. (2020)	•	•								•	•	•	•	
106	Fawy Wang (2022)	•	•	•	•						•	•		•	
107	Ferlisi et al. (2021)	•	•		•			•			•				

No.	Reference	Notes
108	Franceschini et al. (2022)	Geotagged News and Content Analysis
109	Francesco et al. (2022)	TWI
110	Ciccareoe et al. (2021)	Flow Accumulation and TWI
111	Ghorbanzadeh et al. (2022)	NDVI
112	Haken et al. (2022)	
113	He et al. (2022)	
114	Helbert et al. (2022)	
115	Hughes et al. (2020)	ENSO
116	Jain et al. (2021)	
117	Picarelli et al. (2022)	
118	Lee et al. (2020)	TWI
119	Li et al. (2022)	
120	Li et al. (2022)	
121	Lima et al. (2021)	TWI
122	Liu et al. (2021)	
123	Mahallem et al. (2022)	
124	Marino et al. (2020)	
125	Ng et al. (2021)	
126	Nguyen and Kim (2021)	
127	Paulin et al. (2022)	
128	Peres-Cancelliere et al. (2021)	
129	Perrone et al. (2020)	
130	Xu et al. (2022b)	

(Continued)

TABLE 3.1   (*Continued*)

Reference Matrix of Landslide Causative Factors Considered for Landslide Hazard Zonation

		Causative Factors (Layers) Considered for Landslide Hazard Assessment												
S. No.	Reference	Slope	Rainfall	Structure	Lithology	Soil	Land Cover	Land Use	Hydrological Conditions	Drainage Network	Landslides	Relief	Seismicity	Geomorphology	Other Factors
131	Rodrigues et al. (2021)	•									•				
132	Rosa et al. (2022)	•	•	•	•						•				
133	Rosi et al. (2020)	•	•	•							•	•			
134	Roy et al. (2022)	•	•								•	•			
135	Sharma et al. (2021)	•	•	•							•	•			
136	Song et al. (2022)	•	•	•	•						•				
137	Strouth et al. (2020)	•	•	•	•						•		•		
138	Ilinca et al. (2022)	•	•								•	•	•		
139	Vittoria et al. (2022)	•	•	•	•	•	•	•			•	•			
140	Wallace et al. (2022)			•											
141	Wang et al. (2022)	•	•	•	•	•	•	•			•	•		•	
142	Wei et al. (2022)	•									•				
143	Xin et al. (2022)	•									•	•			
144	Haken et al. (2021)	•	•	•	•	•	•	•			•				
145	Jia et al. (2021)	•	•								•	•	•		
146	Zhang et al. (2021)	•									•				
147	Tammaso Baggio et al. (2021)	•									•				

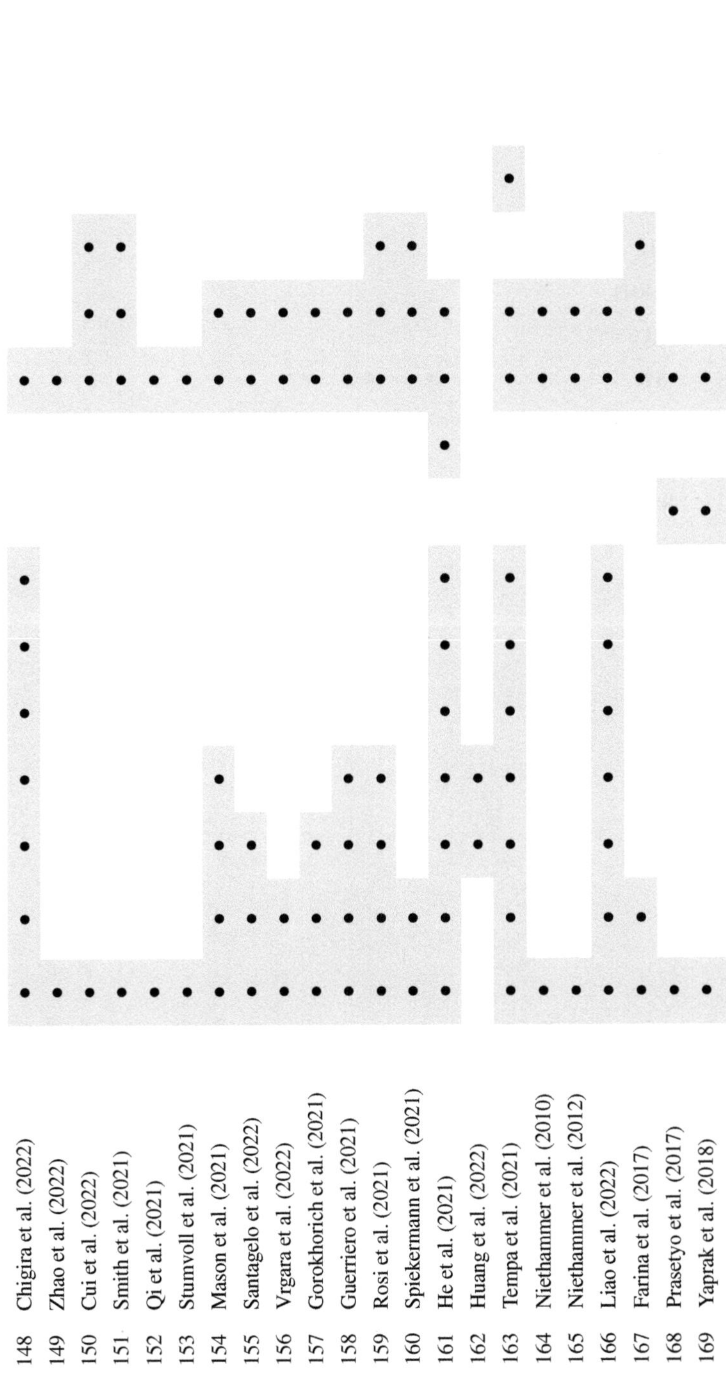

148 Chigira et al. (2022)
149 Zhao et al. (2022)
150 Cui et al. (2022)
151 Smith et al. (2021)
152 Qi et al. (2021)
153 Stumvoll et al. (2021)
154 Mason et al. (2021)
155 Santagelo et al. (2022)
156 Vrgara et al. (2022)
157 Gorokhorich et al. (2021)
158 Guerriero et al. (2021)
159 Rosi et al. (2021)
160 Spiekermann et al. (2021)
161 He et al. (2021)
162 Huang et al. (2022)
163 Tempa et al. (2021)
164 Niethammer et al. (2010)
165 Niethammer et al. (2012)
166 Liao et al. (2022)
167 Farina et al. (2017)
168 Prasetyo et al. (2017)
169 Yaprak et al. (2018)

TABLE 3.2
Reference Matrix to Compare Methods Used for Landslide Hazard Zonation

S. No.	Reference	Study Area	Mapping Unit	LHZ Methods														Other
				BIS	WO	AOR	FR	IVM	WoE	ANN	Prob	LR	MULTI-STAT	THRESLD	AHP	DA	FUZZY	
1	Mantovani et al. (1996)	Europe	Pixel															Deterministic approach
2	Panikkar and Subramaniyan (1997)	Mussourie, India	TMU		●													
3	Rowbotham and Dudycha (1998)	Hong Kong	TMU									●						
4	Atkinson and Massari (1998)	Apennine, Italy	Pixel									●						
5	Guzzetti et al. (1999)	Umbria, C. Italy	Pixel									●				●		
6	Polemio and Sdao (1999)	Apennine, Italy	Pixel								●							
7	Nagarajan et al. (2000)	Goa Highway, India	Pixel		●													
8	Ray (2000)	Himalayas, Garhwal	Pixel					●										
9	Dai and Lee (2001)	Nepal	Pixel		●													
10	Clerici et al. (2002)	Parma basin, Italy	Pixel										●					
11	Parise (2002)	Sele basin, Italy	Slope facet		●													
12	Zêzere (2002)	Lisbon, Portugal	Pixel					●										
13	Cardinali et al. (2002)	Umbria, C. Italy	Pixel				●											

No.	Reference	Location	Mapping unit	Remarks
14	Ercanoglu et al. (2003)	Yenice, Turkey	Pixel	
15	Chau et al. (2004)	Hong Kong	Pixel	
16	Coe et al. (2004)	USA	Pixel	
17	Arrora (2004)	Ganga valley, India	Pixel	
18	Lee (2004)	Korea	Pixel	
19	Gabet et al. (2004)	Khudi basin, Nepal Himalayas	Pixel	
20	Chiang et al.	Haitang, Taiwan	Pixel	LOGIT model
21	Ercanoglu (2005)	Bartin, Turkey	Pixel	
22	Wang and Sassa (2005)	Minamata, Japan	Pixel	
23	Ayalew et al. (2005)	Sado Island, Japan	Pixel	
24	Guzzetti et al. (2005)	Staffora basin, Italy	Pixel	
25	Liu (2005)	Japan	Pixel	
26	Catani et al. (2005)	Arno basin, C. Italy	Pixel	
27	Ayalew and Yamagishi (2005)	Central Japan	Pixel	
28	Jaboyedoff et al. (2005)	Switzerland	Pixel	SWISS CODE
29	Tolga et al. (2005)	Black Sea region, Turkey	UCU	
30	Lee (2005)	Penang Island, Malaysia	Pixel	
31	Sarkar et al. (2006)	Sikkim, India	Pixel	
32	Kanungo (2006)	Himalayas, India	Pixel	
33	Sarkar et al. (2006)	Sikkim, India	Pixel	
34	Guzzetti et al. (2006)	Umbria, C. Italy	Slope facet	
35	Chleborad et al. (2006)	Washington, USA	Pixel	
36	Salciarini et al. (2006)	Umbria, C. Italy	Pixel	TRIGRS
37	Lee and Pradhan (2006)	Penang Island, Malaysia	Pixel	

(Continued)

TABLE 3.2   (*Continued*)
Reference Matrix to Compare Methods Used for Landslide Hazard Zonation

S. No.	Reference	Study Area	Mapping Unit	LHZ Methods														Other
				BIS	WO	AOR	FR	IVM	WoE	ANN	Prob	LR	MULTI-STAT	THRESLD	AHP	DA	FUZZY	
38	Naithani (2007)	Garhwal Himalayas	Slope facet	●														
39	Chang et al. (2007)	Hoshe basin, Taiwan	Pixel									●						
40	Jelínek et al. (2007)	Slovakia, Russia	Pixel															Deterministic approach
41	Ray (2007)	Himalayas, India	Pixel	●													●	
42	Greco et al. (2007)	Calebria, Italy	Pixel									●						
43	Anbalagan et al. (2008)	Nainital, India	Slope facet	●														
44	Floris and Bozzano (2008)	Apennine, Italy	Pixel								●			●				
45	García-Rodríguez et al. (2008)	Salvadev, C. America	Pixel		●							●						
46	Haken et al. (2008)	Kelkit valley, Turkey	Pixel									●						
47	Mathew (2008)	Darjeeling, India	Pixel					●										
48	Mhaskare (2008)	Sahyadris, India	Pixel												●			
49	García-Rodríguez et al. (2008)	El Salvador, C. America	Pixel									●						
50	Ruff and Czurda (2008)	Eastern Alps, Austria	Pixel															Landslide susceptibility index
51	Conoscenti et al. (2008)	NW Sicily, Italy	UCU									●						

				Remarks
52	Dahal and Hasegawa (2008)	Nepal Himalayas, Nepal	Pixel	
53	Lee et al. (2008)	Toraji, CW Taiwan	Pixel	Event-based landslide susceptibility analysis
54	Ghosh et al. (2009)	Darjeeling, India	Slope facet	
55	Ohlmacher and Davis (2009)	Kansas, USA	Pixel	
56	Sharma et al. (2009)	Sikkim, India	Slope facet	
57	Eeckhaut et al. (2009)	Belgium	Slope facet	
58	Pradhan and Lee (2009)	Penang Island, Malaysia	Pixel	
59	Chang and Chiang (2009)	Baichi basin, Taiwan	Pixel	
60	Gunther and Thiel (2009)	Rugen Island, Germany	Pixel	Factor of Safty model, combined LSI
61	Meusburger and Alewell (2009)	Sub-alpine area	Pixel	
62	Pradhan and Lee (2010)	Malaysia	Pixel	
63	Jaiswal et al. (2010)	Nilgiris, India	Pixel	
64	Dahl et al. (2010)	Faeroe Islands	Slope facet	
65	Yeon et al. (2010)	Korea	Pixel	Decision tree
66	Das et al. (2010)	Himalayas, India	Pixel	
67	Mancini et al. (2010)	Daucnia, Italy	Pixel	
68	Erener et al. (2010)	Norway	Pixel	
69	Bălteanu et al. (2010)	Romania	Pixel	Landslide susceptibility index
70	Preuth et al. (2010)	East Belgium, Belgium	Pixel	
71	Blahut et al. (2010)	Lombardy region, C. Italy	Pixel	

(Continued)

TABLE 3.2   (*Continued*)

Reference Matrix to Compare Methods Used for Landslide Hazard Zonation

				LHZ Methods														
S. No.	Reference	Study Area	Mapping Unit	BIS	WO	AOR	FR	IVM	WoE	ANN	Prob	LR	MULTI-STAT	THRESLD	AHP	DA	FUZZY	Other
72	Lee et al. (2010)	Juo Fong region, C. Taiwan	Pixel									●						
73	Kuriakose (2010)	Western Ghats, Kerala, India	Pixel															SHALSTAB, SINMAP, TRIGRS, STARWAR + PROBSTAB
74	Akbar and Ha (2011)	Kaghan valley, Pakistan	Pixel					●										
75	Rotigliano et al. (2011)	Sicily, Italy	Pixel										●					
76	Ghosh (2011)	Darjeeling, India	Pixel								●	●						
77	Das et al. (2011)	North Himalayas	Pixel								●							HSU
78	Akgun (2011)	Izmir City (Turkey)	Pixel									●			●			
79	Kannan et al. (2011)	Bodimetta, Tamil Nadu, India	Slope facet	●														
80	Das (2011)	Garhwal Himalayas	Pixel									●						Physical process-based model
81	Sterlacchini et al. (2011)	Italian Alps, Italy	UCU						●									
82	Atkinson and Massari (2011)	Central Apennines, Italy	Pixel									●						Autologistic regression
83	Goswami et al. (2011)	Sicily	Pixel				●											

No.	Reference	Study area	Mapping unit	Method/model
84	Montrasio et al. (2011)	Emilian Alpine, North Italy	Pixel	TRIGRS
85	Mondal and Maiti (2012)	Darjeeling, India	Pixel	●
86	Saez et al. (2012)	Alps, France	Slope facet	Dendrogeomorphic analysis
87	Xu et al. (2012)	Jiangjiang watershed, China	Pixel	Support vector machine modelling
88	Listo and Vieira (2012)	São Paulo, Brazil	Pixel	SHALSTAB
89	Piacentini et al. (2012)	SE Alps, Italy	Pixel	●
90	Schicker and Moon (2012)	Waikoto region, New Zealand	Pixel	● ●
91	Bui et al. (2012)	Vietnam	Pixel	●
92	Das et al. (2012)	Bhagirathi basin, India	Pixel	Generalised linear modelling
93	Neuhausev et al. (2012)	Vienna Forest, Lower Austria	Pixel	●
94	Pereira et al. (2012)	Santa Marta de Penaguiao Council, Portugal	Pixel	●
95	Balsubramani and Kumaraswamy (2013)	Giri valley, Himachal Pradesh, India	Pixel	● ●
96	Calvello et al. (2013)	Tammaro basin, Italy	Pixel	●
97	Jaiswal and van Westen (2013)	Nilgiri Hills, India	Pixel	●
98	Martha et al. (2013)	NH 109, Rudraprayag, India	Pixel	●
99	Mercogliano et al. (2013)	Italy	Pixel	Global circulation model and HIRESSS model

TABLE 3.3

Recent Trends in Methodologies Adopted for Landslide Hazard Assessment

S. No.	Reference	Method Adopted	S. No.	Reference	Method Adopted	S. No.	Reference	Method Adopted
1	Chao Zhou et al. (2022)	Dynamic hazard mapping, support vector machine	24	Ng et al. (2021)	Machine learning for LS prediction	47	Chigira et al. (2022)	Landslide event–specific analysis
2	Che Ming et al. (2022)	UAV-based landslide event analysis	25	Nguyen and Kim (2021)	Convolutional neural network (CNN) model	48	Chen et al. (2022)	Thermal effect-based debris flow modelling
3	Chi wen Chen (2021)	Mass movement warning system	26	Paulin et al. (2022)	Landslide volume estimation	49	Zhao et al. (2022)	Rainfall threshold modelling
4	Dewttee et al. (2020)	Landslide inventory using historical data	27	Peres-Cancelliere et al. (2021)	Rainfall threshold, intensity duration modelling using MATLAB	50	Cui et al. (2022)	LS inventory using kernel density
5	Fawy Wang (2022)	Penetration test, Doppler array test	28	Perrone et al. (2020)	LS risk assessment using ERT (electrical resistivity tomography) and SRT (seismic refraction tomography)	51	Smith et al. (2021)	Object-based image analysis (comparison of LS data acquisition and susceptibility methods)
6	Ferlisi et al. (2021)	Quantitative risk assessment	29	Xu et al. (2022a)	Deep learning method for prediction of earthquake-induced LS using MFFENet (multi-scale feature fusion with encoder-decoder network and ADANET (adversarial domain adaptation network)	52	Qi et al. (2021)	Landslide inventory in fault zones, Tibet

No.	Reference	Description	No.	Reference	Description	No.	Reference	Description
7	Franceschini et al. (2022)	LS inventory using automatic data mining, sematic engine to classify geotagged news	30	Rodrigues et al. (2021)	LSZ using natural language processing (NLP) and machine learning (support vector machine)	53	Stumvoll et al. (2021)	Slow-moving landslide monitoring system
8	Francesco et al. (2022)	Quantitative risk assessment	31	Rosa et al. (2022)	Fuzzy verification framework	54	Mason et al. (2021)	Landslide event–specific analysis
9	Ciccareoe et al. (2021)	Rainfall threshold model of LSZ	32	Rosi et al. (2020)	3-D rainfall threshold model	55	Santagelo et al. (2022)	Landslide event–specific analysis
10	Ghorbanzadeh et al. (2022)	LS inventory of object-based image analysis	33	Roy et al. (2022)	Monitoring of reactivation of landslides	56	Vrgara et al. (2022)	General linear modelling for LS prediction
11	Haken et al. (2022)	LS inventory of earthquake-triggered landslides	34	Sharma et al. (2021)	Landslide inventory using news archives	57	Gorokhorich et al. (2021)	Cyclone-induced LS prediction using global precipitation model (GPM)
12	He et al. (2022)	Study of structural measures for landslide mitigation	35	Song et al. (2022)	Multiple surge load model for modelling debris flow impact	58	Guerriero et al. (2021)	Finite element stability analysis model
13	Helbert et al. (2022)	Bayesian information criterion (BIC) model for LS risk assessment using public database	36	Strouth et al. (2020)	Landslide social risk assessment – FN criteria	59	Rosi et al. (2021)	Linear regression model
14	Hughes et al. (2020)	Detection of paleo-landslides using bathymetric and sub-bottom data acquisition	37	Ilinca et al. (2022)	Landslide kinematics and change detection (structure form motion model)	60	Spiekermann et al. (2021)	Linear regression model for LSZ at global scale
15	Jain et al. (2021)	Semiautomatic debris flow model	38	Vittoria et al. (2022)	LaRiMiT web-based landslide risk mitigation tools	61	He et al. (2021)	Random forest model for LSZ

(Continued)

TABLE 3.3 (Continued)
Recent Trends in Methodologies Adopted for Landslide Hazard Assessment

S. No.	Reference	Method Adopted	S. No.	Reference	Method Adopted	S. No.	Reference	Method Adopted
16	Picarelli et al. (2022)	Historical LS deformations	39	Wallace et al. (2022)	Prediction of landslide runout distance	62	Huang et al. (2022)	Critical rainfall threshold model for LSZ
17	Lee et al. (2020)	LSZ using extreme value distribution, rainfall threshold, and machine learning	40	Wang et al. (2022)	Landslide event–specific analysis	63	Tempa et al. (2021)	Multi-influencing factor (Knowledge-driven), weighted overlay for LSZ
18	Li et al. (2022)	Object detection algorithm and image segmentation for loess landslide	41	Wei et al. (2022)	LSZ using frequency ratio (FR) and attention constrained neural network machine learning (ACNN)	64	Niethammer et al. (2010)	Landslide monitoring using radio-controlled UAVs
19	Li et al. (2022)	Automatic landslide profile analysis – LS geometry	42	Xin et al. (2022)	Image-based identification of recent landslides	65	Niethammer et al. (2012)	Landslide monitoring using radio-controlled UAVs
20	Lima et al. (2021)	Logistic regression, mixed effect modelling	43	Haken et al. (2021)	PGA model	66	Liao et al. (2022)	Real-time landslide monitoring
21	Liu et al. (2021)	Time series analysis, wavelet analysis	44	Jia et al. (2021)	Non-susceptibility model using cooperative open online landslide repository (COOLR)	67	Farina et al. (2017)	Use of multi-copter drones for landslide investigations
22	Mahallem et al. (2022)	Smooth particle hydrodynamics	45	Zhang et al. (2021)	LS stability analysis using grid-based physical model	68	Prasetyo et al. (2017)	3D modelling of landslides using digital surface model and AHP
23	Marino et al. (2020)	Hydrological monitoring using AWS	46	Tammaso Baggio et al. (2021)	Simulation of debris flow using entertainment model	69	Yaprak et al. (2018)	Monitoring rapidly occurring landslides using UAVs

TABLE 3.4

Reference Matrix to Compare Landslide Causative Factors Considered for LHZ in India

Causative Factors (Layers) Considered for Landslide Hazard Assessment

S. No.	Reference	Slope	Rainfall	Lineament	Lithology	Soil	Land Cover	Land Use	Hydrological Conditions	Drainage	Landslides	Relief	Seismicity	Tectonics	Geomorphology	Weathering	Other Factors
1	Ahmad et al. (2013)	●			●							●					
2	Anbalagan et al. (2008)	●		●	●		●	●	●			●					
3	Anbalagan and Parida (2013)	●	●		●	●											
4	Arrora (2004)	●		●	●		●	●		●	●				●		
5	Arunkumar et al. (2013)	●			●					●							
6	Avinash and Ashamanjari (2010)	●		●		●	●	●									
7	Balsubramani and Kumaraswamy (2013)	●		●	●		●	●			●						
8	Bobade et al. (2012)	●	●			●				●	●						
9	Bodas and Kohli (2009)	●		●		●	●	●	●	●	●				●	●	
10	Bodas and Kohli (2010)	●		●			●			●	●		●				
11	Chakraborty (2008)	●	●	●	●	●	●			●		●			●		
12	Champatiray et al. (2013)	●		●	●						●		●				
13	Champatiray et al. (2013)		●					●			●		●				
14	Champatiray et al. (2008)	●		●	●						●	●	●				
15	Champatiray et al. (2009)	●	●		●						●		●				
16	Chandel et al. (2011)	●			●		●	●		●		●					
17	Chauhan et al. (2010)	●		●	●		●	●		●		●					

(Continued)

TABLE 3.4   *(Continued)*

Reference Matrix to Compare Landslide Causative Factors Considered for LHZ in India

Causative Factors (Layers) Considered for Landslide Hazard Assessment

S. No.	Reference	Slope	Rainfall	Lineament	Lithology	Soil	Land Cover	Land Use	Hydrological Conditions	Drainage	Landslides	Relief	Seismicity	Tectonics	Geomorphology	Weathering	Other Factors
18	Das et al. (2011)			•	•						•						
19	Das et al. (2012)	•		•	•	•	•			•						•	
20	Ganapathy et al. (2010)	•	•		•						•				•		
21	Ghosh (2011)	•		•	•	•	•	•	•	•					•		
22	Ghosh et al. (2009)	•		•	•		•	•				•					
23	Jaiswal and van Westen (2013)		•				•	•			•						
24	Jaiswal et al. (2010)				•						•						LS geometry
25	Jaiswal et al. (2010)	•	•		•						•						
26	Jaiswal (2011)	•	•		•						•						
27	Kanungo (2006)	•		•	•		•	•							•		
28	Kanungo et al. (2008)	•		•	•		•	•		•							
29	Karlekar (2012)	•		•	•	•	•	•		•	•				•		Landslide geometry
30	Kumar et al. (2010)	•		•	•			•		•					•		
31	Kumar et al. (2008)			•	•			•	•	•	•				•		
32	Kuriakose (2010)	•				•	•	•			•						Friction angle
33	Lallianthanga et al. (2013)	•		•	•		•	•				•					

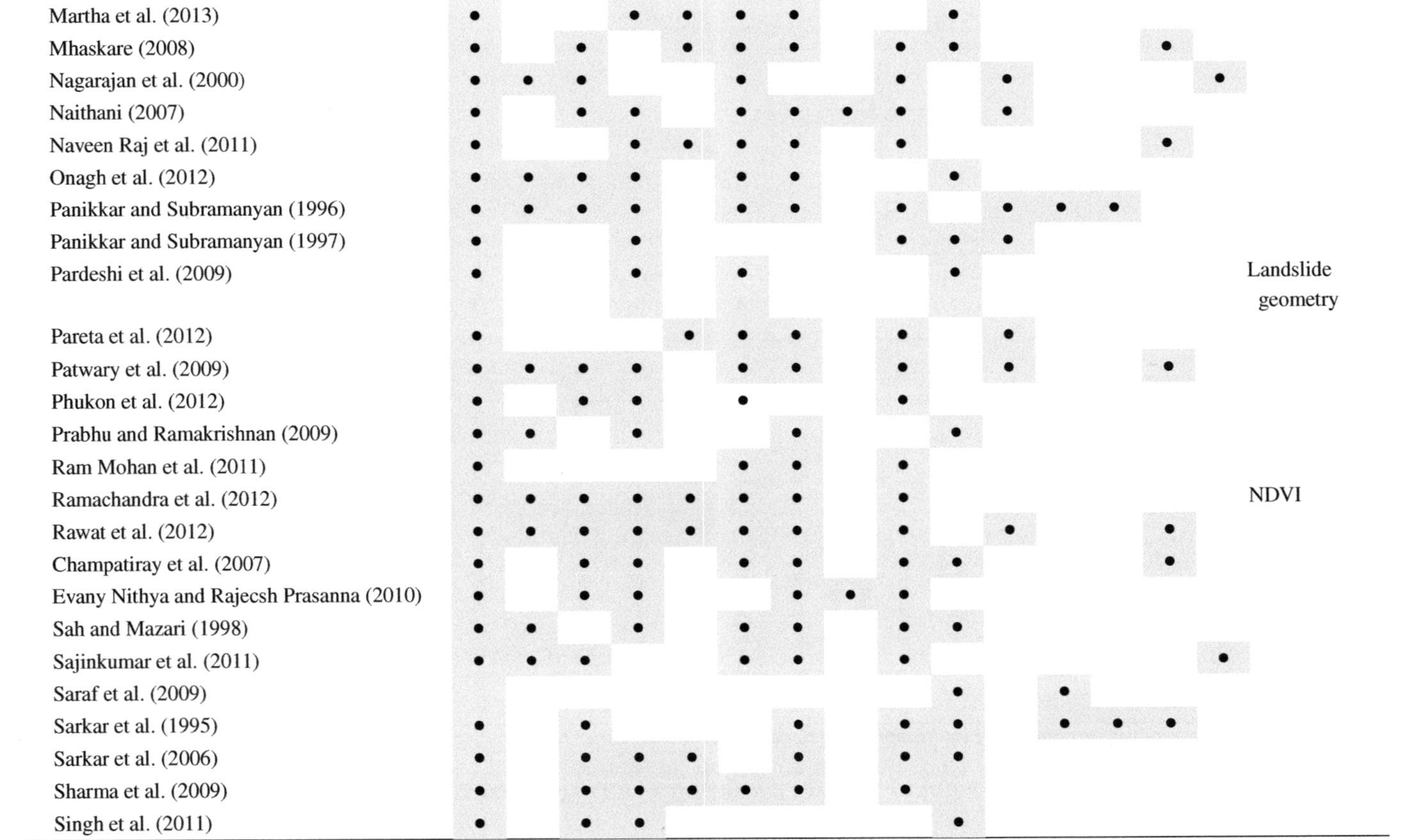

#	Reference
34	Martha et al. (2013)
35	Mhaskare (2008)
36	Nagarajan et al. (2000)
37	Naithani (2007)
38	Naveen Raj et al. (2011)
39	Onagh et al. (2012)
40	Panikkar and Subramanyan (1996)
41	Panikkar and Subramanyan (1997)
42	Pardeshi et al. (2009)
43	Pareta et al. (2012)
44	Patwary et al. (2009)
45	Phukon et al. (2012)
46	Prabhu and Ramakrishnan (2009)
47	Ram Mohan et al. (2011)
48	Ramachandra et al. (2012)
49	Rawat et al. (2012)
50	Champatiray et al. (2007)
51	Evany Nithya and Rajecsh Prasanna (2010)
52	Sah and Mazari (1998)
53	Sajinkumar et al. (2011)
54	Saraf et al. (2009)
55	Sarkar et al. (1995)
56	Sarkar et al. (2006)
57	Sharma et al. (2009)
58	Singh et al. (2011)

(Continued)

TABLE 3.4 (*Continued*)
Reference Matrix to Compare Landslide Causative Factors Considered for LHZ in India

S. No.	Reference	Causative Factors (Layers) Considered for Landslide Hazard Assessment															
		Slope	Rainfall	Lineament	Lithology	Soil	Land Cover	Land Use	Hydrological Conditions	Drainage	Landslides	Relief	Seismicity	Tectonics	Geomorphology	Weathering	Other Factors
59	Singh and Singh (2013)	●						●	●								
60	Singh and Champatiray (2009)	●			●						●						
61	Sriramkumar et al. (2006)	●			●	●	●	●		●					●		
62	Sujatha et al. (2012)	●			●	●	●	●		●		●			●	●	
63	Thigale and Umrikar (2007)	●	●	●				●		●	●		●	●			
64	Vijith et al. (2009)	●		●				●		●					●		Plan curvature
65	Kumar et al. (2010)	●		●	●			●		●					●		
66	Wagh and Deshpande (2013)	●		●	●		●	●									
67	Kannan (2011)	●		●		●		●	●			●					
68	Sarkar (2004)	●		●	●	●	●	●			●						
69	Bodas and Kohli (2012)	●					●	●		●					●	●	
70	De and Jamatia (2011)	●	●		●					●							
71	Basu and De (2003)	●	●		●					●	●						
72	Gomathi et al. (2013)	●				●		●		●	●				●		
73	Tikke et al. (2014)	●		●	●		●	●			●						
74	Despande et al. (2009)	●		●	●		●	●					●				
75	Das (2011)			●	●					●							

TABLE 3.5

Reference Matrix to Compare Landslide Hazard Zonation Methods in India

S. No.	Reference	Study Area	Map Unit	BIS	WO	IVM	Heuristic	LI	FR	LR	ANN	Prob.	AOR	MULTI-STAT	Logit	THRSLD	AHP	HSU	Fuzzy	Geotechnical Investigation	Other
																					Landslide Hazard Methods Adopted
1	Sarkar et al. (1995)	Srinagar Rudraprayag area	Pixel		●																
2	Panikkar and Subramanyan (1996)	Dehradun and Mussourie	Pixel																		Geomorphic analysis
3	Panikkar and Subramanyan (1997)	Mussourie, India	TMU		●																
4	Sah and Mazari (1998)	Kullu valley, Himachal Pradesh	NA					●													
5	Nagarajan et al. (2000)	Goa Highway, India	Pixel																		Discriminant analysis
6	Basu and De (2003)	Sikkim, India	NA																		Site-specific investigation
7	Arrora (2004)	Ganga valley, India	Pixel			●															
8	Sarkar (2004)	Darjeeling Himalayas, India	Slope facet		●																
9	Kanungo (2006)	Darjeeling Himalayas, India	Pixel								●								●		
10	Sarkar et al. (2006)	Sikkim, India	Pixel			●															

(Continued)

TABLE 3.5   (*Continued*)

Reference Matrix to Compare Landslide Hazard Zonation Methods in India

			Map	Landslide Hazard Methods Adopted																		
S. No.	Reference	Study Area	Unit	BIS	WO	IVM	Heuristic	LI	FR	LR	ANN	Prob.	AOR	MULTI-STAT	Logit	THRSLD	AHP	HSU	Fuzzy	Geotechnical Investigation	Other
11	Sriramkumar et al. (2006)	Konkan Railway	Pixel				●														
12	Naithani (2007)	Garhwal Himalayas	Slope facet	●																	
13	Champatiray et al. (2007)	Uttarkashi, Garhwal Himalayas	Pixel	●															●		
14	Thigale and Umrikar (2007)	Dasgaon, Mahad, Raigad	NA																	●	
15	Anbalagan et al. (2008)	Nainital, India	Slope facet	●																	
16	Chakraborty (2008)	NH 108, Uttarkashi	Pixel			●						●									
17	Champatiray et al. (2008)	Kashmir Himalayas	Pixel					●													
18	Kanungo et al. (2008)	Darjeeling, India	Pixel																●		
19	Kumar et al. (2008)	Mumbai-Pune Expressway	NA																	●	
20	Mhaskare (2008)	Sahyadris, India	Pixel														●				
21	Bodas and Kohli (2009)	Western Maharashtra and coastal plains	Slope facet					●													

No.	Reference	Location	Unit							Notes
22	Champatiray et al. (2009)	Pithorgad, Uttarakhand	Pixel		•					
23	Ghosh et al. (2009)	Darjeeling, India	Slope facet	•			•			
24	Pardeshi et al. (2009)	Wada-Khodala Road	NA							Wentworth's method
25	Patwary et al. (2009)	Rishikesh, Garhwal Himalayas	Pixel		•					
26	Prabhu and Ramakrishnan (2009)	Nilgiri district, Tamil Nadu	Pixel			•		•		
27	Saraf et al. (2009)	Chamoli district, Garhwal	Pixel		•					
28	Sharma et al. (2009)	Sikkim, India	Slope facet	•			•			
29	Singh and Champatiray (2009)	Alaknanda catchment	Pixel							SAR interferometry
30	Vijith et al. (2009)	Meenachil catchment, Kerla	Pixel	•						
31	Despande et al. (2009)	Gopeshwar, Chamoli district	Pixel		•					
32	Avinash and Ashamanjari (2010)	Aghnashini basin, Karnataka	Pixel			•				
33	Bodas and Kohli (2010)	Dasgaon, Mahad, Raigad	Slope facet						•	
34	Chauhan et al. (2010)	Chamoli district, Garhwal	Pixel			•				
35	Ganapathy et al. (2010)	Nilgiri district, Tamil Nadu	NA							
36	Jaiswal et al. (2010)	Nilgiris, India	Pixel				•			

(Continued)

TABLE 3.5   (*Continued*)

Reference Matrix to Compare Landslide Hazard Zonation Methods in India

S. No.	Reference	Study Area	Map Unit	Landslide Hazard Methods Adopted																		Other
				BIS	WO	IVM	Heuristic	LI	FR	LR	ANN	Prob.	AOR	MULTI-STAT	Logit	THRSLD	AHP	HSU	Fuzzy	Geotechnical Investigation		
37	Jaiswal et al. (2010)	NH 67, Nilgiri Hills	Pixel									●										
38	Kumar et al. (2010)	Mumbai-Pune Expressway	NA																		Markland test to identify potential slope failures
39	Kuriakose (2010)	Western Ghats, Kerla, India	Pixel																		SHALSTAB, SINMAP, TRIGRS, STARWAR + PROBSTAB
40	Evany Nithya and Rajesh Prasanna (2010)	Nilgiri district, Tamil Nadu	Pixel				●														
41	Kumar et al. (2010)	Mumbai-Pune Expressway	NA																		Stability analysis (Bishop and Peterson method)
42	Chandel et al. (2011)	Kullu district, Himachal Pradesh	Pixel		●																

No.	Reference	Study area	Mapping unit						Method/remarks
43	Das (2011)	Garhwal Himalayas	Pixel	•		•			SSPC
44	Ghosh (2011)	Darjeeling, India	Pixel	•	•				
45	Jaiswal (2011)	Nilgiris, India	Pixel		•				
46	Kannan et al. (2011)	Bodimetta, Tamil Nadu, India	Slope facet •						
47	Naveen Raj et al. (2011)	Nilgiri Hills, India	Pixel						Relative effect method (REM)
48	Ram Mohan et al. (2011)	NH 67, Nilgiri Hills	Pixel			•			
49	Sajinkumar et al. (2011)	Kerala	NA						Chemical index of alteration (role of weathering in slope instability)
50	Singh et al. (2011)	Senapati district, Manipur	Slope facet •						
51	Kannan (2011)	Bodimetta, Tamil Nadu, India	Slope facet •						
52	De and Jamatia (2011)	Darjeeling Himalayas, India	Slope facet •						
53	Das (2011)	Garhwal Himalayas	Pixel			•			SSPC
54	Bobade et al. (2012)	Sawantwadi, Sindhudurg district	NA					•	
55	Das et al. (2012)	Bhagirathi basin, India	Pixel			•			

(Continued)

TABLE 3.5 (*Continued*)

Reference Matrix to Compare Landslide Hazard Zonation Methods in India

S. No.	Reference	Study Area	Map Unit	BIS	WO	IVM	Heuristic	LI	FR	LR	ANN	Prob.	AOR	MULTI-STAT	Logit	THRSLD	AHP	HSU	Fuzzy	Geotechnical Investigation	Other
56	Karlekar (2012)	Raigad district, Maharashtra	Pixel					●						●							Multiple regression equation
57	Onagh et al. (2012)	Bhagirathi basin	Pixel							●											
58	Pareta et al. (2012)	Yamuna basin, Giri River					●														
59	Phukon et al. (2012)	Guwahati, Assam	Pixel														●				
60	Ramachandra et al. (2012)	Sharavati basin, Uttar Kannada	Pixel							●											
61	Rawat et al. (2012)	South district, Sikkim	Pixel				●														
62	Sujatha et al. (2012)	Kodaikanal, Nilgiri Hills	Pixel									●									
63	Bodas and Kohli (2012)	Western Maharashtra and Konkan	NA																		Field investigation
64	Ahmad et al. (2013)	Poladpur-Mahabaleshwar Road	NA																		UDEC Barton-Bandis model
65	Anbalagan and Parida (2013)	Harmony village, Pindar basin	Slope facet																		Circular failure charts method (Hoek and Bray)

66	Arunkumar et al. (2013)	Ooty, Tamil Nadu	Pixel
67	Balsubramani and Kumaraswamy (2013)	Giri valley, Himachal Pradesh, India	Pixel
68	Champatiray et al. (2013)	Yamunotri shrine	Pixel
69	Champatiray et al. (2013)	Mandakini valley	Pixel
70	Jaiswal and van Westen (2013)	Nilgiri Hills, India	Pixel
71	Lallianthanga et al. (2013)	Mamit town, Mizoram	Pixel
72	Martha et al. (2013)	NH 109, Rudraprayag, India	Pixel
73	Singh and Singh (2013)	NE India	NA
74	Wagh and Deshpande (2013)	Mahabaleshwar Ghat (SH 72)	Pixel
75	Gomathi et al. (2013)	Nilgiris, Tamil Nadu	Pixel
76	Tikke et al. (2014)	Sahyadris, India	NA	LRA

TABLE 3.6

Reference Matrix Depicting RS Data Used for Landslide Hazard Zonation in India

S. No.	Reference	Year	Study Area	Major Region	Landslide Type	Data Used
1	Sarkar et al. (1995)	1995	Garhwal Himalayas	Garhwal Himalayas		
2	Panikkar and Subramanyan (1996)	1996	Dehradun and Mussourie	Garhwal Himalayas	Rockslide and debris slide	Aerial photo, IRS LISS II, SPOT, and SOI toposheets
3	Panikkar and Subramanyan (1997)	1997	Mussourie, India	Garhwal Himalayas	Debris slide	IRS LISS II, SPOT
4	Sah and Mazari (1998)	1998	Kullu valley, Himachal Pradesh	Kullu district, Himachal Pradesh Himalayas	Debris slide, slump, debris flow	Geological maps
5	Nagarajan et al. (2000)	2000	Goa Highway, India	Western Ghats, Maharashtra	Rockfall, debris slide	IRS satellite data, toposheet, field investigation
6	Basu and De (2003)	2003	Sikkim, India	NE Himalayas	Debris slide	Geological map
7	Arrora (2004)	2004	Ganga valley, India	Garhwal Himalayas		
8	Sarkar (2004)	2004	Darjeeling Himalayas, India	NE Himalayas	Debris slide	IRS 1C, LISS III, 1D PAN data, Topo DEM, SPT PAN, PAN + PAN, LISS III (merged)
9	Kanungo (2006)	2006	Himalayas, India	Garhwal Himalayas	Debris slide	TOPO Dem, IRS LISS II
10	Sarkar et al. (2006)	2006	Sikkim, India	NE Himalayas	Rock and debris slide	IRS LISS II
11	Sriramkumar et al. (2006)	2006	Konkan Railway	Konkan coastal plains	Boulder fall, debris slide	IRS 1Band, IRS1D, SOI toposheets
12	Champatiray (2007)	2007	Himalayas, India	NE Himalayas	Debris slide	IRS LISS III, FCC, IRS ID, PAN
13	Naithani (2007)	2007	Garhwal Himalayas	Garhwal Himalayas	Debris slide, debris flow	IRS LISS III
14	Thigale and Umrikar (2007)	2007	Dasgaon, Mahad, Raigad	North Konkan	Debris translational slides	Geological maps, SOI toposheets
15	Anbalagan et al. (2008)	2008	Nainital, India	North Himalayas	Rockfall, slump, debris slide	Geological maps

16	Chakraborty (2008)	2008	NH 108, Uttarkashi	Garhwal Himalayas	Debris slide, rockslide, rockfall	Cartosat1, IRS P6, GPS survey
17	Champatiray et al. (2008)	2008	Kashmir Himalayas	Kashmir Himalayas	Rockfall, debris slide	ASTER DEM, SRTM DEM
18	Kanungo et al. (2008)	2008	Darjeeling, India	NE Himalayas	Rockslide	IRS 1C, LISS III, 1D PAN data
19	Kumar et al. (2008)	2008	Mumbai-Pune Expressway	Western Ghats	Rockfall, debris fall	Field data
20	Mhaskare (2008)	2008	Sahyadris, India	Western Ghats	Slump	
21	Bodas and Kohli (2009)	2009	Western Maharashtra and coastal plains	Western Ghats and coastal plains	Debris slide, slump, rockfall, wedge failure	Field data
22	Champatiray et al. (2009)	2009	Pithorgad, Uttarakhand	Garhwal Himalayas	Debris slide, debris flow	Google Earth, BHUVAN (IEOV), LISS III and LISS IV, IRS, PAN, CARTOSAT I
23	Ghosh et al. (2009)	2009	Kursiong, Darjeeling district	NE Himalayas	Debris slide	DEM
24	Pardeshi et al. (2009)	2009	Wada-Khodala Road	North Konkan	Boulder fall, debris slide, slump	Topo maps, PWD records
25	Patwary et al. (2009)	2009	Rishikesh, Garhwal Himalayas	Garhwal Himalayas	Shallow debris slide, rockfall, rockslide	IRS ID, LISS III, PAN (merged)
26	Prabhu and Ramakrishnan (2009)	2009	Nilgiris district, Tamil Nadu	Western Ghats	Debris slide, shallow soil slide	LANDSAT TM, DEM
27	Saraf et al. (2009)	2009	Chamoli district, Garhwal	Garhwal Himalayas	Debris fall	SPOT, ASTER, LANDSAT ETM
28	Sharma et al. (2009)	2009	Sikkim, India	NE Himalayas	Debris slide	ASTER DEM, CARTOSAT 1, QuickBird
29	Singh and Champatiray (2009)	2009	Alaknanda catchment	Garhwal Himalayas	Debris slide	Multiple SAR images, InSAR, DEM, DGPS survey
30	Vijith et al. (2009)	2009	Meenachil catchment, KerlaWestern Ghats		Debris slide	Topo maps, Topo DEM, IRS P6, LISS III

(Continued)

TABLE 3.6 (*Continued*)

Reference Matrix Depicting RS Data Used for Landslide Hazard Zonation in India

S. No.	Reference	Year	Study Area	Major Region	Landslide Type	Data Used
31	Despande et al. (2009)	2009	Gopeshwar, Chamoli district	Garhwal Himalayas	Debris slide	Topo DEM
32	Avinash and Ashamanjari (2010)	2010	Aghnashini basin, Karnataka	Western Ghats	Rockfall, debris slide	Toposheets, soil data (NBSS & LUP), SRTM DEM, LANDSAT ETM
33	Bodas and Kohli (2010)	2010	Dasgaon, Mahad, Raigad	Coastal plains	Debris slide, slump	Toposheets, field data
34	Chauhan et al. (2010)	2010	Chamoli district, Garhwal	Garhwal Himalayas	Debris slide, rockslide	IRS PAN, LISS III and IV, GPS, Google Earth
35	Ganapathy et al. (2010)	2010	Nilgiris district, Tamil Nadu	Western Ghats	Debris slide, boulder fall	NA
36	Jaiswal et al. (2010)	2010	Nilgiris, India	Western Ghats	Debris slide	Historical LS records
37	Jaiswal et al. (2010)	2010	Nilgiri Hills, India	Western Ghats	Debris slide	Historical LS records
38	Jaiswal et al. (2010)	2010	Nilgiri Hills, India	Western Ghats	Debris slide	Historical LS records
39	Kumar et al. (2010)	2010	Mumbai-Pune Expressway	Western Ghats	Rockfall	Field data
40	Kuriakose (2010)	2010	Western Ghats, Kerla, India	Western Ghats	Debris flow	SPOT MS,
41	Evany Nithya and Rajecsh Prasanna (2010)	2010	Nilgiris district, Tamil Nadu	Western Ghats	Rockslide, debris slide	DEM, IRS LISS III, PAN (merged)
42	Kumar et al. (2010)	2010	Mumbai-Pune Expressway	Western Ghats	Rockfall, debris fall	Field data
43	Chandel et al. (2011)	2011	Kullu district, Himachal Pradesh	Himachal Himalayas	Debris slide	LANDSAT ETM+, IRS P6, ASTER DEM, SOI sheets
44	Das (2011)	2011	Garhwal Himalayas	Garhwal Himalayas	Debris slide	CARTOSAT 1, RESOURCESAT 1
45	Ghosh (2011)	2011	Darjeeling, India	NE Himalayas	Debris slide, rockslide	DEM

46	Kannan et al. (2011)	2011	Bodimetta, Tamil Nadu, India	Western Ghats	Debris slide	IRS 1C, LISS III
47	Naveen Raj et al. (2011)	2011	Nilgiri Hills, India	Western Ghats	Debris slide	LANDSAT, DEM (Topo)
48	Ram Mohan et al. (2011)	2011	NH 67, Nilgiri Hills	Western Ghats	Debris slide, circular failure	LANDSAT MSS, DEM
49	Sajinkumar et al. (2011)	2011	Kerala	Western Ghats	Debris slide, debris flow, rockfall	
50	Singh et al. (2011)	2011	Senapati district, Manipur	NE Himalayas	Debris slide	SOI toposheets
51	Kannan (2011)	2011	Bodimetta, Tamil Nadu, India	Western Ghats	Debris slide	IRS 1C, LISS III
52	De and Jamatia (2011)	2011	Darjeeling Himalayas, India	NE Himalayas	Debris slide	LISS III, IRS PAN data
53	Bobade et al. (2012)	2012	Sawantwadi, Sindhudurg district	South Konkan	Slump	Field data, lab analysis (soil)
54	Das (2012)	2012	Bhagirathi basin	Garhwal Himalayas	Debris slide, rockslide	RESOURCESAT, CARTOSAT 1, LANDSAT TM, IRS 1C and 1D
55	Das et al. (2012)	2012	Uttarkashi	Garhwal Himalayas	Debris slide	CARTOSAT 1
56	Karlekar (2012)	2012	Raigad district, Maharashtra	North Konkan	Soil creep, rockfall, debris slide	DTM
57	Onagh et al. (2012)	2012	Bhagirathi basin	Garhwal Himalayas	Debris slide (translational)	DEM, Google Earth, GPS
58	Pareta et al. (2012)	2012	Yamuna basin, Giri River	Himachal Himalayas	Debris slide, rockfall	ASTER DEM, RESOURCE SAT, LISS IV
59	Phukon et al. (2012)	2012	Guwahati, Assam	NE Himalayas	Debris slide	SRTM DEM, GPS, LANDSAT ETM, LISS III
60	Ramachandra et al. (2012)	2012	Sharavati basin, Uttar Kannada	Western Ghats	Debris slide, debris slump	IRS LISS III, Topo DEM, GPS

(Continued)

TABLE 3.6 (*Continued*)
Reference Matrix Depicting RS Data Used for Landslide Hazard Zonation in India

S. No.	Reference	Year	Study Area	Major Region	Landslide Type	Data Used
61	Rawat et al. (2012)	2012	South district, Sikkim	NE Himalayas	Debris slide	IRS LISS III and IV, toposheets, ASTER DEM
62	Sujatha et al. (2012)	2012	Kodaikanal, Nilgiri Hills	Western Ghats	Debris slide, debris flow	IRS LISS III
63	Bodas and Kohli (2012)	2012	Western Maharashtra and Konkan	Western Ghats and Konkan plains	Debris slide, boulder fall, rockslide	Field investigations
64	Ahmad et al. (2013)	2013	Poladpur-Mahabaleshwar Road	Western Ghats	Rockfall, wedge failure	Field data
65	Anbalagan and Parida (2013)	2013	Harmony village, Pindar basin	Garhwal Himalayas	Debris slide (rotational)	GPS, toposheets, field data
66	Arunkumar et al. (2013)	2013	Ooty, Tamil Nadu	Western Ghats	Debris flow	Topographical maps, geological maps
67	Balsubramani and Kumaraswamy (2013)	2013	Giri valley, Himachal Pradesh, India	Western Himalayas	Debris slide	LANDSAT ETM+, ASTER GDEM
68	Champatiray et al. (2013)	2013	Yamunotri shrine	Garhwal Himalayas	Rockfall, debris slide	CARTOSAT 1, IRS LISS IV, SRTM DEM, GeoEye, CARTOSAT DEM
69	Champatiray et al. (2013)	2013	Mandakini valley	Garhwal Himalayas	Rockslide, debris slide	RESOURCESAT, LISS IV, CARTOSAT II, PAN (merged)
70	Jaiswal and van Westen (2013)	2013	Nilgiri Hills, India	Western Ghats	Debris slide	Historical LS records
71	Lallianthanga et al. (2013)	2013	Mamit town, Mizoram	NE Himalayas	Debris slide, slump	QuickBird, IRS P5, CARTOSAT 1

72	Martha et al. (2013)	2013	NH 109, Rudraprayag, India	Garhwal Himalayas	Debris slide, debris flow	CARTOSAT 1, LISS IV
73	Singh and Singh (2013)	2013	NE India	NE Himalayas	Slump (debris)	Field data
74	Wagh and Deshpande (2013)	2013	Mahabaleshwar Ghat (SH 72)	Western Ghats	Debris flow, rockslide, rockfall	CARTOSAT 1, DEM, IKONOS, PAN, Google Earth
75	Gomathi et al. (2013)	2013	Nilgiris district, Tamil Nadu	Nilgiri Hills	Debris slide	Historical LS records
76	Tikke et al. (2014)	2014	Sahyadris, India	Western Ghats	Slump, debris slide, rockslide	Historical LS records

TABLE 3.7

Recent Trends in the Use of Geospatial Information for Landslide Hazard Assessment

S. No.	Reference	RS Data Used	S. No.	Reference	RS Data Used	S. No.	Reference	RS Data Used
1	Chao Zhou et al. (2022)	SAR images	25	Ng et al. (2021)	DEM, DTM, NDVI, MRR	48	Chigira et al. (2022)	ALOS PRISM, ALOS World 3D, Topo-data, ASTER DEM
2	Che Ming et al. (2022)	LiDAR, SPOT, UAV	26	Nguyen and Kim (2021)	Digital aerial photo, LiDAR	49	Chen et al. (2022)	FBG sensor
3	Chi wen Chen (2021)	DEM	27	Paulin et al. (2022)	LiDAR, WS-DNR, Pre-sliding DEM	50	Zhao et al. (2022)	TRMM, GPM-IMERG
4	Dewttee et al. (2020)	SRTM DEM, UAV, Google Earth, DInSAR	28	Peres-Cancelliere et al. (2021)	TRIGIS, MATLAB, DEM	51	Cui et al. (2022)	DEM
5	Fawy Wang (2022)	Field data	29	Perrone et al. (2020)	ERT and SRT	52	Smith et al. (2021)	DEM
6	Ferlisi et al. (2021)	DInSAR	30	Xu et al. (2022b)	Google Earth, UAV	53	Qi et al. (2021)	Google Earth, SRTM DEM
7	Franceschini et al. (2022)	DEM, Web GIS	31	Rodrigues et al. (2021)	DEM	54	Stumvoll et al. (2021)	TLS data, 10 m Austrian DEM
8	Francesco et al. (2022)	DTM	32	Rosa et al. (2022)	SMC radar network, ASTER DEM	55	Mason et al. (2021)	Indonesian National DEM, FLAC 2D analysis
9	Ciccareoe et al. (2021)	DEM	33	Rosi et al. (2020)	DEM	56	Santagelo et al. (2022)	Sentinel 1, SAR
10	Ghorbanzadeh et al. (2022)	ALOS DEM, Sentinel 2	34	Roy et al. (2022)	NASA and JAXA	57	Vrgara et al. (2022)	Suspended sediment concentration (SSC)

No.	Reference	Data/Method	No.	Reference	Data/Method	No.	Reference	Data/Method
11	Haken et al. (2022)	Sentinel 2 Google Earth, TANDEM – X	35	Sharma et al. (2021)	Google Earth	58	Gorokhorich et al. (2021)	SRTM DEM, LiDAR, NOAA
12	He et al. (2022)	DEM	36	Song et al. (2022)	Field data	59	Guerriero et al. (2021)	SAR
13	Helbert et al. (2022)	ENSO	37	Strouth et al. (2020)	Field data	60	Rosi et al. (2021)	UAV, LiDAR
14	Hughes et al. (2020)	GNSS, Bathymetric DEM	38	Ilinca et al. (2022)	SRTM DEM, UAV	61	Spiekermann et al. (2021)	TIMSS, LiDAR
15	Jain et al. (2021)	CARTO DEM, ResourceSat 2A, Pleiades 1A, SPOT, Sentinel 2	39	Vittoria et al. (2022)	Web-based LS mapping tool	62	He et al. (2021)	SRTM, global lithological map (GLiM)
16	Picarelli et al. (2022)	DInSAR, Google Earth, LANDSAT	40	Wallace et al. (2022)	USGS data, DEM, Google Earth	63	Huang et al. (2022)	ASTER, LANDSAT TM
17	Lee et al. (2020)	EnviSat, TRMM	41	Wang et al. (2022)	Google Earth, DSM, UAV	64	Tempa et al. (2021)	High-resolution DEM, UAV
18	Li et al. (2022)	DEM ALPA (automatic landslide profile analysis)	42	Wei et al. (2022)	ALOS, DEM, LANDSAT 8	65	Niethammer et al. (2010)	DSM, UAV
19	Li et al. (2022)	DEM, Google Earth	43	Xin et al. (2022)	UAV, visible infrared imaging radiometric suite (VIIRS)	66	Niethammer et al. (2012)	TLS, LiDAR, orthomosaic DEM, UAV
20	Lima et al. (2021)	DTM	44	Haken et al. (2021)	SRTM DEM, GPM, IMERG, USGS Shakemap system	67	Liao et al. (2022)	Drone-based UAV
21	Liu et al. (2021)	Sentinel 1, InSAR, DEM, Sentinel 2	45	Jia et al. (2021)	SRTM DEM, GFLD	68	Farina et al. (2017)	UAV-multi-copter drones

(Continued)

TABLE 3.7   (*Continued*)

Recent Trends in the Use of Geospatial Information for Landslide Hazard Assessment

S. No.	Reference	RS Data Used	S. No.	Reference	RS Data Used	S. No.	Reference	RS Data Used
22	Mahallem et al. (2022)	Field investigations	46	Zhang et al. (2021)	DEM	69	Prasetyo et al. (2017)	Orthomosaic photo, UAV, DSM, GPS survey
23	Marino et al. (2020)	Geotechnical investigations	47	Tammaso Baggio et al. (2021)	GRASS GIS 7, DTM, LiDAR	70	Yaprak et al. (2018)	DSM, DGPS, UAV, orthomosaic photo
24	Meena et al. (2022)	RapidEye, ALOS, PALS, AR, DEM						

REFERENCES

Akbar, Tahir Ali, and Sung Ryong Ha. 2011. "Landslide Hazard Zoning along Himalayan Kaghan Valley of Pakistan-by Integration of GPS, GIS, and Remote Sensing Technology." *Landslides* 8 (4): 527–540. https://doi.org/10.1007/s10346-011-0260-1.

Akgun, A. 2011. "A Comparison of Landslide Susceptibility Maps Produced by Logistic Regression, Multi-Criteria Decision, and Likelihood Ratio Methods: A Case Study at Lzmir, Turkey." *Landslides (Springer-Verlag).* https://doi.org/10.1007/s10346-011-0283-7.

Alexander, D. 1993. *Natural Disasters*. UCL Press Ltd.

Anbalagan, R., D. Chakraborty, and A. Kohali. 2008. "Landslide Hazard Zonation (LHZ) Maping on Meso-Scale for Systematic Town Planning in Mountainous Terrain." *Journal of Scientific & Industrial Research* 67: 486–497.

Arora, M., A. Dasgupta, and R. Gupta. 2004. "An Artificial Neural Network Approach for Landslide Hazard Zonation in the Bhagirathi (Ganga) Valley, Himalaya." *International Journal of Remote Sensing* 25: 559–572.

Atkinson, P. M., and R. Massari. 2011. "Autologistic Modelling of Susceptibility to Landsliding in the Central Apennines, Italy." *Geomorphology* 130 (1–2): 55–64. https://doi.org/10.1016/j.geomorph.2011.02.001.

Ayalew, L., and H. Yamagishi. 2005. "The Application of GIS-Based Logistic Regression for Landslide Susceptibility Mapping in the Kakuda-Yahiko Mountains, Central Japan." *Geomorphology* 65: 15–31.

Ayalew, L., H. Yamagishi, H. Marui, and T. Kanno. 2005. "Landslides in Sado Island of Japan: Part II. GIS-Based Susceptibility Mapping with Comparisons of Results from Two Methods and Verifications." *Engineering Geology* 81 (4): 432–445. https://doi.org/10.1016/j.enggeo.2005.08.004.

Balsubramani, K., and K. Kumaraswamy. 2013. "Application of Geospatial Technology and Information Value Technique in Landslide Hazard Zonation Mapping: A Case Study of Giri Valley, Himachal Pradesh." *Disaster Advances* 6: 38–47.

Bălteanu, Dan, Viorel Chendeş, Mihaela Sima, and Petru Enciu. 2010. "A Country-Wide Spatial Assessment of Landslide Susceptibility in Romania." *Geomorphology* 124 (3–4): 102–112. https://doi.org/10.1016/j.geomorph.2010.03.005.

Blahut, Jan, Cees J. van Westen, and Simone Sterlacchini. 2010. "Analysis of Landslide Inventories for Accurate Prediction of Debris-Flow Source Areas." *Geomorphology* 119 (1–2): 36–51. https://doi.org/10.1016/j.geomorph.2010.02.017.

Bommer, J., and C. Rodriguez. 2002. "Earthquake-Induced Landslides in Central America." *Engineering Geology* 63 (3–4): 189–220.

Brabb, E. 1993. Proposal for *Worldwide Landslide Hazard Maps*. Proceedings of 7th International Conference and field workshop on landslide in Czech and Slovak Republics, pp. 15–27.

Brusden, D. 1984. "Mudslides." In D. Brusden, D. Prior (Eds.), *Slope Instability* (pp. 363–418). Wiley.

Bureau of Indian Standards. 1998. *Preparation of Landslide Hazard Zonation Maps in Mountainous Terrain – Guidelines (Part2-Macrozonation)*. IS 14496-2, Hill Area Development Engineering (CED 56), New Delhi, pp. 1–20.

Caleca, Francesco, Veronica Tofani, Samuele Segoni, Federico Raspini, Ascanio Rosi, Marco Natali, Filippo Catani, and Nicola Casagli. 2022. "A Methodological Approach of QRA for Slow-Moving Landslides at a Regional Scale." *Landslides* (August 2021): 1539–1561. https://doi.org/10.1007/s10346-022-01875-x.

Calvello, Michele, Leonardo Cascini, and Sabrina Mastroianni. 2013. "Landslide Zoning over Large Areas from a Sample Inventory by Means of Scale-Dependent Terrain Units." *Geomorphology* 182: 33–48. https://doi.org/10.1016/j.geomorph.2012.10.026.

Capobianco, Vittoria, Marco Uzielli, Bjørn Kalsnes, Jung Chan Choi, James Michael Strout, Loretta von der Tann, Ingar Haug Steinholt, Anders Solheim, Farrokh Nadim, and Suzanne Lacasse. 2022. "Recent Innovations in the LaRiMiT Risk Mitigation Tool: Implementing a Novel Methodology for Expert Scoring and Extending the Database to Include Nature-Based Solutions." *Landslides* 19 (7): 1563–1583. https://doi.org/10.1007/s10346-022-01855-1.

Cardinali, M., P. Reichenbach, F. Guzzetti, F. Ardizzone, G. Antonini, M. Galli, M. Cacciano, M. Castellani, and P. Salvati. 2002. "A Geomorphological Approach to the Estimation of Landslide Hazards and Risks in Umbria, Central Italy." *Natural Hazards and Earth System Sciences* 2 (1–2): 57–72. https://doi.org/10.5194/nhess-2-57-2002.

Catani, F., N. Casagli, L. Ermini, G. Righini, and G. Menduni. 2005. "Landslide Hazard and Risk Mapping at Catchment Scale in the Arno River Basin." *Landslides* 2 (4): 329–342. https://doi.org/10.1007/s10346-005-0021-0.

Champatiray, P., S. Dimri, R. Lakhera, and S. Sati. 2007. "Fuzzy Based Methods for Landslide Hazard Assessment in Active Seismic Zone of Himalaya." *Landslides (Springer-Verlag)* 4: 101–110.

Chand, Dipender Singh. 2008. *Landslide Monitoring in Space and Time Using Optical Satellite Imagery and Dem Derived Parameters: Case Study from Garhwal Himalaya, Uttarakhand.* International Institute for Geoinformation Science and Earth Observation, Enschede, Netherlands and Indian Institute of Remote Sensing, National Remote Sensing Agency, Department of Space, Dehradun, India, p. 111.

Chang, K. T., and S. Chiang. 2009. "An Integrated Model for Predicting Rainfall Induced Landslides." *Geomorphology* 105: 366–373.

Chang, K. T., and J. K. Liu. 2004. "Landslide Features Interpreted by Neural Network Method Using a High-Resolution Satellite Image and Digital Topographic Data." *Proceedings of ISPRS XX Congress, Commission VII TS WG VII/5, Istanbul, Turkey* 1: 574–579.

Chau, K. T., Y. L. Sze, M. K. Fung, W. Y. Wong, E. L. Fong, and L. C. P. Chan. 2004. "Landslide Hazard Analysis for Hong Kong Using Landslide Inventory and GIS." *Computers and Geosciences* 30 (4): 429–443. https://doi.org/10.1016/j.cageo.2003.08.013.

Chen, Chi-Wen, Ching Hung, Guan-Wei Lin, Jun-Jih Liou, Shih-Yao Lin, Hsin-Chi Li, Yung-Ming Chen, and Hongery Chen. 2022. "Preliminary Establishment of a Mass Movement Warning System for Taiwan Using the Soil Water Index." *Landslides (Springer-Verlag)* 19: 1779–1789.

Chen, X. L., H. L. Ran, and W. T. Yang. 2012. "Evaluation of Factors Controlling Large Earthquake-Induced Landslides by the Wenchuan Earthquake." *Natural Hazards and Earth System Science* 12 (12): 3645–3657. https://doi.org/10.5194/nhess-12-3645-2012.

Chleborad, Alan F., Rex L. Baum, Jonathan W. Godt, and US Geological Survey. 2006. *Rainfall Thresholds for Forecasting Landslides in the Seattle, Washington, Area-Exceedance and Probability.* http://scholar.google.com/scholar?q=related:94WNYvf6TfEJ:scholar.google.com/&hl=en&num=30&as_sdt=0,5.

Chung, Chang Jo, and Andrea G. Fabbri. 2008. "Predicting Landslides for Risk Analysis – Spatial Models Tested by a Cross-Validation Technique." *Geomorphology* 94 (3–4): 438–452. https://doi.org/10.1016/j.geomorph.2006.12.036.

Clerici, Aldo, Susanna Perego, Claudio Tellini, and Paolo Vescovi. 2002. "A Procedure for Landslide Susceptibility Zonation by the Conditional Analysis Method." *Geomorphology* 48 (4): 349–364. https://doi.org/10.1016/S0169-555X(02)00079-X.

Coe, J, J. Godt, R. Baum, R. Bucknam, and J. Michael. 2004. "Landslide Susceptibility from Topography in Guatemala." *Landslides: Evaluation and Stabilization/Glissement de Terrain: Evaluation et Stabilisation* 2: 69–78. https://doi.org/10.1201/b16816-8.

Colombo, A., L. Lanteri, M. Ramasco, and C. Troisi. 2005. "Systematic GIS Based Landslide Inventory as the First Step for Effective Landslide Hazard Management." *Landslides* 2: 291–301.

Conoscenti, Christian, Cipriano Di Maggio, and Edoardo Rotigliano. 2008. "GIS Analysis to Assess Landslide Susceptibility in a Fluvial Basin of NW Sicily (Italy)." *Geomorphology* 94 (3–4): 325–339. https://doi.org/10.1016/j.geomorph.2006.10.039.

Courture, R. 2011. *Landslide Terminology – National Technical Guidelines and Best Practices on Landslides.* Canada: Open Files 6824; Geological Survey of Canada.

Crozier, M. 1986. *Landslides – Causes, Consequences and Environment.* Croom Helm Ltd.

Cruden, D. M. 1991. "A Simple Definition of a Landslide." *Bulletin of the International Association of Engineering Geology – Bulletin de l'Association Internationale de Géologie de l'Ingénieur* 43 (1): 27–29. https://doi.org/10.1007/BF02590167.

Dahal, Ranjan Kumar, and Shuichi Hasegawa. 2008. "Representative Rainfall Thresholds for Landslides in the Nepal Himalaya." *Geomorphology* 100 (3–4): 429–443. https://doi.org/10.1016/j.geomorph.2008.01.014.

Dahl, M. P. J., L. E. Mortensen, A. Veihe, and N. H. Jensen. 2010. "A Simple Qualitative Approach for Mapping Regional Landslide Susceptibility in the Faroe Islands." *Natural Hazards and Earth System Sciences* 10 (2): 159–170. https://doi.org/10.5194/nhess-10-159-2010.

Dai, F. C., and C. F. Lee. 2001. "Frequency-Volume Relation and Prediction of Rainfall-Induced Landslides." *Engineering Geology* 59 (3–4): 253–266. https://doi.org/10.1016/S0013-7952(00)00077-6.

Dai, F. C., and C. F. Lee. 2002. "Landslide Characteristics and Slope Instability Modeling Using GIS, Lantau Island, Hong Kong." *Geomorphology* 42 (3–4): 213–228. https://doi.org/10.1016/S0169-555X(01)00087-3.

Das, I. 2011. Spatial Statistical Modelling for Assessing Landslide Hazard and Vulnerability. ITC Dissertation number 192, University of Twente, International Institute for Geo-information Science and Earth Observation, Enschede, Netherlands.

Das, I., Alfred Stein, Norman Kerle, and V. K. Dadhwal. 2011. "Probabilistic Landslide Hazard Assessment Using Homogeneous Susceptible Units (HSU) along a National Highway Corridor in the Northern Himalayas, India." *Landslides* 8 (3): 293–308. https://doi.org/10.1007/s10346-011-0257-9.

Das, I., Alfred Stein, Norman Kerle, and Vinay K. Dadhwal. 2012. "Landslide Susceptibility Mapping along Road Corridors in the Indian Himalayas Using Bayesian Logistic Regression Models." *Geomorphology* 179: 116–125. https://doi.org/10.1016/j.geomorph.2012.08.004.

Delgado Garcia, Helbert, Petly David, Bermudez Mouricio, and Sepulveda Sergio. 2022. "Fatal Landslides in Colombia (from Historical Times to 2020) and Their Socio-Economic Impacts." *Landslides (Springer-Verlag)* 19: 1689–1716.

Dewitte, Olivier, Antoine Dille, Arthur Depicker, Désiré Kubwimana, Jean Claude Maki Mateso, Toussaint Mugaruka Bibentyo, Judith Uwihirwe, and Elise Monsieurs. 2021. "Constraining Landslide Timing in a Data-Scarce Context: From Recent to Very Old Processes in the Tropical Environment of the North Tanganyika-Kivu Rift Region." *Landslides* 18 (1): 161–177. https://doi.org/10.1007/s10346-020-01452-0.

Eeckhaut, M., P. Reichenbach, F. Guzzetti, M. Rossi, and J. Poesen. 2009. "Combined Landslide Inventory and Susceptibility Assessment Based on Different Mapping Units: An Example from the Flemish Ardennes, Belgium." *Natural Hazards and Earth System Sciences* 9: 507–521.

Ercanoglu, Murate. 2005. "Landslide Susceptibility Assessment of SE Bartin (West Black Sea Region, Turkey) by Artificial Neural Networks." *Natural Hazards and Earth System Science* 5 (6): 979–992. https://doi.org/10.5194/nhess-5-979-2005.

Ercanoglu, M., C. Gokceoglu, and T. W. J. Van Asch. 2004. "Landslide Susceptibility Zoning North of Yenice (NW Turkey) by Multivariate Statistical Techniques." *Natural Hazards* 32 (1): 1–23. https://doi.org/10.1023/B:NHAZ.0000026786.85589.4a.

Erener, Arzu, H. Sebnem, and B. Düzgün. 2010. "Improvement of Statistical Landslide Susceptibility Mapping by Using Spatial and Global Regression Methods in the Case of More and Romsdal (Norway)." *Landslides* 7 (1): 55–68. https://doi.org/10.1007/s10346-009-0188-x.

Ferlisi, Settimio, Antonio Marchese, and Dario Peduto. 2021. "Quantitative Analysis of the Risk to Road Networks Exposed to Slow-Moving Landslides: A Case Study in the Campania Region (Southern Italy)." *Landslides* 18 (1): 303–319. https://doi.org/10.1007/s10346-020-01482-8.

Floris, Mario, and Francesca Bozzano. 2008. "Evaluation of Landslide Reactivation: A Modified Rainfall Threshold Model Based on Historical Records of Rainfall and Landslides." *Geomorphology* 94 (1–2): 40–57. https://doi.org/10.1016/j.geomorph.2007.04.009.

Franceschini, Rachele, Ascanio Rosi, Filippo Catani, and Nicola Casagli. 2022. "Exploring a Landslide Inventory Created by Automated Web Data Mining: The Case of Italy." *Landslides* 19 (4): 841–853. https://doi.org/10.1007/s10346-021-01799-y.

Froude, M. J., and D. N. Petley. 2018. "Global Fatal Landslide Occurrence from 2004 to 2016." *Natural Hazards and Earth System Sciences* 18: 2161–2181. https://doi.org/10.5194/nhess-18-2161-2018.

Gabet, Emmanuel J., Douglas W. Burbank, Jaakko K. Putkonen, Beth A. Pratt-Sitaula, and Tank Ojha. 2004. "Rainfall Thresholds for Landsliding in the Himalayas of Nepal." *Geomorphology* 63 (3–4): 131–143. https://doi.org/10.1016/j.geomorph.2004.03.011.

Galli, Mirco, Francesca Ardizzone, Mauro Cardinali, Fausto Guzzetti, and Paola Reichenbach. 2008. "Comparing Landslide Inventory Maps." *Geomorphology* 94 (3–4): 268–289. https://doi.org/10.1016/j.geomorph.2006.09.023.

García-Rodríguez, M. J., J. A. Malpica, B. Benito, and M. Díaz. 2008. "Susceptibility Assessment of Earthquake-Triggered Landslides in El Salvador Using Logistic Regression." *Geomorphology* 95 (3–4): 172–191. https://doi.org/10.1016/j.geomorph.2007.06.001.

Ghorbanzadeh, Omid, Hejar Shahabi, Alessandro Crivellari, Saeid Homayouni, Thomas Blaschke, and Pedram Ghamisi. 2022. "Landslide Detection Using Deep Learning and Object-Based Image Analysis." *Landslides* 19 (4): 929–939. https://doi.org/10.1007/s10346-021-01843-x.

Ghosh, Saibal. 2011. Knowledge Guided Empirical Prediction of Landslide Hazard. ITC Dissertation number 190, University of Twente, International Institute for Geo-information Science and Earth Observation, Enschede, Netherlands.

Ghosh, Saibal, C. J. van Westen, E. J. M. Carranza, T. B. Ghoshal, N. K. Sarkar, and M. Surendranath. 2009. "A Quantitative Approach for Improving the BIS (Indian) Method of Medium-Scale Landslide Susceptibility." *Journal of the Geological Society of India* 74 (5): 625–638. https://doi.org/10.1007/s12594-009-0167-9.

Giuseppe, Ciccarese, Mulas Marco, and Corsini Alessandro. 2021. "Combining Spatial Modelling and Regionalization of Rainfall Thresholds for Debris Flows Hazard Mapping in the Emilia-Romagna Apennines (Italy)." *Landslides* 18 (11): 3513–3529. https://doi.org/10.1007/s10346-021-01739-w.

Gokceoglu, C., and E. Sezer. 2009. "A Statistical Assessment on International Landslide Literature (1945–2008)." *Landslides* 6: 345–351.

Gorokhovich, Yuri, and Andrii Vustianiuk. 2021. "Implications of Slope Aspect for Landslide Risk Assessment: A Case Study of Hurricane Maria in Puerto Rico in 2017." *Geomorphology* 391: 107874. https://doi.org/10.1016/j.geomorph.2021.107874.

Goswami, Rajasmita, Neil C. Mitchell, and Simon H. Brocklehurst. 2011. "Distribution and Causes of Landslides in the Eastern Peloritani of NE Sicily and Western Aspromonte of SW Calabria, Italy." *Geomorphology* 132 (3–4): 111–122. https://doi.org/10.1016/j.geomorph.2011.04.036.

Greco, R., M. Sorriso-Valvo, and E. Catalano. 2007. "Logistic Regression Analysis in the Evaluation of Mass Movements Susceptibility: The Aspromonte Case Study, Calabria, Italy." *Engineering Geology* 89 (1–2): 47–66. https://doi.org/10.1016/j.enggeo.2006.09.006.

Gutiérrez, F., M. Soldati, F. Audemard, and D. Bălteanu. 2010. "Recent Advances in Landslide Investigation: Issues and Perspectives." *Geomorphology* 124 (3–4): 95–101. https://doi.org/10.1016/j.geomorph.2010.10.020.

Guzzetti, F. 2000. "Landslide Fatalities and Evaluation of Landslide Risk in Italy." *Engineering Geology* 58: 89–107.

Guzzetti, F. 2003. "Landslide Hazard Assessment and Risk Evaluation: Limits and Prospectives." *Proceedings of the 4th EGS Plinius Conference* (October 2002): 1–4.

Guzzetti, F., P. Aleotti, B. Malamud, and D. L. Turcotte. 2003. "Comparison of Three Landslide Event Inventories in Central and Northern Italy." In *4th EGS Plinius Conference* (pp. 5–9). Mallorca, Spain: Universitat de les Ilies Balears.

Guzzetti, F., A. Carrara, M. Cardinali, and P. Reichenbach. 1999. "Landslide Hazard Evaluation: A Review of Current Techniques and Their Application in a Multi-Study, Central Italy." *Geomorphology* 31: 181–216.

Guzzetti, F., M. Galli, P. Reichenbach, F. Ardizzone, and M. Cardinali. 2006. "Landslide Hazard Assessment in the Collazzone Area, Umbria, Central Italy." *Natural Hazards and Earth System Science* 6 (1): 115–131. https://doi.org/10.5194/nhess-6-115-2006.

Guzzetti, F., P. Reichenbach, M. Cardinali, M. Galli, and F. Ardizzone. 2005. "Probabilistic Landslide Hazard Assessment at the Basin Scale." *Geomorphology* 72 (1–4): 272–299. https://doi.org/10.1016/j.geomorph.2005.06.002.

Hansen, A. 1984. *Slope Instability* (pp. 523–602). D. Brusden, D. Prior (Eds.). New York: John Wiley and Sons.

He, Jian, Limin Zhang, Ruilin Fan, Shengyang Zhou, Hongyu Luo, and Dalei Peng. 2022. "Evaluating Effectiveness of Mitigation Measures for Large Debris Flows in Wenchuan, China." *Landslides* 19 (4): 913–928. https://doi.org/10.1007/s10346-021-01809-z.

Highland, L. M. 2008. "Introduction the Landslide Handbook-a Guide to Understanding Landslides." *The Landslide Handbook – a Guide to Understanding Landslides*, 4–42. http://landslides.usgs.gov/.

Highland, L. M., and P. Bobrowsky. 2008. "The Landslide Handbook – a Guide to Understanding Landslides." *US Geological Survey Circular* 1325: 1–147.

Hughes, K. E., M. Geertsema, E. Kwoll, M. N. Koppes, N. J. Roberts, J. J. Clague, and S. Rohland. 2021. "Previously Undiscovered Landslide Deposits in Harrison Lake, British Columbia, Canada." *Landslides* 18 (2): 529–538. https://doi.org/10.1007/s10346-020-01514-3.

Hutchinson, J. 1988. "Mass movement." In *The Encyclopaedia of Geomorphology* (pp. 688–695). R.W. Fairbridge, Reinold.

Ilinca, Viorel, Ionuț Șandric, Zenaida Chițu, Radu Irimia, and Ion Gheuca. 2022. "UAV Applications to Assess Short-Term Dynamics of Slow-Moving Landslides under Dense Forest Cover." *Landslides* 19 (7): 1717–1734. https://doi.org/10.1007/s10346-022-01877-9.

Jain, Nirmala, Tapas R. Martha, Kirti Khanna, Priyom Roy, and K. Vinod Kumar. 2021. "Major Landslides in Kerala, India, during 2018–2020 Period: An Analysis Using Rainfall Data and Debris Flow Model." *Landslides* 18 (11): 3629–3645. https://doi.org/10.1007/s10346-021-01746-x.

Jaiswal, P. 2011. Landslide Risk Quantification along Transportation Corridors Based on Historical Information. ITC Dissertation number 191, University of Twente, International Institute for Geo-information Science and Earth Observation, Enschede, Netherlands.

Jaiswal, Pankaj, and Cees J. van Westen. 2013. "Use of Quantitative Landslide Hazard and Risk Information for Local Disaster Risk Reduction along a Transportation Corridor: A Case Study from Nilgiri District, India." *Natural Hazards* 65 (1): 887–913. https://doi.org/10.1007/s11069-012-0404-1.

Jaiswal, P., C. J. van Westen, and V. Jetten. 2010. "Quantitative Assessment of Direct and Indirect Landslide Risk along Transportation Lines in Southern India." *Natural Hazards and Earth System Science* 10 (6): 1253–1267. https://doi.org/10.5194/nhess-10-1253-2010.

Jelínek, Róbert, and Peter Wagner. 2007. "Landslide Hazard Zonation by Deterministic Analysis (Veľká Čausa Landslide Area, Slovakia)." *Landslides* 4 (4): 339–350. https://doi.org/10.1007/s10346-007-0089-9.

Kannan, M., E. Saranathan, and R. Anbalagan. 2011. "Macro Landslide Hazard Zonation Mapping – Case Study from Bodi – Bodimettu Ghats Section, Theni District, Tamil Nadu – India." *Journal of the Indian Society of Remote Sensing* 39 (4): 485–496. https://doi.org/10.1007/s12524-011-0112-4.

Kanungo, D., M. Arora, R. Gupta, and S. Sarkar. 2009. "Landslide Risk Assessment Using Concept of Danger Pixel and Fuzzy Set Theory in Darjeeling Himalayas." *Landslides* 5: 407–416.

Kavzoglu, Taskin, Emrehan Kutlug Sahin, and Ismail Colkesen. 2014. "Landslide Susceptibility Mapping Using GIS-Based Multi-Criteria Decision Analysis, Support Vector Machines, and Logistic Regression." *Landslides* 11 (3): 425–439. https://doi.org/10.1007/s10346-013-0391-7.

Kuriakose, S. 2010. *Physically-Based Dynamic Modelling of the Effect of Land Use Changes on Shallow Landslide Initiation in the Western Ghats of Kerala, India.* University of Twente, International Institute for Geo-Information Science and Earth Observation.

Lallianthanga, R. K., and F. Lalbiakmawia. 2013. "Micro-Level Landslide Hazard Zonation of Saitual Town, Mizoram, India Using Remote Sensing and GIS Techniques *1,2." *International Journal of Engineering Sciences & Research Technology* 2 (9): 184–194.

Lee, C. T., C. C. Huang, J. F. Lee, K. L. Pan, M. L. Lin, and J. J. Dong. 2008. "Statistical Approach to Storm Event-Induced Landslides Susceptibility." *Natural Hazards and Earth System Science* 8 (4): 941–960. https://doi.org/10.5194/nhess-8-941-2008.

Lee, Jung Hyun, Hanbeen Kim, Hyuck Jin Park, and Jun Haeng Heo. 2021. "Temporal Prediction Modeling for Rainfall-Induced Shallow Landslide Hazards Using Extreme Value Distribution." *Landslides* 18 (1): 321–338. https://doi.org/10.1007/s10346-020-01502-7.

Lee, S. T. 2005. *Cross-Verification of Spatial Logistic Regression for Landslide Susceptibility Analysis: A Case Study of Korea.* Proceedings, 31st International Symposium on Remote Sensing of Environment, ISRSE 2005: Global Monitoring for Sustainability and Security. St. Petersburg, Russia, 20–24 June 2005. http://www.scopus.com/inward/record.url?eid=2-s2.0-84879728712&partnerID=tZOtx3y1. Accessed in 4 March, 2020.

Lee, S. T., and Biswajeet Pradhan. 2006. "Probabilistic Landslide Hazards and Risk Mapping on Penang Island, Malaysia." *Journal of Earth System Science* 115 (6): 661–672. https://doi.org/10.1007/s12040-006-0004-0.

Lee, S. T., T. T. Yu, W. F. Peng, and C. L. Wang. 2010. "Incorporating the Effects of Topographic Amplification in the Analysis of Earthquake-Induced Landslide Hazards Using Logistic Regression." *Natural Hazards and Earth System Science* 10 (12): 2475–2488. https://doi.org/10.5194/nhess-10-2475-2010.

Li, Huajin, Yusen He, Qiang Xu, Jiahao Deng, Weile Li, and Yong Wei. 2022a. "Detection and Segmentation of Loess Landslides via Satellite Images: A Two-Phase Framework." *Landslides* 19 (3): 673–686. https://doi.org/10.1007/s10346-021-01789-0.

Li, Langping, Hengxing Lan, Alexander Strom, and Renato Macciotta. 2022b. "Landslide Longitudinal Shape: A New Concept for Complementing Landslide Aspect Ratio." *Landslides* 19 (5): 1143–1163. https://doi.org/10.1007/s10346-021-01828-w.

Liang, Xin, Samuele Segoni, Kunlong Yin, Juan Du, Bo Chai, Veronica Tofani, and Nicola Casagli. 2022. "Characteristics of Landslides and Debris Flows Triggered by Extreme Rainfall in Daoshi Town during the 2019 Typhoon Lekima, Zhejiang Province, China." *Landslides* 19 (7): 1735–1749. https://doi.org/10.1007/s10346-022-01889-5.

Lima, Pedro, Stefan Steger, and Thomas Glade. 2021. "Counteracting Flawed Landslide Data in Statistically Based Landslide Susceptibility Modelling for Very Large Areas: A National-Scale Assessment for Austria." *Landslides* 18 (11): 3531–3546. https://doi.org/10.1007/s10346-021-01693-7.

Listo, Fabrizio de Luiz Rosito, and Bianca Carvalho Vieira. 2012. "Mapping of Risk and Susceptibility of Shallow-Landslide in the City of São Paulo, Brazil." *Geomorphology* 169–170: 30–44. https://doi.org/10.1016/j.geomorph.2012.01.010.

Liu, Ya, Haijun Qiu, Dongdong Yang, Zijing Liu, Shuyue Ma, Yanqian Pei, Juanjuan Zhang, and Bingzhe Tang. 2022. "Deformation Responses of Landslides to Seasonal Rainfall Based on InSAR and Wavelet Analysis." *Landslides* 19 (1): 199–210. https://doi.org/10.1007/s10346-021-01785-4.

Ma, Fengshan, Jie Wang, Renmao Yuan, Haijun Zhao, and Jie Guo. 2013. "Application of Analytical Hierarchy Process and Least-Squares Method for Landslide Susceptibility Assessment along the Zhong-Wu Natural Gas Pipeline, China." *Landslides* 10 (4): 481–492. https://doi.org/10.1007/s10346-013-0402-8.

Mancini, F., C. Ceppi, and G. Ritrovato. 2010. "GIS and Statistical Analysis for Landslide Susceptibility Mapping in the Daunia Area, Italy." *Natural Hazards and Earth System Science* 10 (9): 1851–1864. https://doi.org/10.5194/nhess-10-1851-2010.

Mantovani, F., R. Soeters, and C. van Westen. 1996. "Remote Sensing Techniques for Landslide Studies and Hazard Zonationn Europe." *Geomorphology* 15: 213–225.

Marino, Pasquale, Giovanni Francesco Santonastaso, Xuanmei Fan, and Roberto Greco. 2021. "Prediction of Shallow Landslides in Pyroclastic-Covered Slopes by Coupled Modeling of Unsaturated and Saturated Groundwater Flow." *Landslides* 18 (1): 31–41. https://doi.org/10.1007/s10346-020-01484-6.

Martha, Tapas R., Cees J. van Westen, Norman Kerle, Victor Jetten, and K. Vinod Kumar. 2013. "Landslide Hazard and Risk Assessment Using Semi-Automatically Created Landslide Inventories." *Geomorphology* 184: 139–150. https://doi.org/10.1016/j.geomorph.2012.12.001.

Meena, Sansar Raj, Lucas Pedrosa Soares, Carlos H. Grohmann, Cees van Westen, Kushanav Bhuyan, Ramesh P. Singh, Mario Floris, and Filippo Catani. 2022. "Landslide Detection in the Himalayas Using Machine Learning Algorithms and U-Net." *Landslides* 19 (5): 1209–1229. https://doi.org/10.1007/s10346-022-01861-3.

Mercogliano, P., S. Segoni, G. Rossi, B. Sikorsky, V. Tofani, P. Schiano, F. Catani, and N. Casagli. 2013. "Brief Communication a Prototype Forecasting Chain for Rainfall Induced Shallow Landslides." *Natural Hazards and Earth System Science* 13 (3): 771–777. https://doi.org/10.5194/nhess-13-771-2013.

Meusburger, K., and C. Alewell. 2009. "On the Influence of Temporal Change on the Validity of Landslide Susceptibility Maps." *Natural Hazards and Earth System Science* 9 (4): 1495–1507. https://doi.org/10.5194/nhess-9-1495-2009.

Miller, Daniel J., and Kelly M. Burnett. 2007. "Effects of Forest Cover, Topography, and Sampling Extent on the Measured Density of Shallow, Translational Landslides." *Water Resources Research* 43 (3): 1–23. https://doi.org/10.1029/2005WR004807.

Mondal, Sujit, and Ramkrishna Maiti. 2012. "Landslide Susceptibility Analysis of Shiv-Khola Watershed, Darjiling: A Remote Sensing & GIS Based Analytical Hierarchy Process (AHP)." *Journal of the Indian Society of Remote Sensing* 40 (3): 483–496. https://doi.org/10.1007/s12524-011-0160-9.

Monkhouse, F. J. 1970. *Dictionary of Geography (Monkhouse)* (2nd ed.). Edward Arnold Ltd.

Montrasio, L., R. Valentino, and G. Losi. 2011. "Towards a Real-Time Susceptibility Assessment of Rainfall-Induced Shallow Landslides on a Regional Scale." *Natural Hazards and Earth System Sciences* 11: 1927–1947.

Nagarajan, R., A. Roy, R. Vinodkumar, and M. Khire. 2000. "Landslide Hazard Susceptibility Mapping Based on Terrain and Climatic Factors for Tropical Monsoon Region." *Engineering Geology* 58: 275–287.

Naithani, A. 2007. "Macro Landslide Hazard Zonation Mapping Using Univariate Statistical Analysis in Parts of Garhwal Himalaya." *Journal Geological Society of India* 70: 353–368.

National Disaster Management Authority of India. 2009. *National Disaster Management Guidelines: Management of Landslides and Snow Avalanches.* National Disaster Management Authority, Government of India.

National Disaster Management Authority of India. 2019, September. *National Landslide Risk Management Strategy* (pp. 1–68). New Delhi: Government of India.

Nefeslioglu, Hakan A., Tamer Y. Duman, and Serap Durmaz. 2008. "Landslide Susceptibility Mapping for a Part of Tectonic Kelkit Valley (Eastern Black Sea Region of Turkey)." *Geomorphology* 94 (3–4): 401–418. https://doi.org/10.1016/j.geomorph.2006.10.036.

Ng, C. W. W., B. Yang, Z. Q. Liu, J. S. H. Kwan, and L. Chen. 2021. "Spatiotemporal Modelling of Rainfall-Induced Landslides Using Machine Learning." *Landslides* 18 (7): 2499–2514. https://doi.org/10.1007/s10346-021-01662-0.

Nguyen, Ba Quang Vinh, and Yun Tae Kim. 2021. "Regional-Scale Landslide Risk Assessment on Mt. Umyeon Using Risk Index Estimation." *Landslides* 18 (7): 2547–2564. https://doi.org/10.1007/s10346-021-01622-8.

Niethammer, U., S. Rothmund, M. R. James, J. Travelletti, M. Joswig, and Lancaster Environment Centre. 2010. *Uav-Based Remote Sensing of Landslides.* International Archives of Photogrammetry, Remote Sensing and Spatial Information Sciences, Vol. XXXVIII, Part 5 Commission V Symposium, Newcastle upon Tyne, UK.

Ohlmacher, Gregory C., and John C. Davis. 2003. "Using Multiple Logistic Regression and GIS Technology to Predict Landslide Hazard in Northeast Kansas, USA." *Engineering Geology* 69 (3–4): 331–343. https://doi.org/10.1016/S0013-7952(03)00069-3.

Panikkar, S., and V. Subramaniyan. 1997a. "4 SV Pannikar DDun(1997).Pdf." *Current Science* 72 (12): 1117–1121.

Panikkar, S., and V. Subramaniyan. 1997b. "Landslide Hazard Analysis of the Area Around Dehra Dun and Mussoorie, Uttar Pradesh." *Current Science* 73: 1117–1123.

Pardeshi, S., S. Pardeshi, and S. Autade. 2013. "Landslide Hazard Assessment: Recent Trends and Techniques." *SpringerPlus* 2 (523). https://doi.org/10.1186/2193-1801-2-253.

Pardeshi, S., S. Pardeshi, and V. Nagare. 2009. "A Study of Effect of Landslides on Human Environment in Thane District, Maharshtra State." *The Deccan Geographer* 47 (1): 45–56.

Parise, M. 2002. "Landslide Hazard Zonation of Slopes Susceptible to Rock Falls and Topples." *Natural Hazards and Earth System Sciences* 2 (1–2): 37–49. https://doi.org/10.5194/nhess-2-37-2002.

Paulin, Gabriel Legorreta, Katherine A. Mickelson, Trevor A. Contreras, William Gallin, Kara E. Jacobacci, and Marcus Bursik. 2022. "Assessing Landslide Volume Using Two Generic Models: Application to Landslides in Whatcom County, Washington, USA." *Landslides* 19 (4): 901–912. https://doi.org/10.1007/s10346-021-01825-z.

Pelty, D. 2010. "Landslide Hazard." In A. Ayala, I. Goudie (Eds.), *Geomorphological Hazards and Disaster Prevention* (pp. 63–73). Cambridge University Press.

Pereira, S., J. Zezere, and C. Bateira. 2012. "Assessing Predictive Capacity and Conditional Independence of Landslide Predisposing Factors for Shallow Landslide Susceptibility Models." *Natural Hazards and Earth System Sciences* 12: 979–988.

Peres, David J., and Antonino Cancelliere. 2021. "Comparing Methods for Determining Landslide Early Warning Thresholds: Potential Use of Non-Triggering Rainfall for Locations with Scarce Landslide Data Availability." *Landslides* 18 (9): 3135–3147. https://doi.org/10.1007/s10346-021-01704-7.

Perrone, Angela, Filomena Canora, Giuseppe Calamita, Jessica Bellanova, Vincenzo Serlenga, Serena Panebianco, Nicola Tragni, et al. 2021. "A Multidisciplinary Approach for Landslide Residual Risk Assessment: The Pomarico Landslide (Basilicata Region, Southern Italy) Case Study." *Landslides* 18 (1): 353–365. https://doi.org/10.1007/s10346-020-01526-z.

Piacentini, Daniela, Francesco Troiani, Mauro Soldati, Claudia Notarnicola, Daniele Savelli, Stefan Schneiderbauer, and Claudia Strada. 2012. "Statistical Analysis for Assessing Shallow-Landslide Susceptibility in South Tyrol (South-Eastern Alps, Italy)." *Geomorphology* 151–152: 196–206. https://doi.org/10.1016/j.geomorph.2012.02.003.

Polemio, Maurizio, and Francesco Sdao. 1999. "The Role of Rainfall in the Landslide Hazard: The Case of the Avigliano Urban Area (Southern Apennines, Italy)." *Engineering Geology* 53 (3–4): 297–309. https://doi.org/10.1016/S0013-7952(98)00083-0.

Pradhan, Biswajeet, and Saro Lee. 2009. "Landslide Risk Analysis Using Artificial Neural Network Model Focussing on Different Training Sites." *International Journal of Physical Sciences* 4 (1): 01–015.

Pradhan, Biswajeet, and Saro Lee. 2010. "Regional Landslide Susceptibility Analysis Using Back-Propagation Neural Network Model at Cameron Highland, Malaysia." *Landslides* 7 (1): 13–30. https://doi.org/10.1007/s10346-009-0183-2.

Preuth, Thomas, Thomas Glade, and Alain Demoulin. 2010. "Stability Analysis of a Human-Influenced Landslide in Eastern Belgium." *Geomorphology* 120 (1–2): 38–47. https://doi.org/10.1016/j.geomorph.2009.09.013.

Rezig, S., J. Favre, and E. Leroi. 1996a. *Landslides* (pp. 351–355). I. Sennset (Ed.). Balkema.

Rezig, S., J. Favre, and E. Leroi. 1996b. "The Probabilistic Evaluation of Landslide Risk." In I. Sennset (Ed.), *Landslides* (pp. 351–355). Rotterdam: Balkema.

Rodrigues, Saulo Guilherme, Maisa Mendonça Silva, and Marcelo Hazin Alencar. 2021. "A Proposal for an Approach to Mapping Susceptibility to Landslides Using Natural Language Processing and Machine Learning." *Landslides* 18 (7): 2515–2529. https://doi.org/10.1007/s10346-021-01643-3.

Rosi, Ascanio, Samuele Segoni, Vanessa Canavesi, Antonio Monni, Angela Gallucci, and Nicola Casagli. 2021. "Definition of 3D Rainfall Thresholds to Increase Operative Landslide Early Warning System Performances." *Landslides* 18 (3): 1045–1057. https://doi.org/10.1007/s10346-020-01523-2.

Rotigliano, E., V. Agnesi, C. Cappadonia, and C. Conoscenti. 2011. "The Role of Diagnostic Areas in the Assessment of Landslide Susceptibility Models: A Test in the Sicilian Chain." *Natural Hazards* 58 (NA): 981–999.

Rowbotham, D., and D. Dudycha. 1998. "GIS Modelling of Slope Stability in Phewa Watershed, Nepal." *Geomorphology* 26: 151–170.

Roy, Priyom, Nirmala Jain, Tapas R. Martha, and K. Vinod Kumar. 2022. "Reactivating Balia Nala Landslide, Nainital, India – a Disaster in Waiting." *Landslides* 19 (6): 1531–1535. https://doi.org/10.1007/s10346-022-01881-z.

Saaty, T. 2008. "Decision Making with the Analytic Hierarchy Process." *International Journal of Services Sciences* 1 (1): 83–98.

Salciarini, Diana, Jonathan W. Godt, William Z. Savage, Pietro Conversini, Rex L. Baum, and John A. Michael. 2006. "Modeling Regional Initiation of Rainfall-Induced Shallow Landslides in the Eastern Umbria Region of Central Italy." *Landslides* 3 (3): 181–194. https://doi.org/10.1007/s10346-006-0037-0.

Sarkar, S., D. Kanungo, and G. Mehrotra. 1995. "Landslide Hazard Zonation: A Case Study of Garhwal Himalaya, India." *Mountain Research and Development* 15: 301–309.

Sarkar, S., D. Kanungo, A. Patra, and P. Kumar. 2006. "GIS Based Landslide Susceptibility Mapping- A Case Study in Indian Himalaya." In *Disaster Mitigation of Debris Flows, Slope Failures and Landslides* (pp. 617–624). Tokyo, Japan: Universal Academy Press.

Schicker, Renée, and Vicki Moon. 2012. "Comparison of Bivariate and Multivariate Statistical Approaches in Landslide Susceptibility Mapping at a Regional Scale." *Geomorphology* 161–162: 40–57. https://doi.org/10.1016/j.geomorph.2012.03.036.

Selby, M. J. 1982. *Hillslope Materials and Processes*. Oxford University Press.

Sharma, Amol, Chander Prakash, and Anuj Nautiyal. 2022. "Recent Landslide News of Himachal Pradesh, India." *Landslides* 19 (1): 221–224. https://doi.org/10.1007/s10346-021-01802-6.

Sharma, L., N. Patel, M. Ghosh, and P. Debnath. 2009. "Geographical Information System Based Landslide Probabilistic Model with Trivariate Approach – A Case Study in Sikkim Himalaya." In *Eighteenth United Nations Regional Cartographic Conference for Asia and the Pacific*. Bangkok: Economic and Social Council, United Nations.

Sharpe, C. 1938. *Landslides and Related Phenomenon*. Columbia University Press.

Singh, C., K. Behra, and W. Rocky. 2011. "Landslide Susceptibility Along NH-39 between Karong and Mao, Senapati District, Manipur." *Journal Geological Society of India* 78: 559–570.

Song, Dongri, Yitong Bai, Xiao Qing Chen, Gordon G. D. Zhou, Clarence E. Choi, Alessandro Pasuto, and Peng Peng. 2022. "Assessment of Debris Flow Multiple-Surge Load Model Based on the Physical Process of Debris-Barrier Interaction." *Landslides* 19 (5): 1165–1177. https://doi.org/10.1007/s10346-021-01778-3.

Sterlacchini, S., C. Ballabio, J. Blahut, M. Masetti, and A. Sorichetta. 2011. "Spatial Agreement of Predicted Patterns in Landslide Susceptibility Maps." *Geomorphology* 125 (1): 51–61. https://doi.org/10.1016/j.geomorph.2010.09.004.

Strouth, Alex, and Scott McDougall. 2021. "Societal Risk Evaluation for Landslides: Historical Synthesis and Proposed Tools." *Landslides* 18 (3): 1071–1085. https://doi.org/10.1007/s10346-020-01547-8.

Tanyaş, Hakan, Dalia Kirschbaum, Tolga Görüm, Cees J. van Westen, Chenxiao Tang, and Luigi Lombardo. 2021. "A Closer Look at Factors Governing Landslide Recovery Time in Post-Seismic Periods." *Geomorphology* 391. https://doi.org/10.1016/j.geomorph.2021.107912.

Tempa, Karma, Kinley Peljor, Sangay Wangdi, Rupesh Ghalley, Kelzang Jamtsho, Samir Ghalley, and Pratima Pradhan. 2021. "UAV Technique to Localize Landslide Susceptibility and Mitigation Proposal: A Case of Rinchending Goenpa Landslide in Bhutan." *Natural Hazards Research* 1 (4): 171–186. https://doi.org/10.1016/j.nhres.2021.09.001.

Tien Bui, Dieu, Biswajeet Pradhan, Owe Lofman, Inge Revhaug, and Oystein B. Dick. 2012. "Landslide Susceptibility Assessment in the Hoa Binh Province of Vietnam: A Comparison of the Levenberg-Marquardt and Bayesian Regularized Neural Networks." *Geomorphology* 171–172: 12–29. https://doi.org/10.1016/j.geomorph.2012.04.023.

Tofani, V., S. Segoni, A. Agostini, F. Catani, and N. Casagli. 2013. "Technical Note: Use of Remote Sensing for Landslide Studies in Europe." *Natural Hazards and Earth System Science* 13 (2): 299–309. https://doi.org/10.5194/nhess-13-299-2013.

Tolga, Can, Hakan A. Nefeslioglu, Candan Gokceoglu, Harun Sonmez, and Tamer Y. Duman. 2005. "Susceptibility Assessments of Shallow Earthflows Triggered by Heavy Rainfall at Three Catchments by Logistic Regression Analyses." *Geomorphology* 72 (1–4): 250–271. https://doi.org/10.1016/j.geomorph.2005.05.011.

United Nations International Strategy for Disaster Reduction. 2004. *Terminology of Disaster Risk Reduction.* www.unisdr.org/eng/library/lib-terminology-eng home.htm.

van Westen, C. 2010. "GIS for the Assessment of Risk from Geomorphological Hazards." In I. Alcántara-Ayala, A. S. Goudie (Eds.), *Geomorphological Hazards and Disaster Prevention* (pp. 205–219). Cambridge University Press.

Varnes, D. I. 1984. *Landslide Hazard Zonation: A Review of Principles and Practice.* Paris: United Nations Scientific and Cultural Organization.

Wang, Fawu, Zixin Zhao, Ye Chen, Guolong Zhu, Kounghoon Nam, and Zhenhua Ye. 2022. "Guocun Landslide in a Slate Slope with Dip Structure on 27 March 2021 in Tonglu County, Zhejiang Province, China." *Landslides* 19 (6): 1435–1447. https://doi.org/10.1007/s10346-022-01846-2.

Wang, H. B., and K. Sassa. 2005. "Comparative Evaluation of Landslide Susceptibility in Minamata Area, Japan." *Environmental Geology* 47 (7): 956–966. https://doi.org/10.1007/s00254-005-1225-2.

Wu, W., and R. C. Sidle. 1995. "A Distributed Slope Stability Model for Steep Forested Basins." *Water Resources* 31: 2097–2110. https://doi.org/10.1029/95WR01136.

Xu, Chong, Fuchu Dai, Xiwei Xu, and Yuan Hsi Lee. 2012. "GIS-Based Support Vector Machine Modeling of Earthquake-Triggered Landslide Susceptibility in the Jianjiang River Watershed, China." *Geomorphology* 145–146: 70–80. https://doi.org/10.1016/j.geomorph.2011.12.040.

Xu, Ling, Dongdong Yan, and Tengyuan Zhao. 2021a. "Probabilistic Evaluation of Loess Landslide Impact Using Multivariate Model." *Landslides* 18 (3): 1011–1023. https://doi.org/10.1007/s10346-020-01521-4.

Xu, Yuankun, David L. George, Jinwoo Kim, Zhong Lu, Mark Riley, Todd Griffin, and Juan de la Fuente. 2021b. "Landslide Monitoring and Runout Hazard Assessment by Integrating Multi-Source Remote Sensing and Numerical Models: An Application to the Gold Basin Landslide Complex, Northern Washington." *Landslides* 18 (3): 1131–1141. https://doi.org/10.1007/s10346-020-01533-0.

Yang, Che-Ming, Wei-An Chao, Meng-Chia Weng, Yu-Yao Fu, Jui-Ming Chang, and Wei-Kai Huang. 2022. "Che-Ming Yang 2022.Pdf." *Landslides (Springer-Verlag)* 19: 1807–1811.

Yeon, Y., J. Han, and K. Ryu. 2010. "Landslide Susceptibility Mapping in Injae, Korea Using a Decision Tree." *Engineering Geology (Elsevier)* 166: 274–283.

Yoshimatsu, H., and S. Abe. 2006. "A Review of Landslide Hazards in Japan and Assessment of Their Susceptibility Using an Analytical Hierarchic Process (AHP) Method." *Landslides* 3 (2): 149–158. https://doi.org/10.1007/s10346-005-0031-y.

Zêzere, J. L. 2002. "Landslide Susceptibility Assessment Considering Landslide Typology. A Case Study in the Area North of Lisbon (Portugal)." *Natural Hazards and Earth System Sciences* 2 (1–2): 73–82. https://doi.org/10.5194/nhess-2-73-2002.

Zhou, Chao, Ying Cao, Xie Hu, Kunlong Yin, Yue Wang, and Filippo Catani. 2022. "Enhanced Dynamic Landslide Hazard Mapping Using MT-InSAR Method in the Three Gorges Reservoir Area." *Landslides* 19 (7): 1585–1597. https://doi.org/10.1007/s10346-021-01796-1.

Zizioli, D., C. Meisina, R. Valentino, and L. Montrasio. 2013. "Comparison between Different Approaches to Modeling Shallow Landslide Susceptibility: A Case History in Oltrepo Pavese, Northern Italy." *Natural Hazards and Earth System Sciences* 13 (3): 559–573. https://doi.org/10.5194/nhess-13-559-2013.

4 Spatial Patterns of Landslides and Their Characteristics
Case Studies

4.1 INTRODUCTION

Preparation of a complete landslide inventory with considerable details is a primary requirement for effective landslide hazard assessment. A detailed landslide inventory mainly requires data inputs such as the location of the landslide, date of the event, frequency of landslides, causes, and type of mass movement. Past landslide records, multi-temporal satellite images, aerial photographs, and newspaper archives are proven databases for the preparation of landslide inventories. Landslide inventory provides the necessary database for landslide hazard zonation (LHZ) mapping. Complete and reliable landslide inventories provide a necessary database for the identification of landslide-prone areas and evaluation of the validity of LHZ models.

This chapter describes the methods of studying spatiotemporal patterns of landslide distribution and landslide characteristics with the help of appropriate case studies. Besides, methods of analysis of landslide geometry parameters, landslide volume estimation, estimation of runout distance, identification of landslides using past landslide records, satellite image interpretation, field investigations, etc. are demonstrated through case studies. The relationship between the distribution of landslides and slope, elevation, drainage, geology, and rainfall data is established to determine the role of geo-environmental parameters in slope instability in the study area.

4.2 IDENTIFICATION AND DELINEATION OF LANDSLIDE-AFFECTED REGIONS: AN OVERVIEW OF TECHNIQUES

This section of the study focuses on the assessment of spatial and temporal patterns of landslide distribution in North Konkan using past landslide records and field investigations.

Past landslide data were obtained from the public works department, government of Maharashtra, disaster management cell, and National Highways Authority of India (NHAI) subdivisions and *Times of India* archives from 2005

DOI: 10.1201/9781003323488-4

to 2014. The sites of slope failures along major roads were visited, and data about landslide location, slope angle, and landslide type have been recorded in the field using global positioning system (GPS), laser distance metre, and measuring tape. Besides slope failures along the roads, a few landslides in the interior areas have been identified using data from newspaper archives. The identified landslide sites were visited, and a discussion with local people was held for confirmation.

Subsequently, the landslide data are plotted using Google Earth Pro and road development plans (2001–2020). The annual and monthly landslide frequency is obtained using temporal data of landslides. It is compared with rainfall data to establish the relationship between rainfall and landslide occurrences.

Based on the data obtained from government offices and observations recorded during field investigations, major zones of landslide concentrations are identified and mapped. The spatial distribution of landslides in the study area is compared with lithology, slope, aspect, elevation, and zones of structural discontinuities (faults and lineaments).

Besides, an attempt has been made to identify the role of human interventions in the slope instability process along major communication routes. Land use and land cover are also compared with the spatial distribution of slope failures to know the effect of land use. Land cover is also compared to understand any effect of it on the occurrence of slope failures or vice versa. The details of the methodology adopted are given in Figure 4.1.

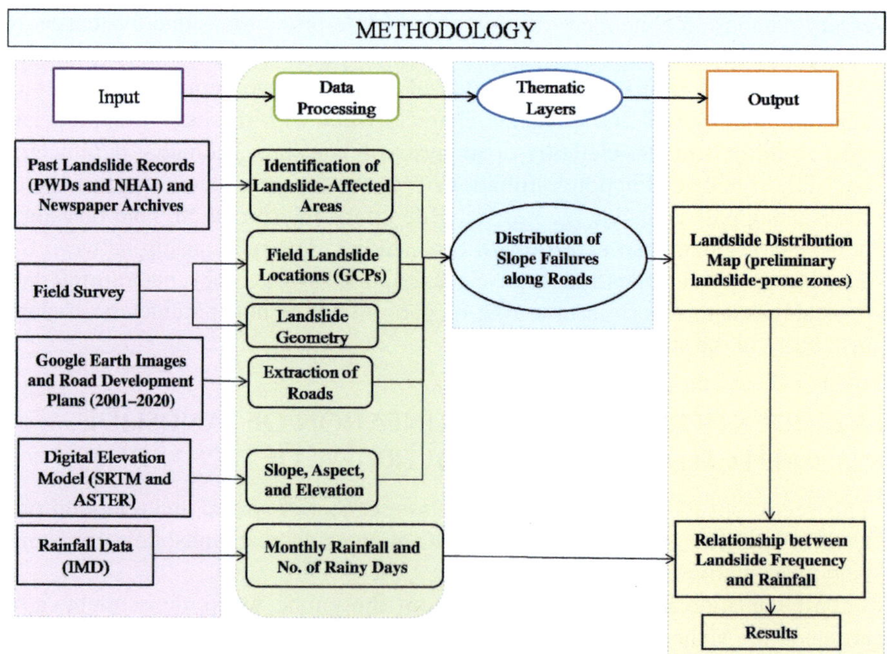

FIGURE 4.1   Methodology Flowchart.

4.3 CASE STUDY 1: LANDSLIDES IN THE WESTERN GHATS MATERIALS

Assessment of spatial patterns of actual and potential landslides is an important step in landslide hazard assessment (LHA). Past landslide records and field surveys are proven data sources for landslide studies (Jaiswal et al., 2010). The North Konkan region of Maharashtra, India, has been selected for this case study (Figure 4.2).

For the present study, landslide-affected areas have been identified using past landslide records of the technical survey register (TS register) of the public works department (PWD), the natural disaster report of the NHAI, newspaper archives, and disaster management cell (DMC of Thane Municipal Corporation). The data obtained from these government offices have been supplemented with the field survey. The details of landslide locations provided by PWD have been identified and confirmed by using GPS locations and road chainage. A total of 199 actual landslide-affected locations were visited during fieldwork, and data related to landslide scar geometry were recorded in the field (Table 4.1).

According to State Disaster Management Authority of India, North Konkan is a part of Konkan and is one of the three major landslide-prone areas in South Asia and is susceptible to landslides of different magnitudes.

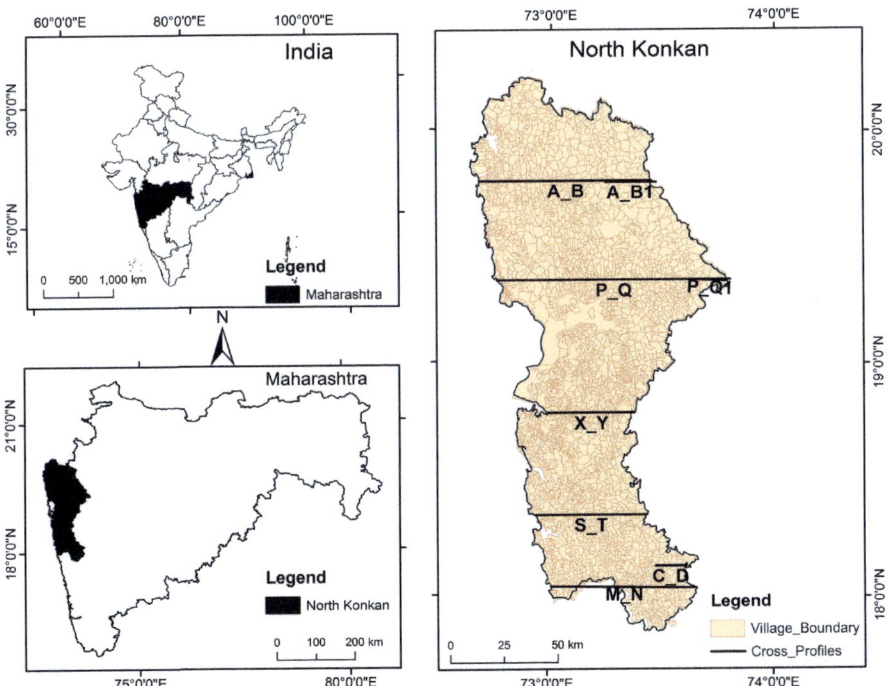

FIGURE 4.2 Location Map of the Study Area: North Konkan, Maharashtra, India.

TABLE 4.1

Landslides in North Konkan (2004–2014)

						Areas/Roads				
Year	Jawhar Division	Thane Division	NH-222	TMC	NH-3	Mahad	Alibag	NH-17	Expressway	Total
2004	08	11	DNA	DNA	10	DNA	10	DNA	33	72
2005	48	DNA	14	DNA	DNA	32	15	91	47	247
2006	0	07	14	DNA	DNA	13	06	45	11	96
2007	0	02	46	DNA	01	100	07	46	19	221
2008	0	03	36	DNA	02	18	0	14	04	77
2009	01	01	11	01	01	35	04	15	02	71
2010	01	0	35	01	0	26	64	10	0	137
2011	0	0	09	05	01	19	22	25	01	82
2012	04	0	04	04	01	17	03	0	0	33
2013	01	01	08	01	0	18	01	0	02	32
2014	0	01	01	0	03	01	0	02	01	09
Total	63	26	178	12	19	279	132	248	120	1077

(*Source*: Public Works Departments, National Highways Authority of India, India Meteorological Department, *Times of India* archives)

North Konkan is susceptible to slope failures of different magnitudes. Most of the reported landslides in the study area are confined to the western slopes of the Western Ghat escarpment, slopes of isolated hillocks in the central uplands, and along the edges of low-level plateaus (Figure 4.3). The records of past landslide events show significant variation in their distribution pattern (Table 4.1). Major concentrations of the landslide-affected areas in North Konkan are the following:

1 Daberi-Behedpada Circle
2 Khodala area (Devbandhar)
3 Kasara Ghat
4 Parali-Ujjaini Road
5 Malshej Ghat
6 Kalwa-Mumbra
7 Pune-Mumbai Expressway (Khandala section)
8 Neral-Matheran Road
9 Sukeli Pass
10 Shrivardhan-Shekhadi Road
11 Varandh Ghat
12 Mahad Circle
13 Poladpur Circle

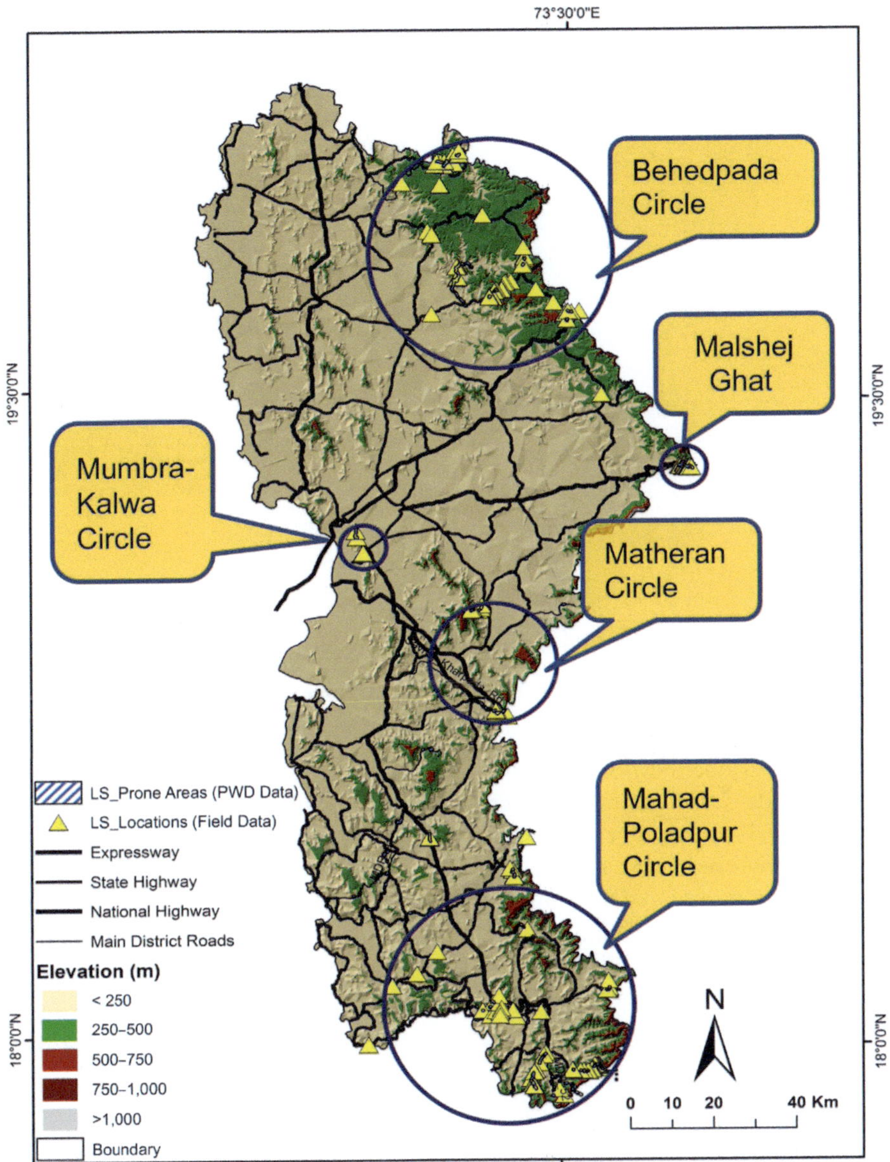

FIGURE 4.3 Distribution of Slope Failures in the North Konkan Region.

4.3.1 Dhaberi-Behedpada Circle

This area is located in the north-east corner of the study area. Administratively, the area is in Mokhada tehsil. This area is marked by highly dissected topography;

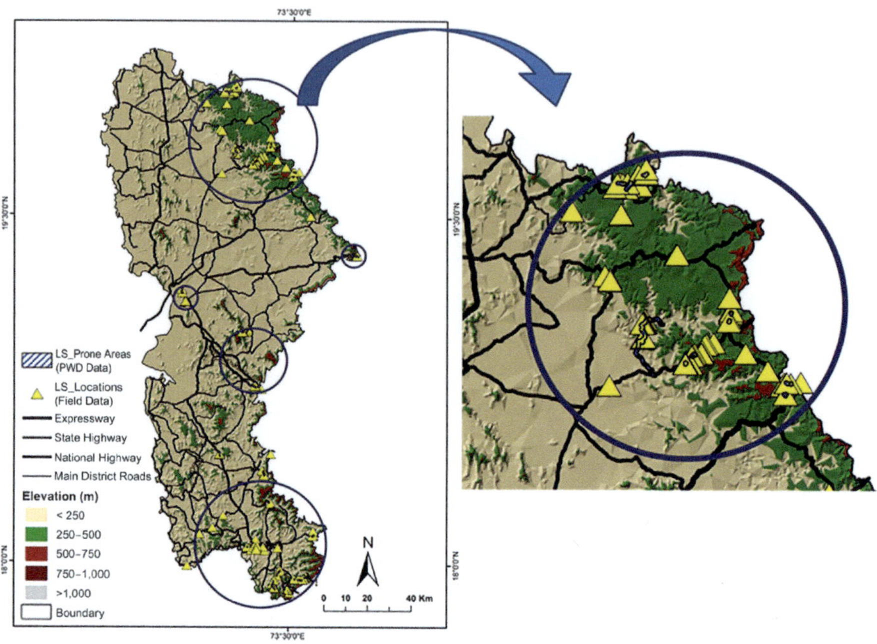

FIGURE 4.4　An Expanded View of the Behedpada Circle in North Konkan.

several landslide locations have been identified along SH 28. Debris slump and debris slide are common types of slope failures, but wedge failures are also observed. This area is covered with the oldest basaltic formation (Salher) and classified as megacryst flow. A major landslide site with a runout distance of 500 metres is observed near Behedpada village, which destroyed an agricultural field, a well, and a few hamlets in July 2005. Steep slopes $(25°)$, heavy rainfall, barren upper slopes, and slope cutting for road construction are major landslide factors in this area (Figure 4.4).

4.3.2　KHODALA CIRCLE

This is another major landslide-affected area, which includes slope failures along the Wada-Khodala Road, SH 37 (in the Khodala-Parali section), Khodala-Mokhada Road, and Kasara-Khodala Road.

Pardeshi et al. (2009) identified 31 actual and potential landslide sites along the Khodala-Wada Road using slope analysis and field investigation. The study revealed that besides disruption to traffic, damage to vegetation cover and geomorphic damage are major effects of landslide events in this area.

This area is characterised by a lower megacryst lava flow horizon with 60 km thickness near Suryamal (Pardeshi et al., 2009). A total of 21 slope failures have been identified along SH 34 (Khodala-Wada), most of which are identified as debris slides, whereas few rotational slides are observed at places. These landslides are comparatively small in extent (scar length ranges between 19 and 25 m). General slopes are between 15° and 30° along the road stretch. The area around Suryamal is intruded by north-south trending dykes and has steep slopes, coarse material, and dense drainage networks; the existence of fracture joints and lineament, along with heavy rainfall, causes landslides in the 8 km long section of Khodala-Wada Road (Pardeshi et al., 2009).

Besides the Khodala-Wada Road, the Khodala-Mokhada Road also experiences landslides. In July 2005, landslides near Dolara and Devbandh caused natural damming on the Pinjal River. The burst of the landslide dam caused severe damage to a hostel, temple, and settlement downstream. A huge mass, 85 m long and 59 m wide, went into the stream for about 600 metres, damaging the road connecting Khodala and Mokhada (Figure 4.4).

Besides these two areas, a few landslide-affected areas are also observed along the road connecting Kasara and Mokhada (SH 37), near Kiniste village. Steep natural slope (29°–38°); proximity to major drainage networks; joints; heavy rainfall; and loose, coarse slope-forming materials are major causes for landslide occurrence in this area.

4.3.3 KASARA GHAT

The Mumbai-Agra Highway (NH 3) is one of the busiest communication routes which get affected by landslides. During the rainy season in the Kasara Ghat section, seven landslide locations have been identified in the field. Besides steep natural slopes and rainfall, slope cutting and blasting for road widening also caused slope failures in this section.

Recently (30 July 2014), landslides were reported at five places in the old Kasara Ghat section, which affected road traffic for the entire day (*Times of India*, 31 July 2014). Frequent slope failure incidents are also reported along the New Kasara Ghat Road near Igatpuri diversion. Translational debris slide is a common type of slope failure in this section.

Another major landslide event occurred in the Igatpuri section along the railway line near Hotel Manas. A 43.2 m wide and 65 m long debris slump disrupted rail traffic for three to five hours on 30 July 2014 (*Times of India*, 31 July 2014) after receiving 234 mm of rainfall within 24 hours on 29 July 2014.

Kasara Ghat starts from km 465/000 near Manas Hotel to km 474/000 at Latifwadi for about 9 km. In the peak monsoon season of 2004, on 2 and 3 August, a heavy spell of about 540 mm of rain caused a disastrous landslide episode (PWD, Nashik). Three major and seven minor landslide events were reported, of which two were fatal landslides. This landslide episode caused traffic disruption, damage to vehicles, and damage to roadside vegetation.

4.3.4 PARALI-UJJAINI ROAD

Parali-Ujjaini Road near Wada is also an important landslide-affected area. A total of ten fresh landslides are identified along just a 4 km road between Parali and Ujjaini village. Translational debris slides are a common type of slope failure in this section. A steep natural slope is a major cause of slope failure in this area. Few landslide scars are as long as 200 m, and displaced material are transported for over 500 m downslope. The traffic along Parali-Ujjaini Road was disrupted by landslide events during the peak rainy season. High-gradient streams cause slope undercutting and slope instability in this area (Figure 4.4).

A recent landslide on the bank of the Gargai River has been observed as the largest event in this section with a scar length of 119 m and width of 26.16 m; the material displaced from the scar has been deposited in the river channel.

4.3.5 Malshej Ghat Section

This landslide-affected area is located along NH 222, one of the major communication routes connecting Ahmednagar district to the Konkan coastal strip (Figure 4.5). As per the record of the NHAI (NH Division 3), a total of 178 landslides have been reported in Malshej Ghat during 2005–2013. The length of the ghat section is about 10 km between chainage 84 / 250 km and 94 / 500 km. Maximum landslide events are observed between 90 / 000 km and 93 / 000 km section. Translational debris slide is the predominant slope failure type in this area, whereas rockslides and wedge failure are also observed near the 92 / 000 km mark. Almost all landslide events in Malshej Ghat occur during the peak monsoon period, indicating rainfall is a major trigger in landslide occurrence.

Besides rain, steep slopes (30°–45° and up to 80°), the existence of fractured rocks, and sparsely vegetated upslope areas are major preparatory factors associated with landslides in the Malshej Ghat area. Anthropogenic interventions like slope cutting, road widening, and blasting also contribute to slope instability. While clearing the debris material from the road, it is dumped downslope from the site. This dumping of material also causes loss of vegetation and secondary landslides on the debris resting in the downslope direction. Road traffic gets disturbed due to landslides in Malshej Ghat almost every year, leading to huge losses in terms of time (delay in traffic flow), fuel loss, loss of working hours, damage to man-made structures, etc. Recently, a major landslide event on 15 September 2014 in the Malshej Ghat area interrupted road traffic for almost eight days. Another landslide event on 24 July 2013 near the tunnel of Malshej Ghat claimed two lives, after which the ghat was closed for almost a week.

4.3.6 Mumbra Circle

This is one of the most important landslide-affected areas in the study area because of the high risk to human settlements. Most of the slum areas are at the foot of Parsik Hill, which get damaged during landslide events. The 4 km stretch of the Mumbra-Kalwa bypass road is vulnerable to slope failure mainly due to the steep natural slope ($> 30°$) and slope cutting for road construction. Wedge failure and translational debris slide are common slope failure types in this section (Figure 4.3).

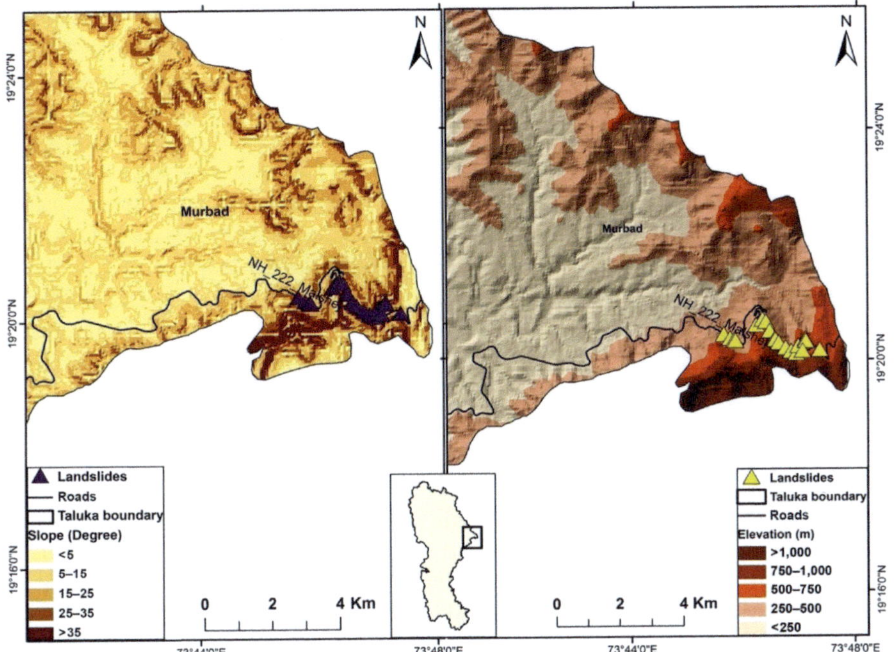

FIGURE 4.5   Slope Failures in the Malshej Ghat Section (NH 222).

Four fatal landslides have been reported in different localities of Kalwa and Mumbra, such as Gholai Nagar (Kalwa), Sainik Nagar (Mumbra), Shanti Nagar (Mumbra), and the foothills of the Mumbra Devi Temple during 2010–2012 (DMC, Thane Municipal Corporation).

Uncontrolled human encroachment on the slopes of Parsik Hill and slope modification are mainly responsible for slope instability in this area. A recent landslide event on 18 June 2013 occurred in a stone quarry in Turbhe and claimed the lives of five workers (*Times of India*, 2013).

Besides steep natural slopes and heavy rainfall, reckless mining operations near Parsik Hill are responsible for slope instability in this area.

4.3.7 PUNE-MUMBAI EXPRESSWAY

The Mumbai-Pune Expressway is one of the busiest communication roads in India opened for traffic in March 2002. This is probably the first expressway connecting the eastern plateau of Maharashtra to the Konkan coastal region. The ghat section of the expressway is susceptible to landslides and slope failure and has experienced several landslide events in 2003–2004.

Rockfall, toppling, and wedge failure are common on this road. In this section of the ghat, 13 compound pahoehoe flows are observed (Figure 4.3).

Earlier studies on the ghat section of this expressway (Kumar et al., 2010) indicated that the weak layer along open joints and intersecting joints of rock facilitated rapid weathering and erosion due to seepage of rainwater. Four landslide-affected sites are identified during fieldwork: viz., Amrutanjan Bridge, either end of Adoshi Tunnel, and a site below Duke Retreat Resort. One of the most frequent landslide-affected sites is located near the 41 km mark on the expressway. The slope failure at this site interrupts traffic almost every year during monsoon season. A major fatal landslide at this site, on 26 June 2009, claimed the life of one person, and seven others were injured (*Times of India*, 2009).

Mitigation measures, such as fixing metal nets, bolting the rock, and construction of culverts and retaining walls, have been taken up by MSRDC (Maharashtra State Road Development Corporation) for stabilising slopes along the expressway at sites vulnerable to landslides.

4.3.8 NERAL-MATHERAN ROAD

The Neral-Matheran Road is 11.50 km and is a major district road (MDR 9) that connects Neral railway station and Matheran hill station in Raigad district (Figure 4.6). Matheran is situated at 750 m ASL (above sea level). A total of 73 landslides have been reported from 2010 to 2012. Landslide hazards have been assessed using historical landslide records from the PWD and via field investigations.

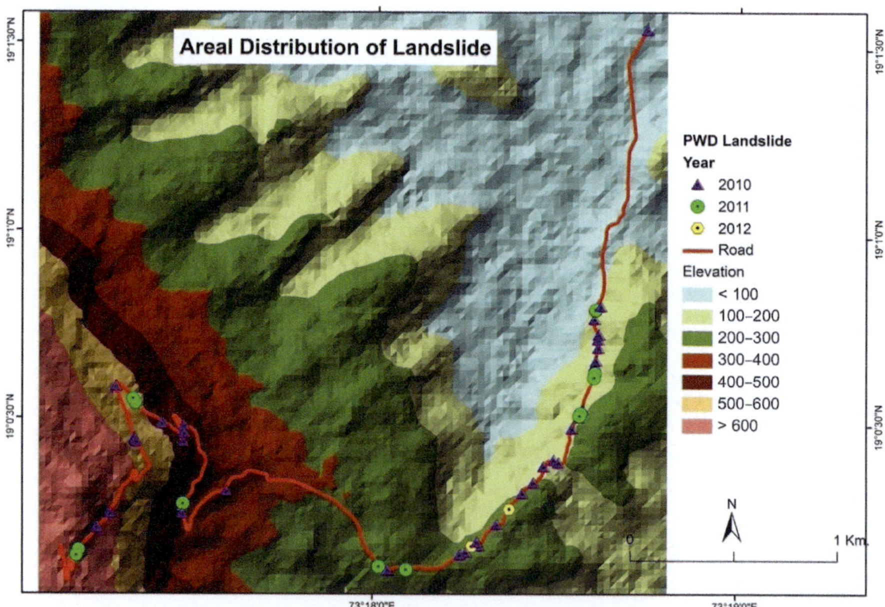

FIGURE 4.6   Distribution of Slope Failures along Neral-Matheran Road.

Occurrences of landslides on this road are mainly associated with rainfall during the monsoon season. Intense monsoon rain is a major landslide triggering factor responsible for slope failure in the hilly tracks of Matheran. After the analysis of landslide occurrences and rainfall, it is observed that there is a positive correlation between rainfall occurrence and landslide events. It is also observed that rainfall occurrence before landslide events is a major cause of landslides in the study area. From the field study, it is observed that landslides in the study area are often responsible for traffic disruption, loss of topsoil cover, and loss of vegetation.

4.3.9 SUKELI PASS

This landslide-affected area is located along the busiest section of the Mumbai-Goa National Highway (NH 17) between road chainage 64/000 and 66/000. Steep slopes (35°–40°), combined with heavy rainfall, often cause slope failure in this road section. A recent landslide event in Sukeli Pass blocked the highway for two to three hours. A total of 14 landslide events have been reported in this section during 2005–2013. Translational debris slide is a common landslide type in this area.

The area around Sukeli Pass is covered with compound and simple flows, classified under elephant formation, at an elevation of 90–113 m.

4.3.10 SHRIVARDHAN CIRCLE

This zone includes minor slope failures along Mangaon-Mhasla Road, Bagmandla-Shrivardhan Road, and Shrivardhan-Shekhadi Road. A total of seven slope failure sites have been identified in the field, of which four are located along the Mangaon-Mhasla Road and three along the coastal road connecting to Shrivardhan and Shekhadi. Mangaon-Mhasla Road passes through Sai village. This area exhibits highly rugged topography with a general slope between 25° and 35° and even more at some places. Four landslide sites are located at different places: one near Dongaroli village, one in Sai Ghat, and two sites near Mhasla Creek and Bagmandla. All landslides in this zone are classified as translational debris slides.

Geologically, the area is covered with simple lava flow (aa type), classified under the Diveghat subgroup, capped with laterite at places. The area around Shrivardhan is intruded by north-west to south-east trending lineaments.

Another major concentration of landslides, running parallel to the coastline between Shrivardhan and Shekhadi, is covered with compound and simple basaltic flows with a total absence of aluminium. Steeper slopes in the upslope direction (35°–55°) cause slope failures in this road section. Translational debris slides and boulder falls are common slope failure types in this area. The upslope is sparsely vegetated and devoid of vegetation at a few places, facilitating easy movement of loose, weathered material downslope. However, no major fatal landslide events have been reported in this section.

4.3.11 Varandh Ghat

Varandh Ghat, located along the state highway (SH 70) connecting Bhor tehsil of Pune district in the east and Mahad tehsil (Raigad district) in the west is an important landslide-affected area. The road passes through a narrow Varandh pass for about 10 km in the ghat section. Six landslide sites have been identified during the field survey. The Waghjai Temple area is the most vulnerable to landslides. Near-vertical slopes and fractured rock frequently cause landslides at this site. Several times, landslides caused damage to the Waghjai Temple and also disruption to road traffic in this section of the ghat.

The Parmachi landslide episode of 1992 claimed the lives of 12 people near Parmachi village. Few minor landslide scars have also been identified along this road between Parmachi village and Barasgaon. A recent landslide event on 20 July 2014 near Waghjai Temple interrupted road traffic for eight days. Debris slide is a common type of slope failure in this area, whereas boulder fall and toppling are also observed in some places. Heavy rainfall, steep to near-vertical slopes, slope cutting, blasting, and loosening of slope-forming material are the main causes of landslide initiation along this section of Varandh Ghat.

4.3.12 Mahad Circle

This is the largest landslide-affected zone with the maximum number of fatal landslides in the landslide episode of 2005 (Figure 4.7). This zone includes landslides around Mahad town, which can be subcategorised into two:

 i Dasgaon area
 ii Tol-Ambet Road

This area is located on the banks of the Savitri River. Geologically, the lowlands around Savitri and its tributaries are covered with essentially pahoehoe flows of Karla formation up to 100 m above mean sea level, whereas the hilly area around Savitri River is dominated by simple aa flows, classified under the Diveghat formation of the Sahyadri group. Relatively closely spaced north-south trending lineaments are observed 3–5 km west of Dasgaon village. This area presents highly rugged topography due to the existence of the foothills of the Western Ghat escarpment and isolated hillocks around Mahad. The regolith saturated with rainwater caused slope failure during heavy spells of rainfall in July 2005.

Debris flow is a common slope failure type observed in this area. The observed landslide scar shows concave upslopes and convex downslopes. From 23 to 26 July 2005, intense rainfall of over 87.8 mm was recorded within three days, which caused slope failure in this area. Jui, Dasgaon, Rohan, Tudil, Muthavli, Gothe, Kondvite, and Sav are the major landslide-affected areas in this zone (Figure 4.8). The details of the damage caused by this landslide episode are given in Table 4.2.

Thigale and Umrikar (2007) studied hydrothermal anomalies generated at six places (Dasgaon, Jui, Rohan, Tudil, Gothe, and Kondvite villages) as a unique

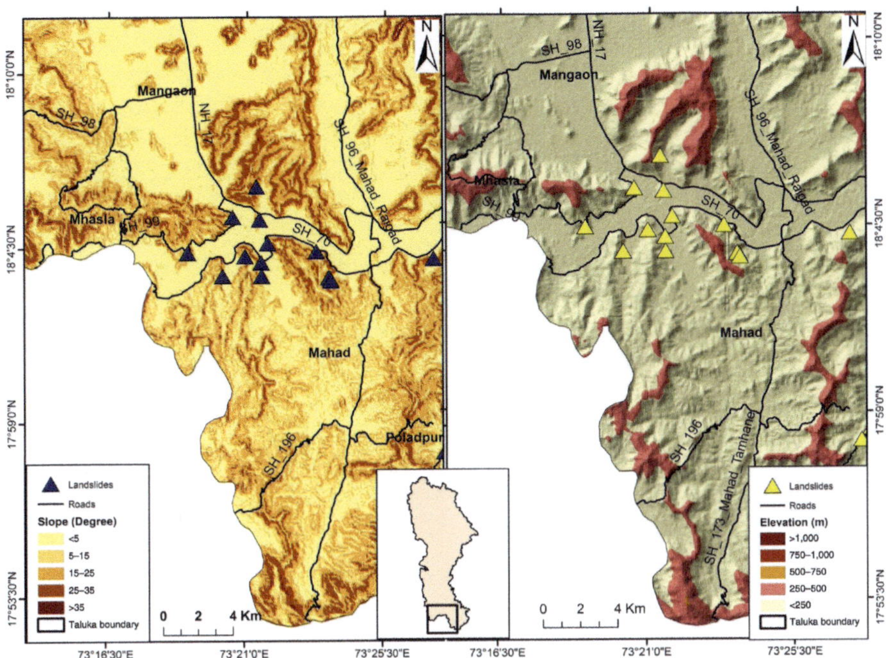

FIGURE 4.7   Distribution of Landslides in Mahad Circle.

TABLE 4.2
Details of Losses Caused by the July 2005 Landslide Episode around Mahad

S. No.	Landslide Name/Location	No. of Fatalities	No. of Injured Persons	Other Damages
1	Kondvite	36	100+	35–40 houses damaged
2	Rohan	14	01	44 houses completely damaged
3	Dasgaon	48	–	45+ downslope hamlets
4	Dasgaon_1	–	100+	40+ hamlets damaged
5	Jui	98	4	43 houses and 200+ cattle, 28 cowsheds
6	Tudil	–	–	Partially damaged school building
7	Muthavli	–	–	Damage to cropland

Source: *Times of India* archives and field survey

phenomenon during this landslide episode. According to them, groundwater is forced through pores and cracks due to increased sheer stress, which leads to the generation of electric currents along subsurface flows, increasing the temperature

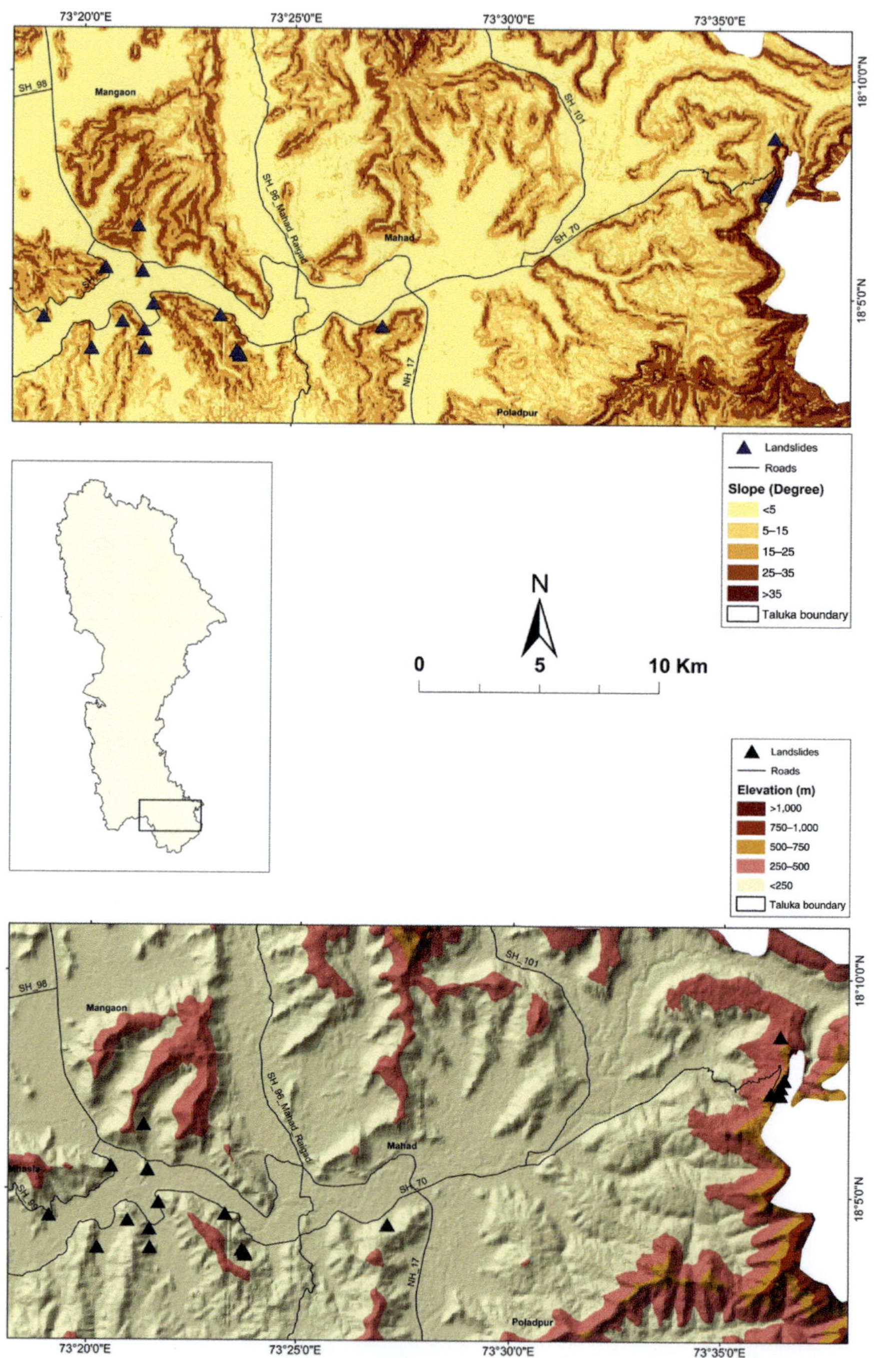

FIGURE 4.8   Slope Failures around Mahad Town.

and giving rise to the formation of dust and steam. The residents have reported that many dead bodies of people were found burnt and even a few people were injured by this hot, muddy water during the rescue operation.

Tol-Ambet Road, running roughly parallel to the bank of the Savitri River, is another important landslide-affected area. Two landslide events have been identified during the field survey: one near Sape-Tarfe village and another near Dabhol mohalla. The former landslide event of July 2005 caused total blockage of road traffic and also destroyed cultivated land in a downslope direction. Besides natural slopes and rainfall, anthropogenic activities such as road widening, blasting, cultivation along the slopes, and removal of natural vegetation also contribute to slope instability in this area.

4.3.13 Poladpur Circle

This is the southernmost landslide-prone area in North Konkan (Figure 4.9). It includes slope failures along the following communication routes.

   i Poladpur-Mahabaleshwar Road (Ambenali Ghat)
  ii Mumbai-Goa National Highway (NH 17) (Kashedi Ghat section)
 iii Poladpur-Devpur-Kotwal Road

   i Poladpur-Mahabaleshwar Road (Ambenali Ghat)

The area around Poladpur is covered with simple (aa type) lava flows up to 700 metres above mean sea level, classified under Diveghat formation. The uplands (700–900 metres ASL) are characterised by aa flows of Purandargarh formation. The area to the south and south-east of Poladpur town exhibits highly rugged terrain marked by the Western Ghat escarpment.

The communication routes across the ghat section are highly vulnerable to slope failures. Poladpur-Wai Road (SH 72) is an important communication route connecting the places of Konkan lowlands and eastern plateau areas. The ghat section (Ambenali Ghat) of this road is about 22 km long and is vulnerable to slope failures, especially during the peak monsoon season. Twenty-two landslide locations have been identified during a field survey in the entire stretch of Ambenali Ghat.

Translational debris slide is a predominant type of slope failure in this area. Steep natural slope (25°–45°) proximity to minor streams, weathered rock, and heavy rainfall are the major factors responsible for the initiation of landslides in this area. Besides, anthropogenic activities such as slope cutting, road widening, blasting, and deforestation also contribute to slope instability. One of the characteristic features of landslide-affected locations in Ambenali Ghat is its proximity to the high-gradient intermittent streams, indicating the role of subsurface water movement in the process of slope instability.

A recent landslide event, on 4 August 2014, near Chirekhind, blocked the road traffic in Ambenali Ghat for two days. Few landslide events have also been reported in 2008 and 2013 (*Times of India*, 2013).

ii) Mumbai-Goa National Highway (NH 17) (Kashedi Ghat section)

This is the most important landslide-prone area in Poladpur circle. Kashedi Ghat is located to the south of Poladpur town on the Mumbai-Goa National Highway (NH 17), the busiest communication route in the Konkan region. It is the main corridor to enter South Konkan. The total length of the ghat is about 14 km.

As per the records available at the NHAI, over 248 landslide events were reported from 2004 to 2014. Several incidents of landslides have been reported in the news, revealing that landslides in Kasedi Ghat interrupt road traffic during monsoon season almost every year. Steep natural slope, moderate to highly weathered rock, and loose slope-forming material, combined with intense monsoonal rain, cause slope failures in this area. There was a major landslide episode at 15 locations in Kashedi Ghat occurred on 9 September 2015 causing traffic blockage for three days.

A total of 11 landslide sites have been identified during the field survey. Most of the slope failures are debris slides, but rotational (slump) slides and debris flow

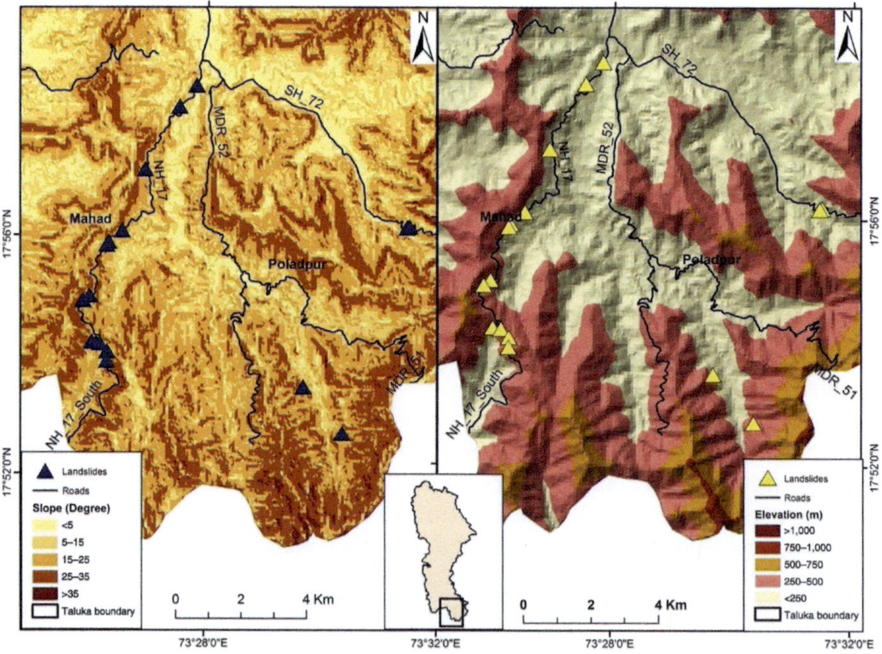

FIGURE 4.9　Distribution of Slope Failures in the Kashedi Ghat Section.

(especially at those sites located near a stream) are also observed at a few places. It is observed that slope cutting and blasting for road widening also contribute to slope instability, although a few retaining walls are constructed to stabilise the slope along the ghat road section.

iii) Poladpur-Devpur-Kotwal Road

This zone is located along the Devpur-Golegani-Kotwal Road, about 21 km south-east of Poladpur town. Two landslide sites have been identified during the field survey. The residents reported that both these fatal landslides occurred during the spell of heavy rainfall in July 2005.

The landslide near Kotwal Bk village was debris flow. A huge mass of debris was transported for about 1 km downslope, leaving behind a narrow and concave landslide scar. This caused the death of two persons with total damage to the cultivated land in downslope areas. Another event was observed near Kotwal Bk village, where a debris flow washed away one house, claiming seven lives in a family.

4.4 TEMPORAL PATTERNS OF LANDSLIDE DISTRIBUTION

The assessment of temporal patterns of landslide distribution is an important step in LHA. It also helps in understanding the relationships between landslide occurrence and its relationship with geo-environmental parameters associated with slope instability processes. This section discusses temporal patterns of landslide distribution in North Konkan.

Past landslide records are a proven database for temporal LHA (Jaiswal et al., 2010). In the present study, data about past landslide records have been obtained from the divisional offices of PWDs, NHAI subdivisions, disaster management cells, and newspaper archives for the period from 2004 to 2014. However, continuous data for the entire period could not be obtained for a few areas due to the unavailability of records in the concerned departments.

The landslide data obtained from the government offices have been compiled to obtain annual and monthly landslide frequencies, which have been then compared with corresponding rainfall data.

4.4.1 SLOPE FAILURES IN 2004–2014

The overall annual frequency of landslides in North Konkan is higher in the years 2005 (251 events), 2007 (221 events), and 2010 (137 events). The fact is that most landslides remained unreported; it is difficult to establish a concrete relation between annual landslide frequency and landslide causative factors. The assessment of annual landslide frequency based on past trends would help in a better understanding of slope instability in the area under investigation. However, the lack of complete and consistent landslide records puts limitations on further analysis.

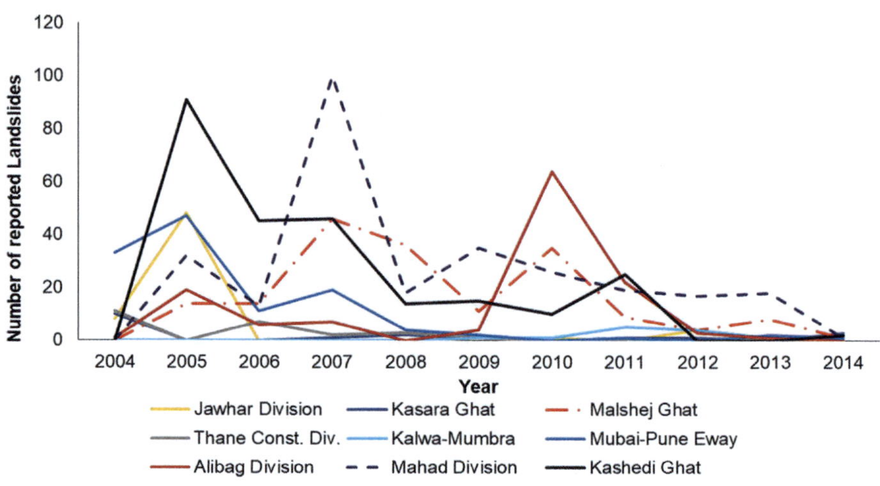

FIGURE 4.10   Landslide Occurrences in North Konkan during 2004–2014.

However, there is significant variation in the annual frequency of landslides in this region. The highest frequency of 279 landslide events was recorded in Mahad PWD jurisdiction during 2004–2014, followed by Kashedi Ghat (248), Malshej Ghat (178), and Mumbai-Pune Expressway, which are the major communication routes in the North Konkan region (Figure 4.10).

Although 2005, 2007, and 2010 show higher landslide frequency, there is a significant spatial variation in annual rainfall and the corresponding number of landslides. However, in almost all landslide-prone areas, landslide frequency was higher in 2005, which was the year of good rainfall (Figure 4.11).

4.4.2   Landslide Frequency and Monthly Rainfall (2004–2014)

Since most of the recorded landslide occurrences are attributed to the monsoon season, it shows a relationship between landslide events and rainfall. It was therefore important to establish a relationship between rainfall and landslide events recorded in the study area. The monthly rainfall data for the period between 2004 and 2014 was obtained from India Meteorological Department. An attempt has been made to establish a relation between monthly rainfall amount and landslide frequency for the respective period.

It is observed that landslide occurrence in the North Konkan region of Maharashtra is attributed to the monsoon season because almost all reported landslides occurred from June to September. The maximum frequency of 522 (48.69%) landslide events occurred in July, followed by August (29.66%) and June (12.97%), which are the rainiest months of the year (Figure 4.12). It reveals that rainfall is an important landslide triggering factor for landslide initiation in this area, as supported by earlier studies (Nagarajan et al., 2000; Thigale and Umrikar, 2007; Karlekar, 2012).

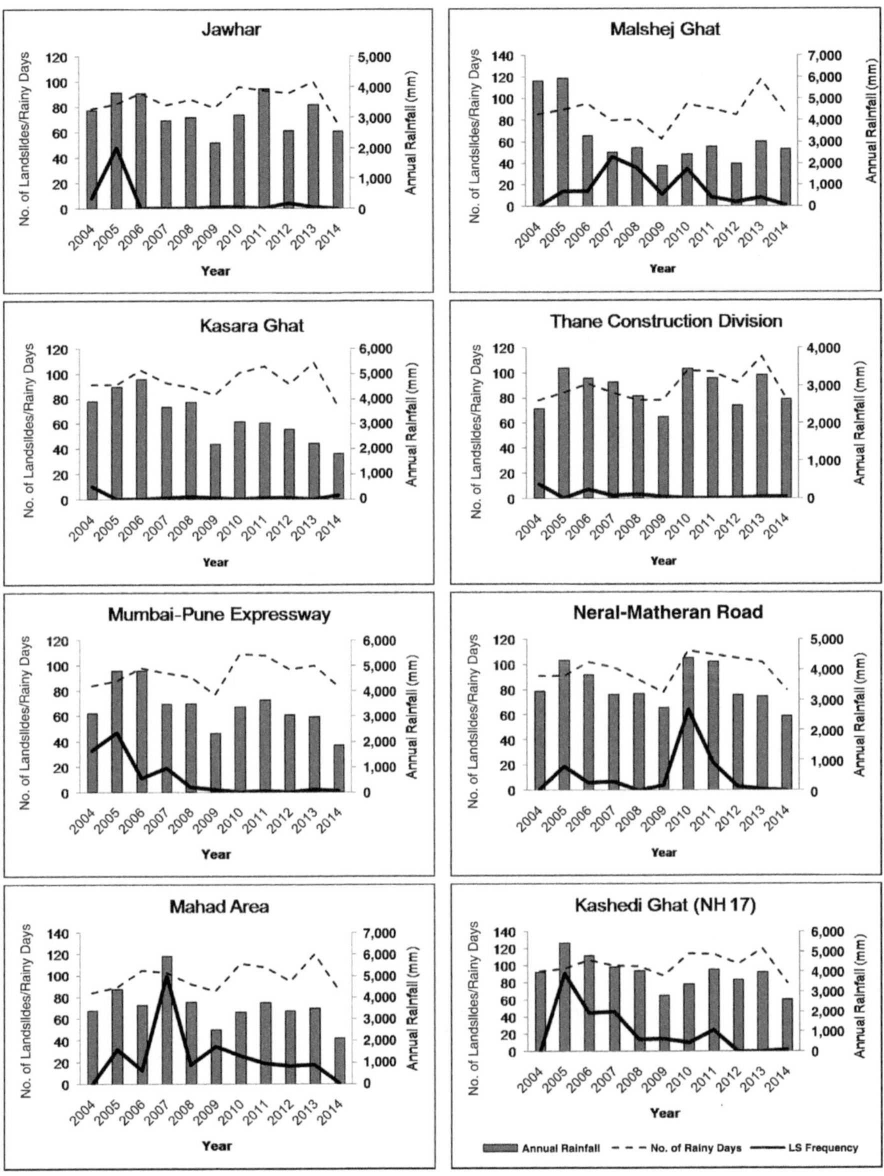

FIGURE 4.11   Annual Rainfall and Landslide Frequency in North Konkan (2004–2014).

(Source: Public Works Department, National Highways Authority of India, India Meteorological Department)

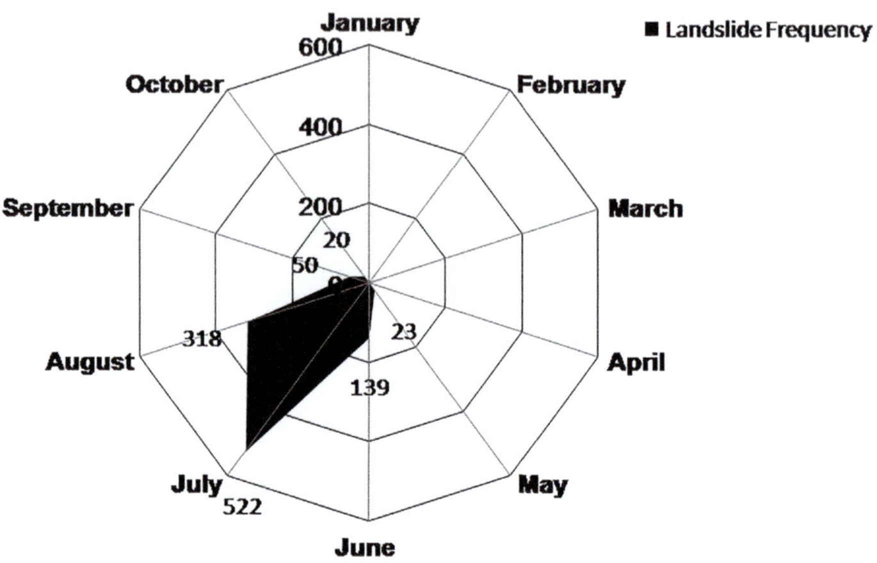

FIGURE 4.12   Monthly Landslide Frequency (2004–2014).

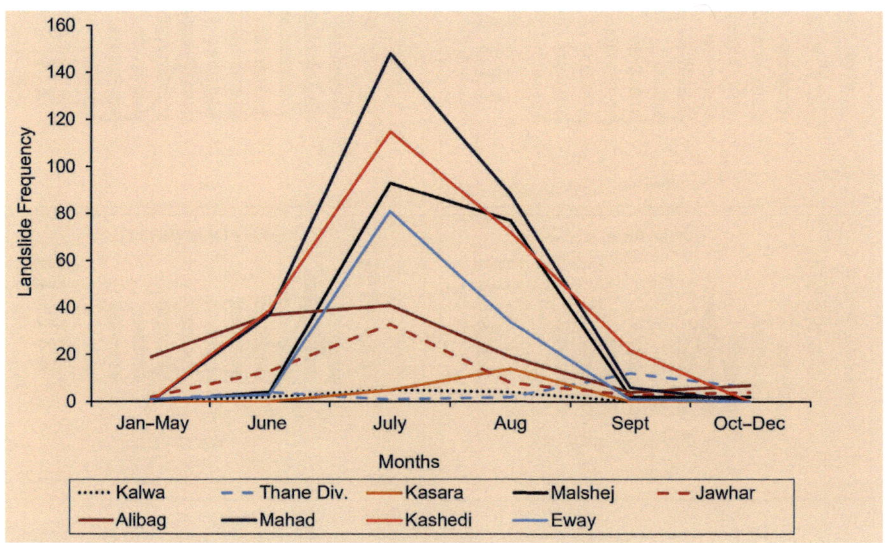

FIGURE 4.13   Area-Wise Monthly Landslide Frequency (2004–2014).

So far as the spatial pattern of monthly landslide frequency in the study area is concerned, a similar pattern is observed in almost all parts of North Konkan (Figure 4.13). Maximum landslide frequency is found to be in July, followed by August and September, which is the period of above average rainfall.

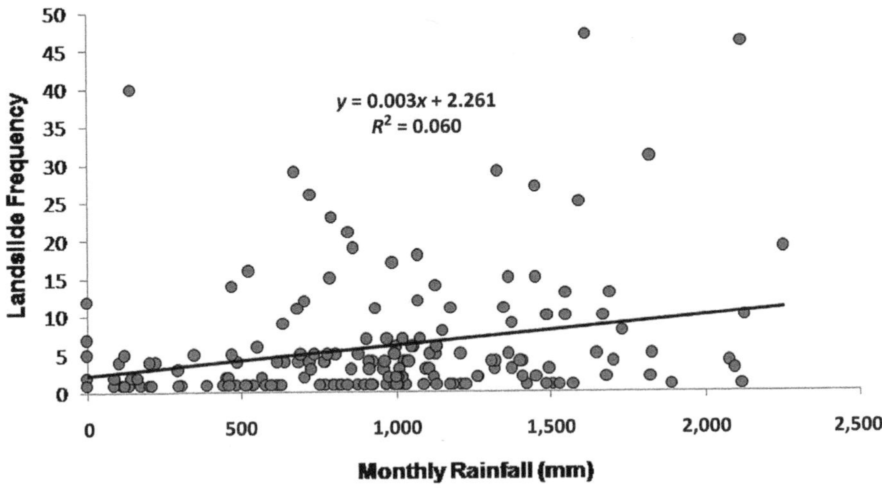

FIGURE 4.14   Monthly Landslide Frequency and Rainfall (2004–2014).

There is a positive correlation (r = 0.2451) between monthly rainfall and land-slide frequency and is found to be at a 0.001 level of significance (Figure 4.14). Since the rainfall data are considered on a monthly basis, the correlation is pos-itive but not very significant. It indicates that landslide frequency increases with increasing rainfall but rainfall is not the only factor that causes landslides in this area (Figure 4.15).

4.5   LANDSLIDE TYPES AND GEOMETRY

Consideration of landslide scar geometry is of fundamental importance in land-slide hazard and risk assessment. However, very little attention is given to this except for a few recent studies on landslide scar geometry in South Asia and East Europe (Bhandari and Kotuwegoda, 1996; Pardeshi et al., 2009; Nikoleeva et al., 2014). This section of the study deals with the application of landslide geometry parameters in LHA.

4.5.1   METHODOLOGY

For assessment of landslide morphology, the landslide scar geometry param-eters such as landslide length, scar width, slope angle, and toe extent have been recorded using a laser distance meter and measuring tape. The landslide scar geometry parameters for 161 landslide sites in North Konkan have been recorded during the field survey. The landslide geometry parameters such as length-to-width ratio (L/W), form factor (F), shape factor (S), and elongation ratio (ER) have been calculated. Landslide area and volume have also been

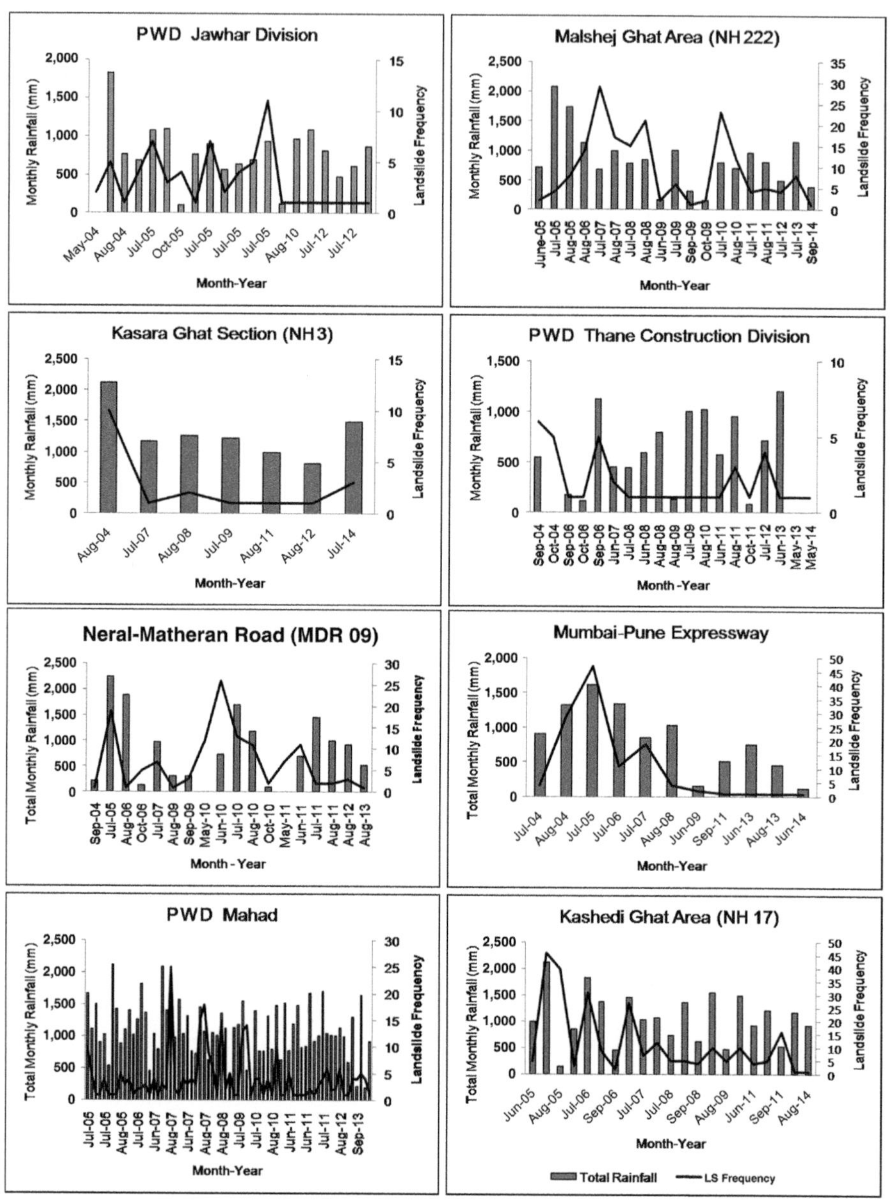

FIGURE 4.15 Monthly Landslide Frequency and Rainfall.

estimated using landslide scar geometry parameters. Landslides are also classified into different types using Varnes's (1984) scheme of classification of mass movement.

4.5.2 SLOPE FAILURE TYPE

The type of slope failure reflects the lithological and geomorphic characteristics of a given area and the nature of slope-forming material. The types of slope failure have been identified during the field survey using Varnes's (1984) scheme of classification of mass movements (Figure 4.16).

The observation reveals that the translational debris slide is the most common type of slope failure in North Konkan, followed by rotational debris slump, rockslides, and debris flow (Figure 4.16). Translational debris slides are common in the north and north-east hill complex of Palghar district (Sai-Vavar Road, Behedpada area of Jawhar-Mokhada), Neral-Matheran hill complex, Parsik Hill, and central Raigad district. Debris slumps (rotational slides) are observed along Wada-Khodala Road (Palghar district), at places along Jawhar-Sakharshet Road, and at some localities in south Raigad district. However, debris flow is associated mainly with the slope failure sites around Mahad town (Raigad district) and one major debris flow near Behedpada village in the north-east part of Palghar district. Rockfall and wedge failure sites are observed at places along the ghat sections of Malshej Ghat, Mumbai-Pune Expressway (Khandala section), Tamhini Ghat, and Varandh Ghat (Waghjai Temple area).

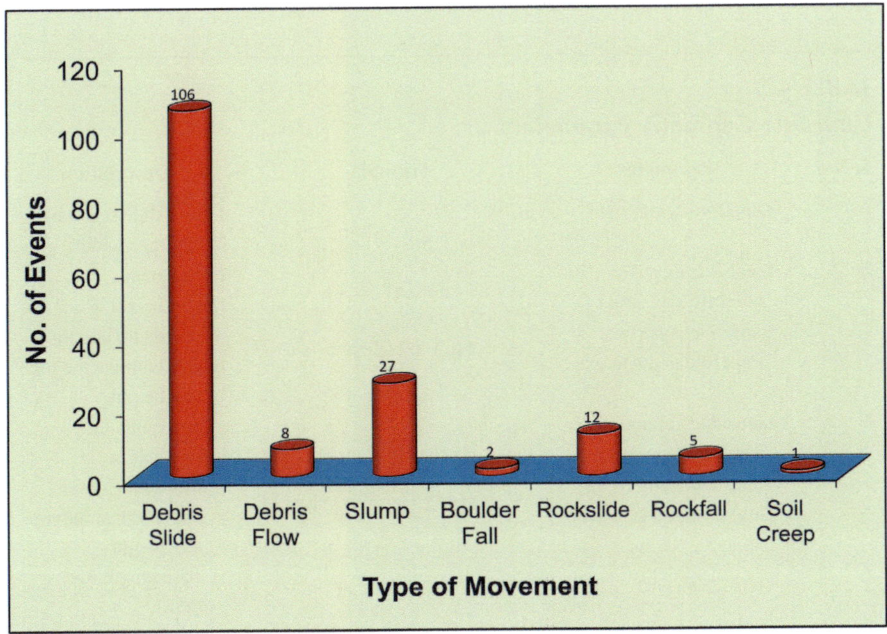

FIGURE 4.16   Classification of Slope Failures in North Konkan.

4.5.3 Landslide Geometry

Landslide size, volume, and shape have been estimated using landslide geometry parameters (scar length, width, slope angle, and toe extent) recorded during the field survey. The following parameters have been calculated for analysis (Table 4.3).

4.5.3.1 Length-to-Width Ratio

The landslide scar length-to-width ratio determines the shape of the landslide. The most common L/W ratio in North Konkan is 1–2. The majority (76.4%) of the landslides in North Konkan have L/W ratio below 2, and most of these are translational debris slides, whereas debris flows around Mahad and at places in the north-east parts of Palghar district attain lengths three to four times greater than the widths. Higher values of the L/W ratio represent elongated landslide scars (Figure 4.17). The L/W ratio of landslide scar determines the coverage of road width by displaced material and hence can help prioritise rescue operations after the slope failure event.

There is a positive correlation between landslide length and average scar width ($r = 0.61$) and is statistically at a 0.001 level of significance (Figure 4.18).

Lower values of landslide L/W ratio in North Konkan reveal that the runout distance of most slope failure sites is limited to the road width. In many cases, a landslide event causes partial road blockage because the displaced material is

---

TABLE 4.3

Landslide Geometry Parameters

S. No.	Parameter	Formula	Description
1	Length-to-width ratio	L/W	L, landslide scar length W, average width
2	Landslide area (for conical-shape scar)	$A = \Pi * r * \sqrt{r^2} = +h^2$	r, toe extent h, scar height, P = 3.14
3	Landslide area (for trapezoidal-shape scar)	$A = (a + b)/2 * h$	a, scar width at base b, scar width at the top h, scar height
4	Landslide volume (for conical-shape scar)	$V = (1/3 * \Pi * r^2 * h)/2$	r, toe extent h, scar height
5	Landslide volume (for trapezoidal-shape scar)	$V = ((1/2 * (a + b) * h) * H$	a, scar width at base b, scar width at the top h, scar height H, base length (toe extent)

(*Source*: For S. Nos. 1, and 3, Bhandari R. and Kotuwegoda W. (1996) and for S. Nos. 2 and 4, Pardeshi et al. (2009))

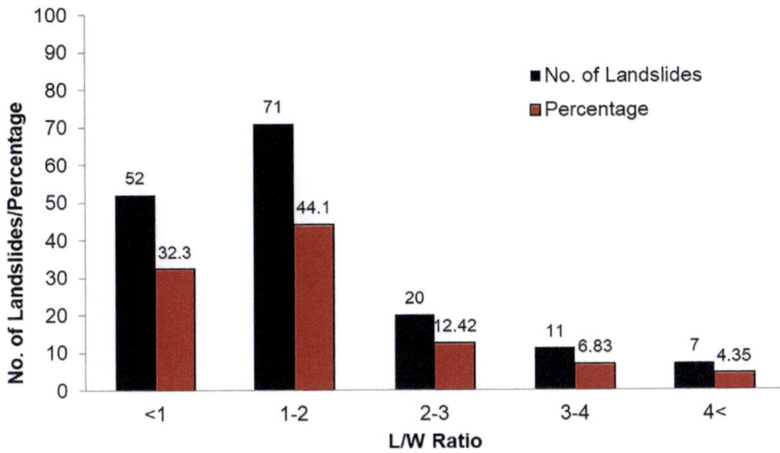

FIGURE 4.17   Frequency of Landslide Length-to-Width Ratio.

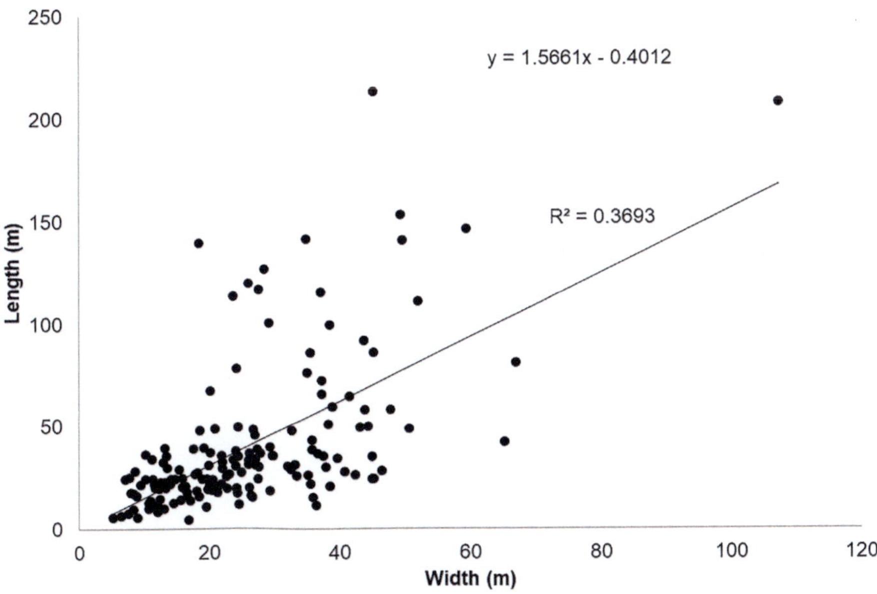

FIGURE 4.18   Relationship between Landslide Length and Width.

transported just up to the base of the slope. However, in the case of narrow and long landslide scars, there is more possibility to cover the entire road width, and subsequently, this may cause total traffic blockage, which is the case of Malshej Ghat (NH 222) and Ambenali Ghat (SH 72).

4.5.3.2 Landslide Area and Volume

Consideration of landslide scar area and volume of debris is instrumental in LHA because it helps in determining the scale of damage and cost of removal of displaced material. The landslide area and volume of displaced material have been estimated using the geometry of regular geometric shapes such as cones and trapezoids for landslides in the study area (Figures 4.19 and 4.20). Since the depth of

TABLE 4.4

Estimated Landslide Area and Volume

Parameters	Landslide Area $\left(\mathbf{m}^2\right)$	Landslide Volume $\left(\mathbf{m}^3\right)$
Maximum	64952.86	549896.4
Minimum	135.964	50.597
Average	2649.057	15235.12
Standard deviation	6365.406	48751.99

(*Source*: Field survey)

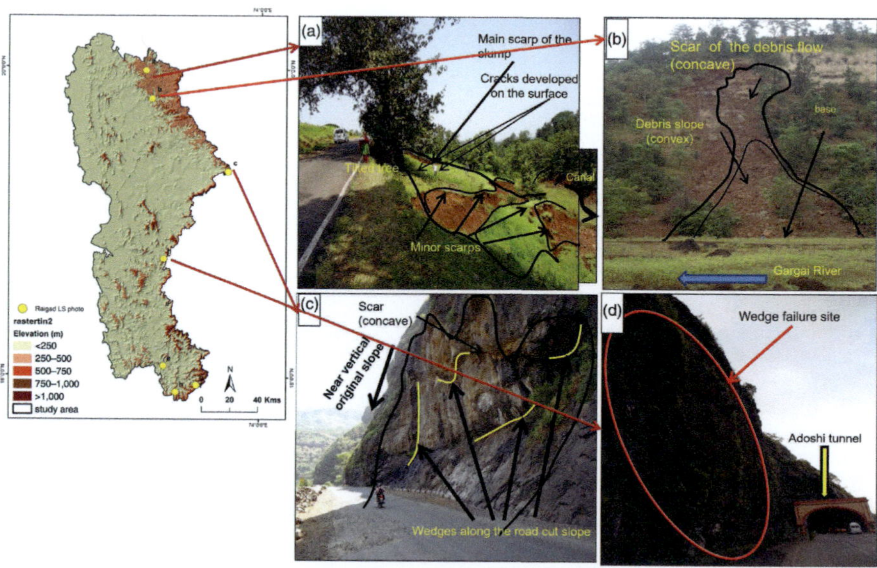

FIGURE 4.19 General Characteristics of Selected Landslides in the North Konkan Region. (a) Slump near Khodala Village. (b) Debris Slide caused by Undercutting by Gargai River. (c) Rock Wedges Formed along Road Cut Slope in Malshej Ghat. (d) Rockfall Zone near Adoshi Tunnel of Mumbai – Pune Expressway.

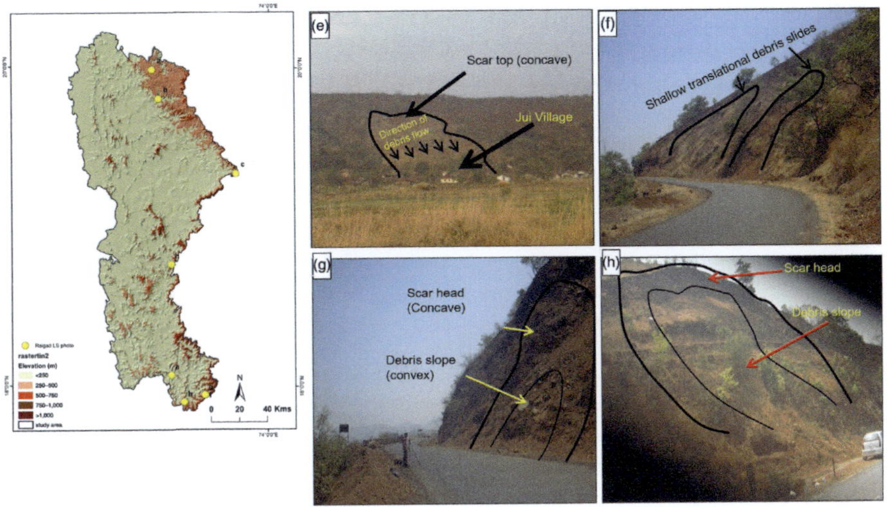

FIGURE 4.20 General Characteristics of Selected Landslides in the North Konkan Region. (e) Jui Village Landslide. (f) Shallow Translational Landslides along Poladpur-Mahabaleshwar Road. (g) Debris Slides along Poladpur-Mahabaleshwar Road. (h) Debris Slide in Carandh Ghat.

the landslide scar has not been determined scientifically, the values for landslide area and volume are approximate.

The area and volume of landslides in North Konkan are moderate to low as compared to the slope failures reported in the Himalayan region. Dasgaon (DAS1) is the largest landslide in North Konkan with an area and volume of 64,952.86 m^2 and 5, 49,896.4 m^3, respectively.

The average landslide area and volume in this area are 2,649.86 m^2 ($\sigma = 6,265.4$) and 15235.12 m^3 $\left(\sigma = 48751.99\right)$, respectively (Table 4.4).

There is a great variation in the landslide area and volume throughout the study area. The size and volume of the landslide in most of the North Konkan area are smaller than average except for a few slope failures located around Mahad town and those in Jawhar plateau (Behedpada landslide and Ujjaini landslide). The information about the landslide area will help to determine the road extent affected by slope failure, and estimation of the volume of displaced material after the landslide will help to understand the amount of work to be done in debris removal and so help with the preparation of machinery.

Consideration of landslide morphology is an important step in landslide hazard and risk assessment. Landslide size, volume estimation, and shape analysis help in determining potential losses caused due to slope failures and also in prioritising landslide mitigation activities.

4.6 CASE STUDY 2: IDENTIFICATION OF LANDSLIDE-PRONE ZONES IN PARTS OF THE NEPAL HIMALAYAS USING GEOSPATIAL INFORMATION

The case study of the identification of landslide-prone zones using geospatial information in parts of the Nepal Himalayas is aimed at demonstrating a different approach to identifying and mapping landslide-affected areas using remotely sensed data through visual image interpretation techniques.

4.6.1 THE STUDY AREA

Administratively, the study area lies in the Sagarmatha and Koshi provinces of Nepal located in the eastern parts of the country. Identification and mapping of landslides are done in the hilly areas around Chhintang village near the confluence of the Tomar and Koshi Rivers (Figure 4.21). The region is characterised by steep slopes of the southern mountainous terrain of the Nepal Himalayas.

The region is covered with younger Himalayan sediments such as sandstones, siltstones, and conglomerates. This region is drained by numerous perennial Himalayan rivers and streams. Streams and rivers flowing through this region

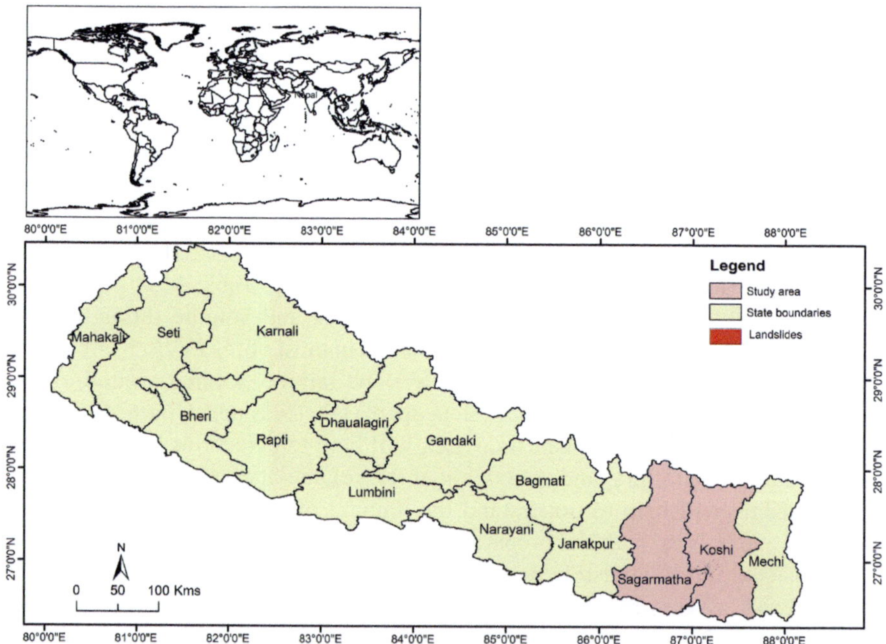

FIGURE 4.21 Study Area: Location Map.

play a significant role in the mass movement or movement of slope-forming materials, particularly those at the proximate distance from streams.

4.6.2   MATERIALS AND METHODS

This study was aimed at identifying and delineating landslide-affected areas in parts of the Nepal Himalayas. An attempt has been made to demonstrate how multi-date Google Earth images can be used to identify and map past landslide events using visual image interpretation techniques. Landslide scars in parts of Sagarmatha and Koshi provinces of Nepal have been delineated using multi-date satellite data (LANDSAT/Copernicus) in Google Earth Pro.

A total of 69 landslide polygons have been digitised, and landslide geometry parameters (such as slope angle, scar length, width, average scar depth) are recorded (Figure 4.22). Further, landslide area, landslide scar length-to-width ratio, the volume of displaced material, etc. are estimated using appropriate statistical techniques.

4.6.3   RESULTS

Based on the visual interpretation of LANDSAT/Copernicus satellite imageries of the study area, it is observed that the majority of landslides identified in this area are attributed to foothills of Nepal-Himalayan Shiwaliks, particularly those in proximity to drainage. It is observed that the initiation of landslide processes in this area is dominated by strong slope undercutting by high-energy rivers and

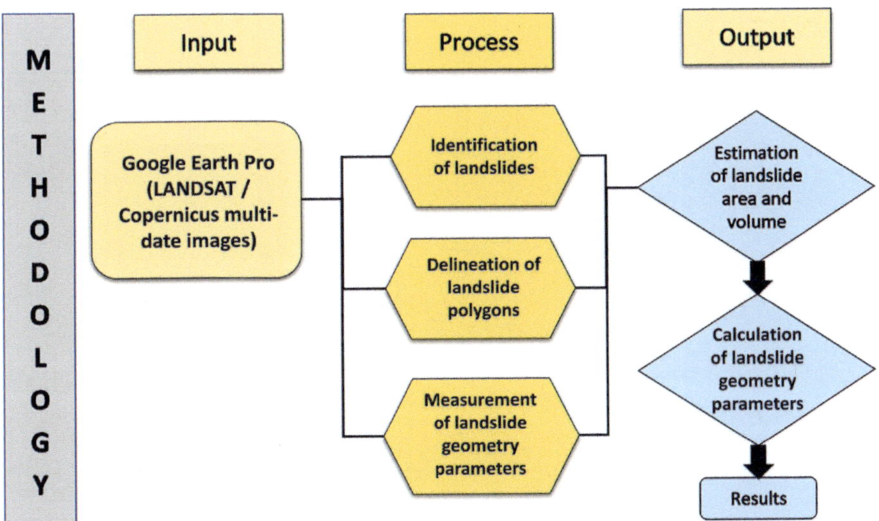

FIGURE 4.22   Methodology.

streams flowing through this region. Loose, unconsolidated materials on hillsides easily move in a downslope direction under the influence of gravity (Figure 4.23).

4.6.3.1 Landslide Morphology

Like other parts of the world, landslides in Nepal are observed in the areas of steep slopes. Based on the observation of 68 landslides in the Sagarmatha and Koshi provinces of Nepal, it is found that 89.29% of the observed landslides are attributed to slopes greater than 35° (Table 4.5).

TABLE 4.5
Slope Angle and Landslide Frequency

Slope (Degrees)	No. of Landslides	Percentage of Landslides
< 15	0	0.00
15–25	2	2.94
25–35	5	7.35
35–45	28	41.18
45–55	24	35.29
> 55	9	13.24
Total	68	100.00

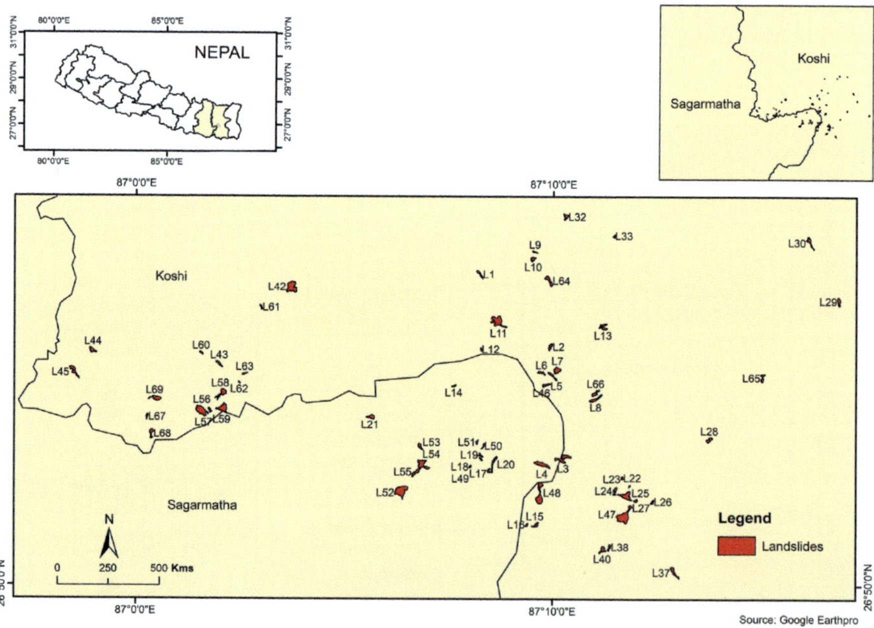

FIGURE 4.23   Spatial Distribution of Landslides in Sagarmatha and Koshi Provinces, Nepal.

4.6.3.2 Landslide Geometry Parameters

Landslide scar length, width, runout distance, average depth of the scar, and other landslide geometry parameters have been used to estimate the area of the landslide scar and volume of landslide displaced material. The details of the method adopted for the calculation of the estimated landslide scar area and volume of displaced material are given below.

Based on the geometric parameters of observed landslide polygons, the landslide length and width ratio has been computed. The classification of landslide scar geometry based on the ratio of landslide scar length and width indicates that the majority of the landslides are of L/W ratio greater than 2 (Figure 4.24). It indicates that the majority of landslides are elongated in shape. Generally, elongated landslides are characterised by a more saturated mass of slope-forming materials which moves in a downslope direction with greater velocity due to step slopes, combined with unconsolidated and saturated slope-forming materials. A high L/W ratio also indicates that these landslides have had a greater runout distance and therefore may cause much damage if they occur in inhabited areas (Figure 4.25).

Estimation of landslide scar area and volume of debris is important in LHA because it helps in determining the scale of damage and cost of removal of displaced material. The landslide area and volume of displaced material have been estimated using the geometry of regular geometric shapes such as cones and trapezoids for landslides in the study area (Figure 4.26).

The average landslide area and volume in this area are 1,00,183 m^2 (σ = 1,79,970) and 28,66,156 m^3 (σ = 85,99,647), respectively.

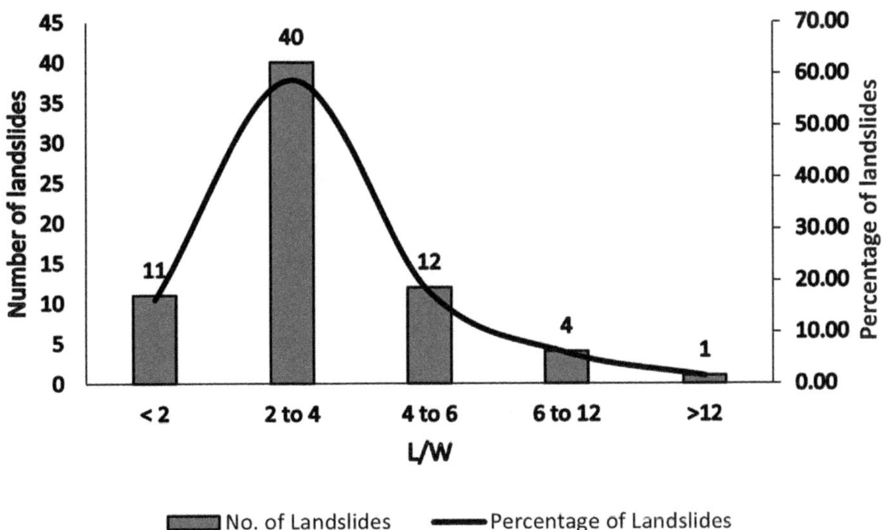

FIGURE 4.24   Frequency of Landslide Length-to-Width Ratio.

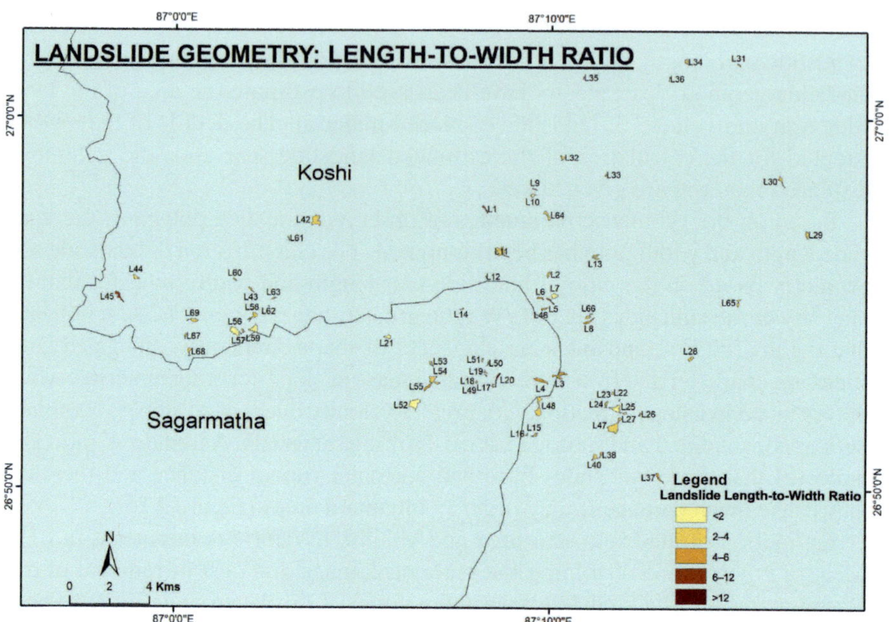

FIGURE 4.25　Frequency of Landslide Length-to-Width Ratio.

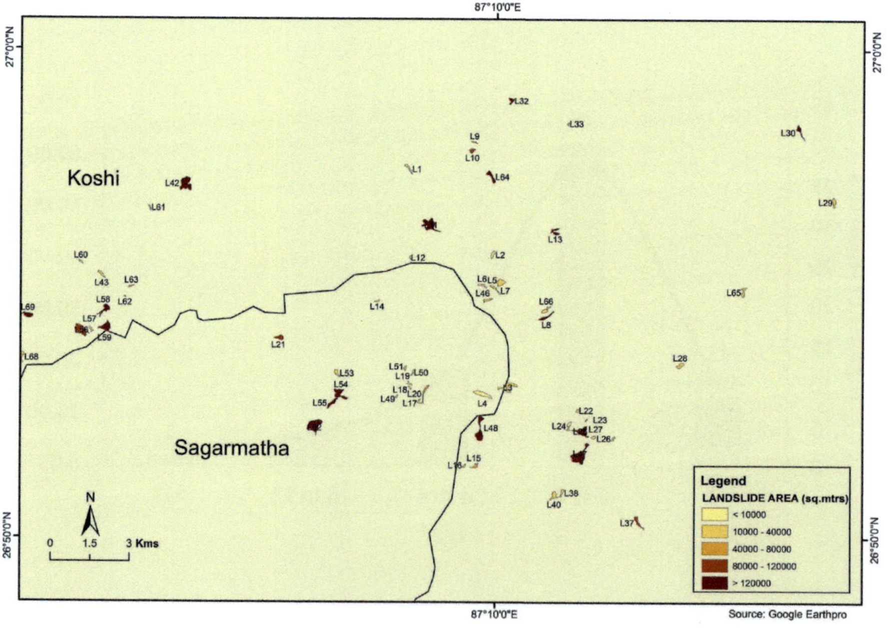

FIGURE 4.26　Estimated Landslide Area.

TABLE 4.6
Estimated Landslide Area and Volume

Parameters	Landslide Area $\left(\mathbf{m}^2\right)$	Landslide Volume $\left(\mathbf{m}^3\right)$
Maximum	1381.66	352.73
Minimum	1028373	60617700
Average	100183	2866156
Standard deviation	179970	8599647

(*Source*: Author)

There is a great variation in the landslide area and volume throughout the study area. The size and volume of the landslides in most of the study area are smaller than average (Table 4.6).

4.7 LIMITATIONS

The data about landslide events have been obtained from the PWD of the government of Maharashtra, the NHAI, and the railway department for the period of 11 years from 2004 to 2014. Landslide locations have been identified, and parameters of landslide scar geometry have been recorded in the field. However, one of the major limitations of identification of the slope failures in the field was that older landslide events could not be identified in the field, possibly due to the development of vegetation cover on the scar and modification of the slope due to slope cutting and construction of retaining structures along the road.

Moreover, PWD and NHAI offices have not maintained complete details of landslide events in their records, such as the exact date of the event, damage details, the volume of the displaced material, duration of traffic blockage, and expenditure incurred towards repairing the damage. Therefore, it was difficult to prepare the complete landslide inventory for the study area. To maximise the completeness of the landslide inventory, extensive fieldwork has been carried out for recording the details of landslide events through the assessment of scar geometry. Besides, multidate images through Google Earth are used for visual interpretation and to identify landslide locations. Additionally, the details of landslide events have also been obtained from newspaper archives of the *Times of India* for the period from 2004 to 2014. To obtain information regarding major landslide events that occurred off the road, local people have been interviewed, and field locations have been visited.

4.8 CONCLUSION

A complete and detailed landslide inventory is a prerequisite for effective landslide hazard assessment. Historical landslide records are proven data sources for complete landslide inventory. This study attempted to prepare a landslide distribution map using past landslide records from the PWD, NHAI, and newspaper archives. The lack of complete landslide records is an important limitation in preparing a reliable and comprehensive landslide inventory for the North Konkan region. Unfortunately, the information regarding the actual date of the event, exact location, and landslide dimensions are not recorded properly. Moreover, it was difficult to identify the landslide scar in the field, mainly due to relict landslides and also due to anthropogenic modifications of the slopes. It is therefore difficult to apply multivariate statistical models of landslide susceptibility. Hence, there is a scope for the application of indirect methods of LHA in areas with a poor database.

REFERENCES

Bhandari, R., and W. Kotuwegoda. 1996. "Consideration of Landslide Geometry and Runout in a Landslide Inventory." In K. Sennset (Ed.) *Landslide Glissements de Terrain. 3, Proceedings of 7th International Symposium on Landslides* (pp. 1859–1864). Balkema Rotterdam.

Jaiswal, P., C. J. van Westen, and V. Jetten. 2010. "Quantitative Assessment of Direct and Indirect Landslide Risk along Transportation Lines in Southern India." *Natural Hazards and Earth System Science* 10 (6): 1253–1267. https://doi.org/10.5194/nhess-10-1253-2010.

Karlekar, S. 2012. "Landslide Hazard Zonation in Raigad District of Maharashtra: A Multivariate Approach." *Journal of Indian Geomorphology* 1: 75–82.

Kumar, K., P. Prasad, and S. Kimothi. 2010. "Rock Fall and Subsidence on Mumbai-Pune Expressway." *Journal of Geo-Engineering Case Histories* 2 (1): 24–39.

Nagarajan, R., A. Roy, R. Vinodkumar, and M. Khire. 2000. "Landslide Hazard Susceptibility Mapping Based on Terrain and Climatic Factors for Tropical Monsoon Region." *Engineering Geology* 58: 275–287.

Nikoleeva, E., T. Walter, M. Shirzaei, and J. Zschau. 2014. "Landslide Observation and Volume Estimation in Central Georgia Based on L – Band InSAR." *Natural Hazards and Earth System Sciences* 14: 675–688.

Pardeshi, S., S. Pardeshi, and V. Nagare. 2009. "A Study of Effect of Landslides on Human Environment in Thane District, Maharshtra State." *The Deccan Geographer* 47 (1): 45–56.

Thigale, S., and B. Umrikar. 2007. "Disastrous Landslide Episode of July 2005 in the Konkan Plains of Maharashtra, India with Special Reference to Tectonic Control and Hydrothermal Anomaly." *Current Science.* 92 (3): 383–385.

Varnes, D. I. 1984. *Landslide Hazard Zonation: A Review of Principles and Practice.* Paris: United Nations Scientific and Cultural Organization.

5 Landslide Susceptibility Zonation

Case Studies

5.1 INTRODUCTION

Landslide susceptibility zonation (LSZ) is an important step in landslide susceptibility assessment (LHA). LSZ refers to the delineation of a region into zones based on its actual and potential degree of susceptibility to landslide occurrence (NRDMS, 1984). 'Landslide susceptibility zonation is the process of division of land surface into areas and ranking of these areas according to the degree of actual or potential susceptibility from landslides or other mass movements' (Varnes, 1984). According to Courture (2011), landslide susceptibility zones represent homogeneous areas with respect to the degrees of actual or potential landslide susceptibility. Landslide susceptibility zonation maps represent areas with a degree of landslide susceptibility based on several causative (preparatory) and triggering factors responsible for landslide occurrence in the given area.

There are several approaches to LSZ, developed and applied in different parts of the world. These approaches are broadly categorised as heuristic, statistical (bivariate and multivariate), probabilistic, physical process-based models, and multi-criteria decision-making (MCDA) methods. The applicability of LSZ methods in a given area depends upon the availability of complete landslide database, scale of mapping, mapping unit, knowledge and experience of the investigators, etc. (Pardeshi et al., 2013).

5.2 LANDSLIDE SUSCEPTIBILITY ZONATION USING THE HEURISTIC APPROACH IN PARTS OF THE WESTERN GHATS AND WESTERN COASTAL AREAS OF INDIA: CASE STUDY 1

Data-driven (statistical) methods of landslide susceptibility zonation are proven to be more accurate and reliable for the prediction of slope failures. However, they require a huge landslide database (date and time of the slope failure event, volume of displaced material, annual frequency of slope failure events, details of damaged caused by slope failures, etc.) for a considerable period. In India, particularly in the Konkan region of Maharashtra, the lack of a complete landslide database is a major limitation in adopting such data-driven LSZ methods. Therefore,

DOI: 10.1201/9781003323488-5

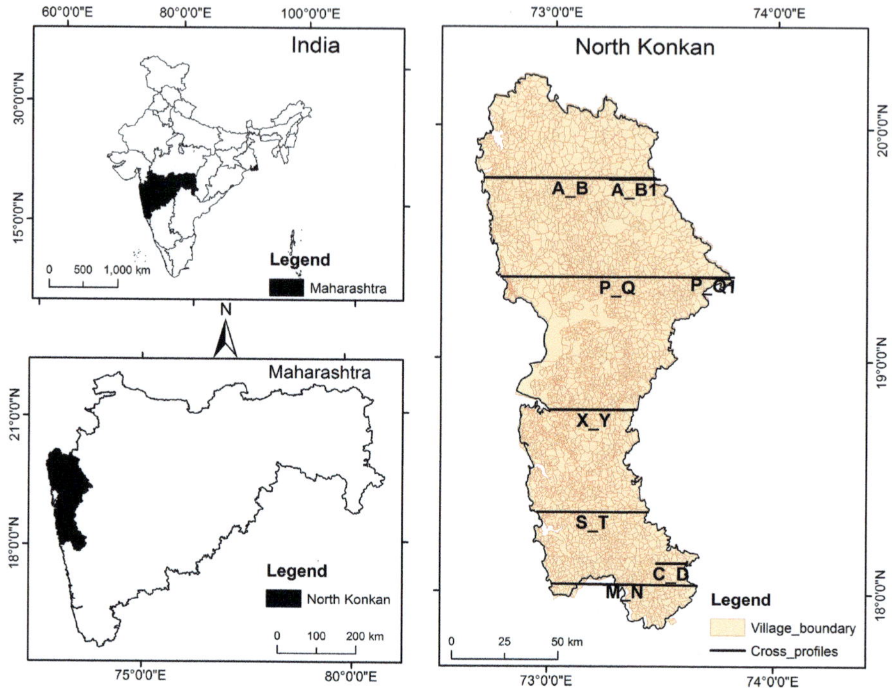

FIGURE 5.1 Location Map of the Study Area.

indirect methods (heuristic, physical process-based models, and MCDA) for LSZ are largely adopted in a data-scarce environment like India. This section deals with landslide susceptibility zonation in the North Konkan region of Maharashtra using a modified form of Bureau of Indian Standard (BIS) based LHEF (landslide susceptibility evaluation factors) rating method and analytic hierarchy process (AHP)-based model.

LSZ mapping has been carried out in the North Konkan region using a modified form of BIS-based LHEF rating scheme and AHP-based model (Figure 5.1). To carry out LSZ in the study area, landslide-related data layers have been generated through different data sources. The details of these data sources are given below.

5.2.1 DATA SOURCES

Slope failures are caused by several geo-environmental and triggering parameters. Based on the review of earlier works carried out in the Western Ghat region, ten landslide causative factors have been determined for LSZ in the study area: viz., slope angle, slope aspect, relief, distance from drainage, distance from structural discontinuities (lineaments, faults, and dykes), lithology, vegetation cover (NDVI), land use and land cover, distance from road, and rainfall. The details of the sources of information of these parameters are given in Table 5.1.

5.2.2 METHODOLOGY

To identify and delineate landslide susceptibility zones in North Konkan, all layers of landslide causative factors are preprocessed in a GIS environment. The LHEF rating scheme (modified after Bureau of Indian Standards, 1998) is used to determine the weights for all input parameters. These thematic data layers are then reclassified using the LHEF rating scheme in ArcGIS 9.3. Finally, these thematic data layers are integrated using 'weighted sum' operation in GIS to derive

TABLE 5.1

Data Sources of Landslide Causative Factors for Landslide Susceptibility Zonation

S. No.	Parameter	Source	Scale/ Spatial Resolution	Particulars	Format
1	Slope gradient (degrees)	SRTM DEM	30 m	–	Raster dataset
2	Slope aspect	SRTM DEM	30 m	–	Raster dataset
3	Elevation (m)	SRTM DEM	30 m	–	Raster dataset
4	Drainage	SRTM DEM	30 m	–	Raster dataset
5	Land use and land cover	LANDSAT 8 images	30 m	Date: 27/03/2013 Band combination: 5:4:3	Raster dataset
6	NDVI	LANDSAT 8 images	30 m	Date: 27/03/2013 Band combination: 5:4:3	Raster dataset
7	Lithology	Geological quadrangles District resource maps	1 : 125000 _	47 F&B, 47A, 47 E Thane and Raigad	– –
8	Structure (lineaments, dykes, and faults)	SRTM DEM LANDSAT 8 images Geological quadrangles District resource maps	30 m 30 m 1 : 125000 –	– – 47 F&B, 47A, 47 E Thane and Raigad	Raster dataset Raster dataset – –
9	Roads	Google Earth Pro Road development plans (2001–2020)	– –	– Thane, Palghar, and Raigad	KMZ –
10	Rainfall (mm)	IMD	–	Mean annual rainfall (mm)	Raster dataset

TEHD (total estimated susceptibility) for each mapping unit. The output map is then classified using the natural breaks method to delineate the region into five landslide susceptibility classes: viz., very high susceptibility, high susceptibility, moderate susceptibility, low susceptibility, and very low susceptibility. The BIS-based LSZ map is compared with the landslide distribution map to evaluate the results (Figure 5.2).

The analytic hierarchy process is another approach adopted for LSZ in the study area. The weights for individual landslide causative factors are derived by pairwise comparison matrices using the scale of absolute numbers given by Saaty (2008). Both inter-parameter and intra-parameter comparisons are done through pairwise comparison matrices. The judgements used in paired comparison are evaluated by calculation of consistency ratio (CR). The weights derived from pairwise comparisons are assigned to the thematic data layers. Finally, these data layers are integrated in a GIS environment using weighted sum operation, and five LSZ classes have been identified. The results of the AHP-based LSZ map has been validated by comparing it with actual landslide distribution. Finally, LSZ maps based on BIS and AHP methods have been compared to finalise landslide susceptibility zones for North Konkan.

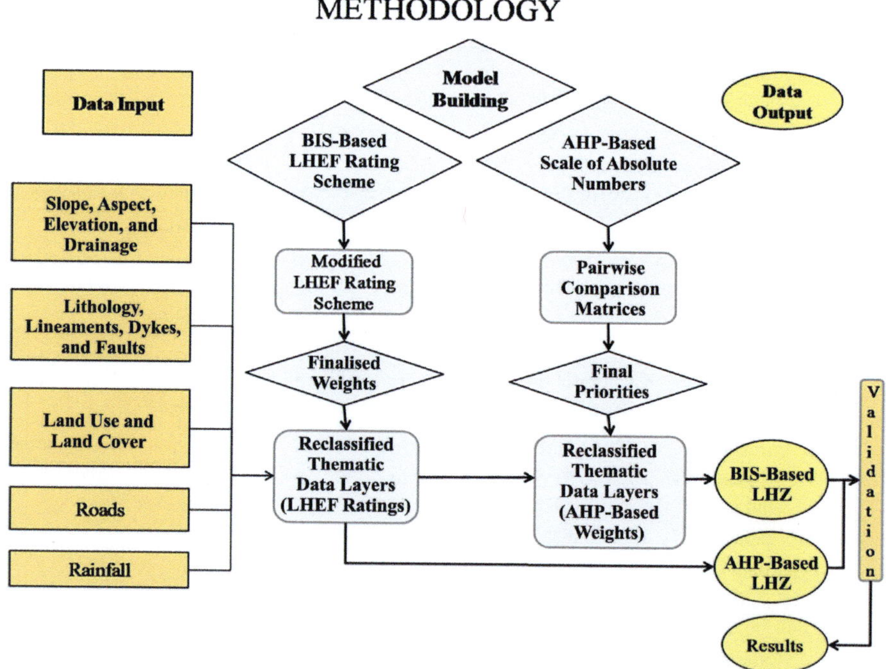

FIGURE 5.2   Flowchart of Methodology.

5.2.3 LANDSLIDE CAUSATIVE FACTORS/THEMATIC DATA LAYERS

Slope failures are caused by both preparatory and triggering factors. Landslide preparatory or causative factors are inherent properties of slope-forming material, such as slope, aspect, lithology, soil type, structural discontinuities, vegetation cover, porosity, land use and land cover, and relative relief. Landslide triggering factors, on the other hand, are external events or processes affecting slope instability, such as seismicity, rainfall, and anthropogenic activities. Delineation of areas into landslide susceptibility zones requires consideration of landslide causative and triggering factors. The assessment of the relative importance of these parameters in the initiation of slope failure is essential for accurate and reliable landslide susceptibility assessment. To carry out landslide susceptibility zonation in the North Konkan region, ten geo-environmental parameters have been used as input data layers. These parameters are listed here:

1 Slope angle
2 Lithology
3 Structure
4 Relief
5 Drainage
6 Land use and land cover
7 Roads
8 NDVI
9 Slope aspect
10 Rainfall

5.2.3.1 Slope

Slope inclination is one of the most important landslide causative factors. The consideration of steepness of slope in relation to the strength of slope-forming material is very important in LSZ (Varnes, 1984). Generally, steeper slopes are more likely to fail than gentler ones. However, it is important to understand that the strength of the slope-forming material is important in determining the role of slope gradient in slope instability processes. In many cases, poorly jointed, hard and massive rocks, with almost vertical slopes, are more stable than the loose, fragmented slope material along comparatively gentler slopes.

The review of existing literature on LSZ reveals that slope gradient is the most important factor considered for landslide susceptibility zonation. In the present study, slope map has been derived from SRTM DEM (30 m spatial resolution), and slope has been categorised into five slope classes: viz., $> 45°$, $35 - 45°$, $25 - 35°$, $15 - 25°$, and $< 15°$ (Figure 5.3) The spatial patterns of slope gradient in the North Konkan region reveal that the western slopes of the Western Ghat region of the study area, slopes of isolated hillocks, and coastal hill ranges in the coastal plains are characterised by steep slopes. Another concentration of steep-sloping areas is along the Jawhar plateau margins located in the north-east corner of the study area.

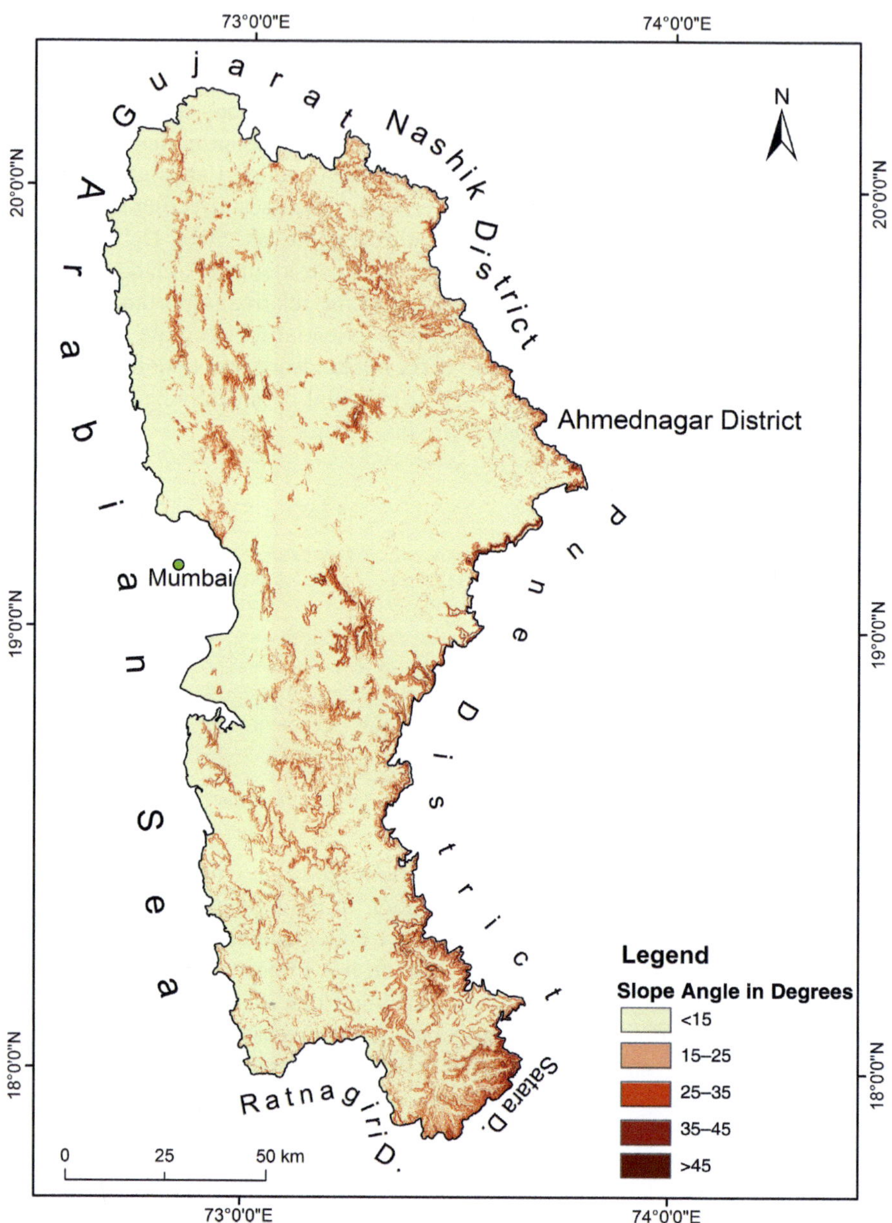

FIGURE 5.3   Slope Map of North Konkan, Maharashtra.

5.2.3.2   Lithology

The composition, texture, and permeability of rock and soils are inherent properties of slope-forming materials, which determine the strength of the material. Therefore, lithological characteristics of the slope-forming material are important factors in determining landslide susceptibility. Loose, fragmented, and weathered rocks and engineering soils are more unstable than compact, massive, and unweathered surface material. Lithology has been considered an important parameter in many of the earlier published works on LSZ.

Since, most of the slope failures in the North Konkan region are associated with the western slopes of the Western Ghat escarpment, composed of basaltic lava, the lithological properties of the series of lava flows in this region are considered for LSZ mapping. The thematic layer of lithology has been extracted from geological quadrangles (Sheet 47 A, E, F, and B) at 1:125000 scale. The lithology layer is then categorised into ten classes on the basis of the age of the lava formation: viz., Salher, Indrayani, Lower Ratangarh, Upper Ratangarh, Megacryst flows (M1, M2, M3), Diveghat formation, Elephanta, Purandargarh, laterites, and alluvium (Figure 5.4). Older lava formations are more stable than the younger ones. Weights are assigned to these formations on the basis of the age of the lava formations.

5.2.3.3   Structure

Structure includes the linear features of the earth's surface, representing discontinuities in homogeneous rocks. The areas with structural discontinuities are marked by dykes, lineaments, faults, and joints. Distance from lineaments and lineament density have been used in many of the earlier studies. The degree of fracturing, jointing, and proximity to tectonically active zones (e.g., lineaments) are important criteria in determining slope stability.

Distance from dykes, faults, and lineaments have been considered for LSZ in the present study. Lineaments, dykes, and faults are extracted from geological quadrangles at 1:125000 scale (Sheet 47 A, E, F, and B). Few details of dykes and lineaments have been digitised from district resource maps of Thane and Raigad districts. Besides, LANDSAT 8 images and SRTM DEM have been used to extract information about these linear features with more details (Figure 5.5). The entire North Konkan is intruded by numerous dykes, faults, and lineaments. The density of these features is more in the north, whereas structural discontinuities are less in Raigad district. All the dykes in this region are dolerite dykes and are aligned generally in the north-south direction (Figure 5.5). However, the faults and lineaments show a north-west to south-east trend, with varying lengths from hundreds of metres to tens of kilometres. The buffer of 50 m from lineaments and faults has been taken for assigning weights to the thematic layer of lineaments.

5.2.3.4   Relief

Absolute and relative relief play a significant role in slope stability. The relief of the region portrays its topography and helps us to understand the geomorphic

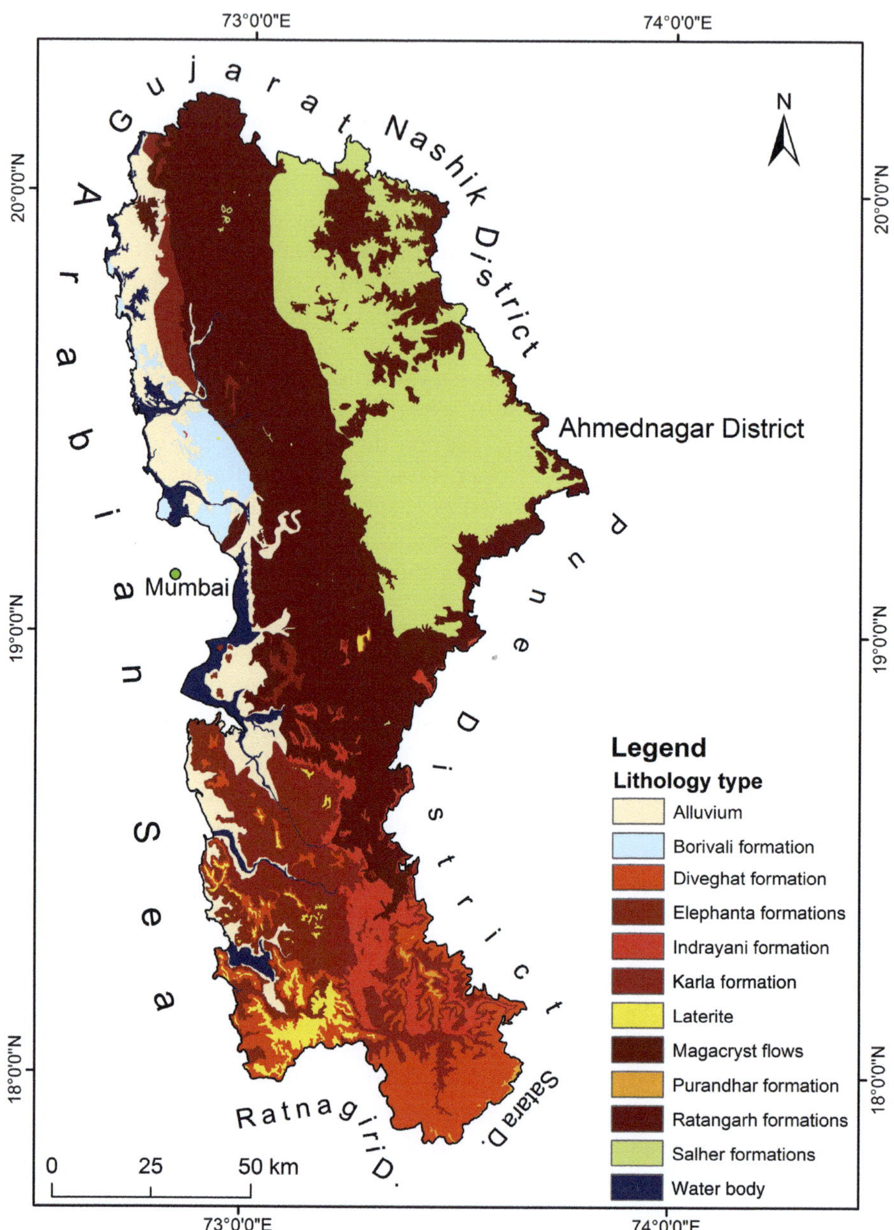

FIGURE 5.4　Lithology Map of North Konkan, Maharashtra.

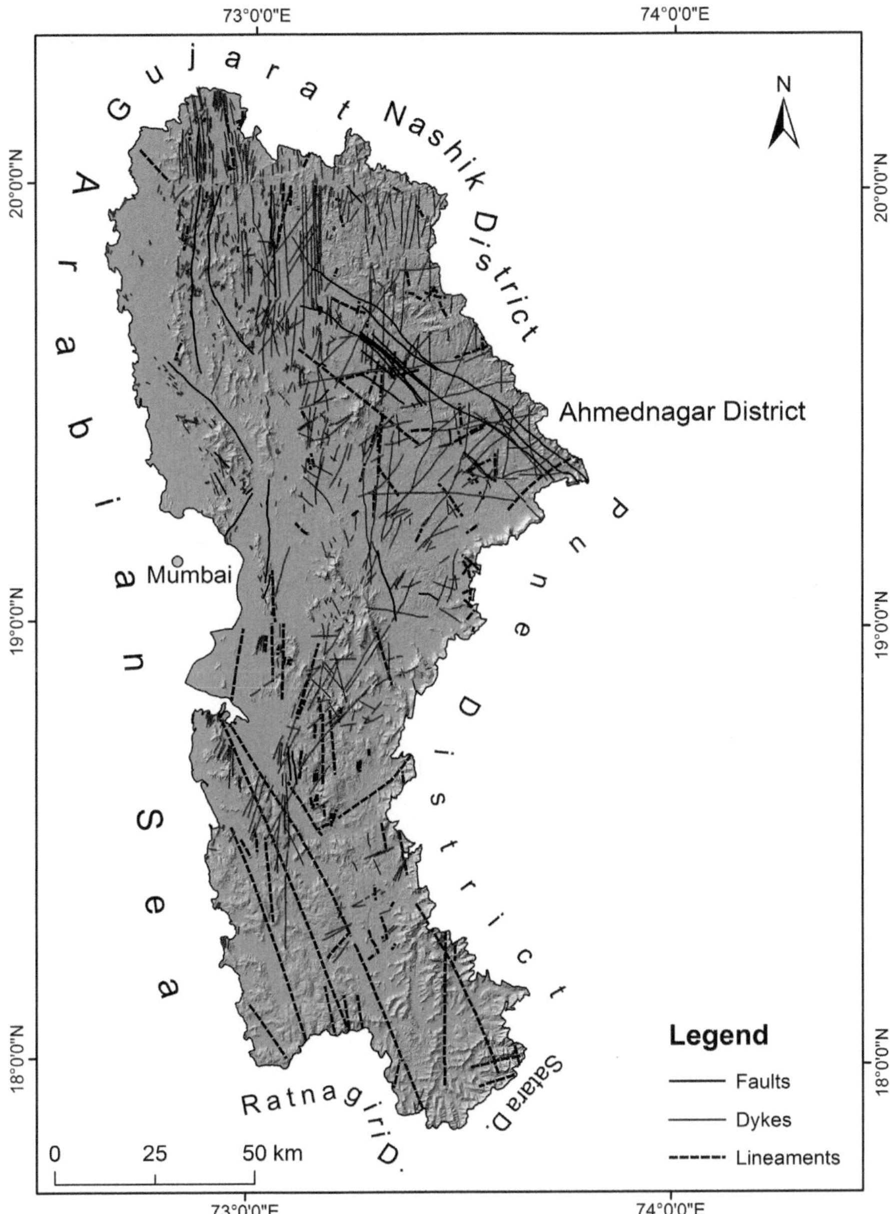

FIGURE 5.5 Map Depicting the Geological Structure of North Konkan, Maharashtra.

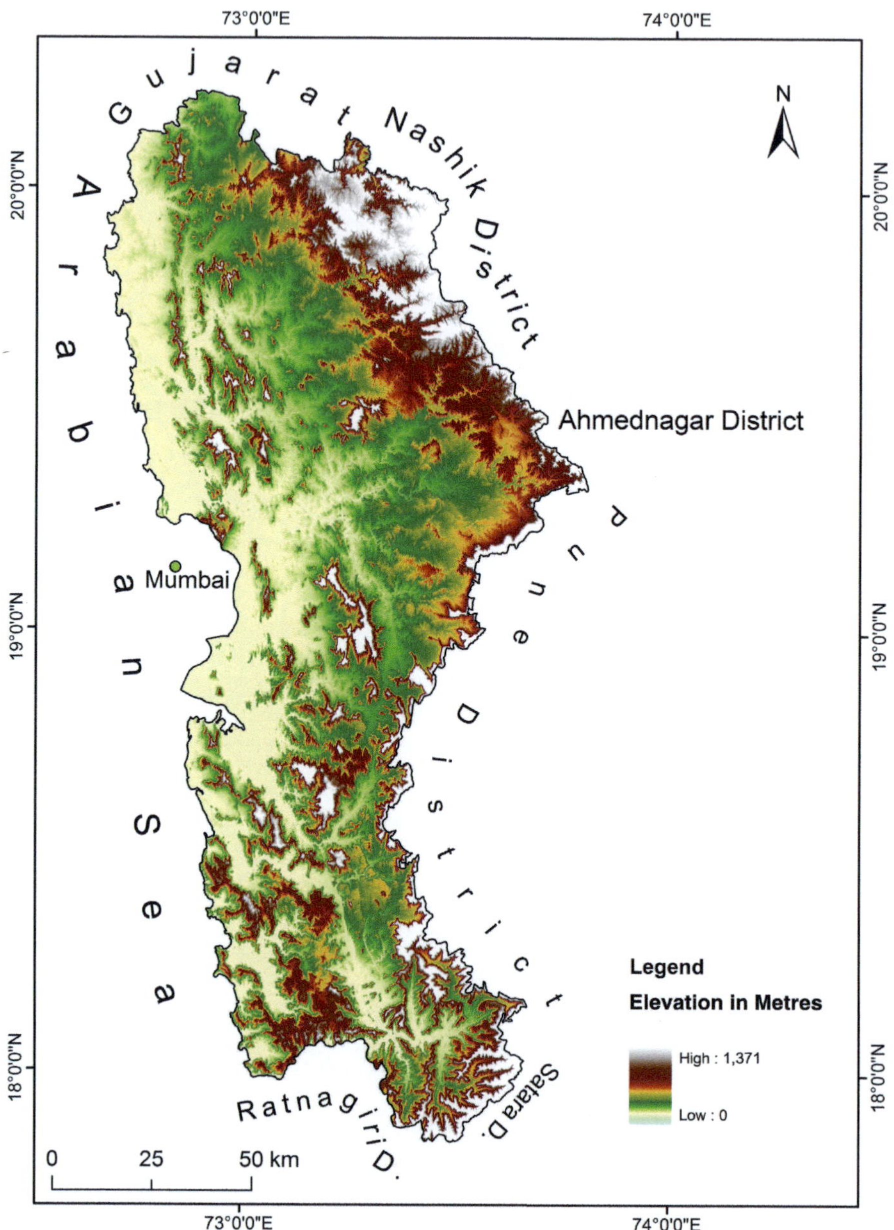

FIGURE 5.6   Relief Map of North Konkan, Maharashtra.

processes operating on these surfaces. Slope failure events are more frequent at higher elevations as compared to low-lying areas. Relief not only causes slope instability but also determines the extent to which the displaced material is moved downslope. Therefore, relief has also been considered as an important input parameter for LSZ.

The relief map of the study area (Figure 5.6) shows that the coastal plains in Thane and Palghar districts have elevation less than 100 m above sea level (ASL), except for a few coastal hill ranges and isolated hills. The north-east corner of the study area is characterised by the dissected Jawhar plateau with an elevation of 300 – 350 m. The central uplands of Raigad district represent more rugged topography, with the elevation ranging from 150 to 350 m. The eastern part of the entire study area, bordered by the Western Ghat escarpment, has an elevation above 300 m, and at a few places (like Igatpuri, Malshej, Khandala, Varandh, Ambenali Ghat), the elevation exceeds 500 m. The elevation data have been obtained from SRTM DEM and are categorised into three broad elevation classes: viz., < 100 m., 100 – 300 m, and > 300 m.

5.2.3.5 Drainage

Drainage plays an important role in slope instability. Proximity to drainage lines promotes saturation of the adjacent slope, and sometimes leads to slope failures. High-gradient streams in mountainous areas often lead to toe cutting, which further leads to landslides. Occasionally, landslides along major streams also result in the formation of temporary landslide dams. Proximity to drainage and drainage density have been used as data input for LSZ in many earlier studies.

In the present work, the drainage layer has been extracted from SRTM DEM, and a 100 m buffer from the drainage line has been created to categorise this layer into two: viz., areas within the 100 m buffer from drainage lines and areas beyond the buffer of 100 m from drainage lines. These categories have been used to assign weights for LSZ (Figure 5.7).

5.2.3.6 Land Use and Land Cover

Land use (agriculture, built-up areas, roads, and railway lines) and land cover (forest areas, barren land, water bodies, sand deposits, rocky outcrops, etc.) also play important roles in slope instability. Anthropogenic interventions in the natural environment sometimes make slopes unstable. Anthropogenic activities like the construction of roads and railway lines in hilly terrains lead to the modification of slopes. Toe cutting, blasting, and road widening may pose serious threats to the stability of natural slopes. Besides these, anthropogenic activities such as the removal of upslope vegetation, cultivation, and construction may also lead to slope instability in the areas where other landslide causative factors reach near the threshold of slope instability. Barren lands and rocky outcrops are considered to be more susceptible to slope failures. The brief review of earlier literature on LSZ indicates that land use and land cover are important data inputs for LSZ mapping.

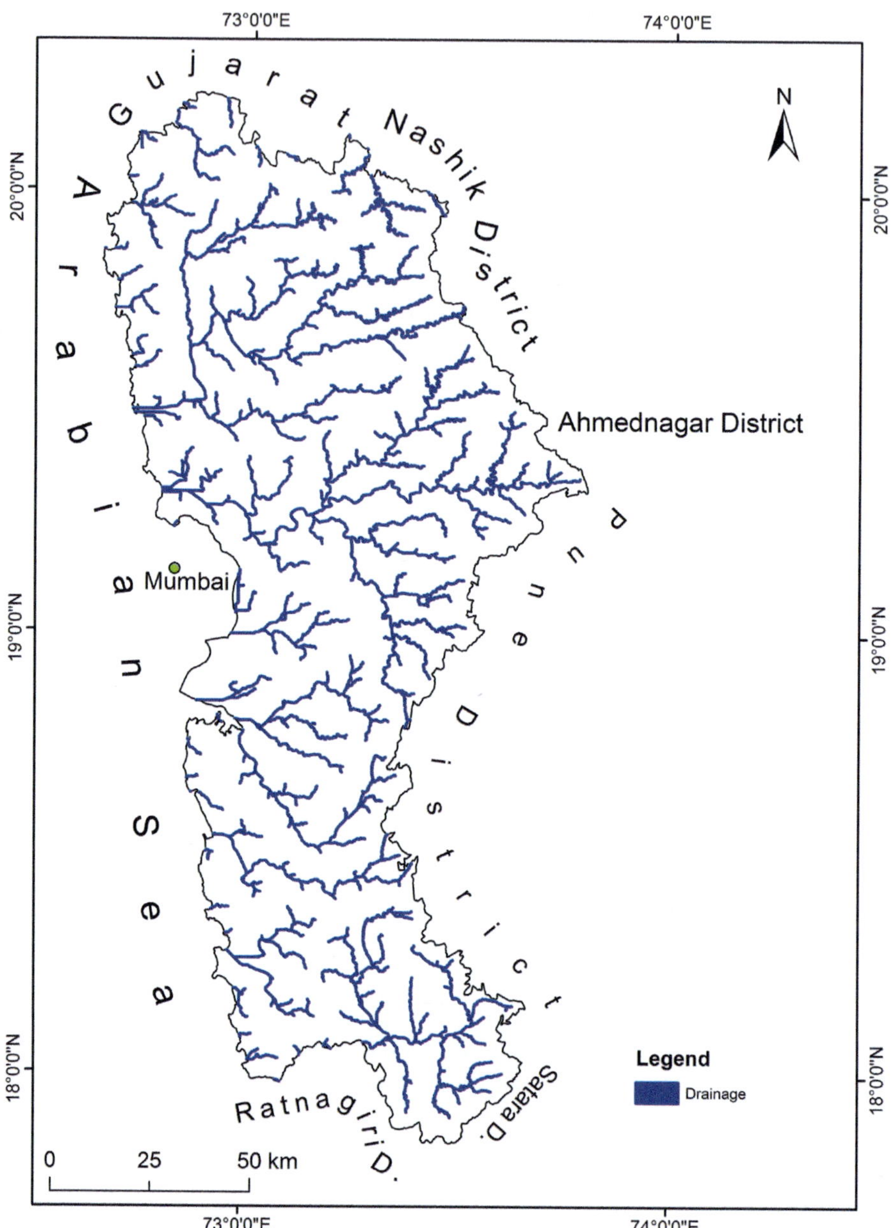

FIGURE 5.7   Drainage Map of North Konkan, Maharashtra.

TABLE 5.2

Information of Satellite Image used for Land Use and Land Cover

Spacecraft ID	LANDSAT 8
Sensor	OLI_TIRS
Path and rows	147_46 and 48
	148_46 and 47
Output format	GEOTIFF
Image acquisition date	18 and 25 April 2013
No. of bands	07
Spatial resolution	30 m

(*Source*: Earth Explorer, https://glovis.usgs.gov/)

To extract information about land use and land cover in the North Konkan region, LANDSAT 8 images have been used (Table 5.2). Supervised classification has been carried out in Erdas Imagine 9.2 software, and land use and land cover classes are identified using maximum likelihood algorithm. Seven land use and land cover classes have been identified, such as barren land, rocky outcrops, coastal sand deposits, natural vegetation, cultivated land, built-up areas, and water bodies (Figure 5.8) (Table 5.3). The classified image has been evaluated using kappa statistics.

5.2.3.7 Distance from Road

Removal of lateral support by human activities like slope cutting for road construction and widening is an important cause of slope failure (Varnes, 1984). Construction of roads on steep slopes often results in slope failure events of varying magnitudes. Therefore, roads are considered as the most dynamic zones representing human interventions in rugged terrain. Proximity to major roads is considered an important causative factor for LSZ mapping. Slope modification for road construction and road widening often results in loss of strength of the slope-forming material in the upslope direction. Secondary slope failures in the downslope direction also affect natural and human structures down slope. The thematic data layer of roads can be used in both LSZ mapping and landslide risk assessment.

Most slope failures reported in North Konkan are located along major communication routes passing through the Western Ghats section of the study area. Therefore, distance from roads has been considered for LSZ mapping in the present study. The road layer has been digitised from road development plans (2001–2020) and Google Earth Pro (Figure 5.9). A buffer of 100 m distance from roads has been considered for classifying this layer. Two categories – viz., distance from road < 100 m and > 100 m – have been identified and weights assigned accordingly.

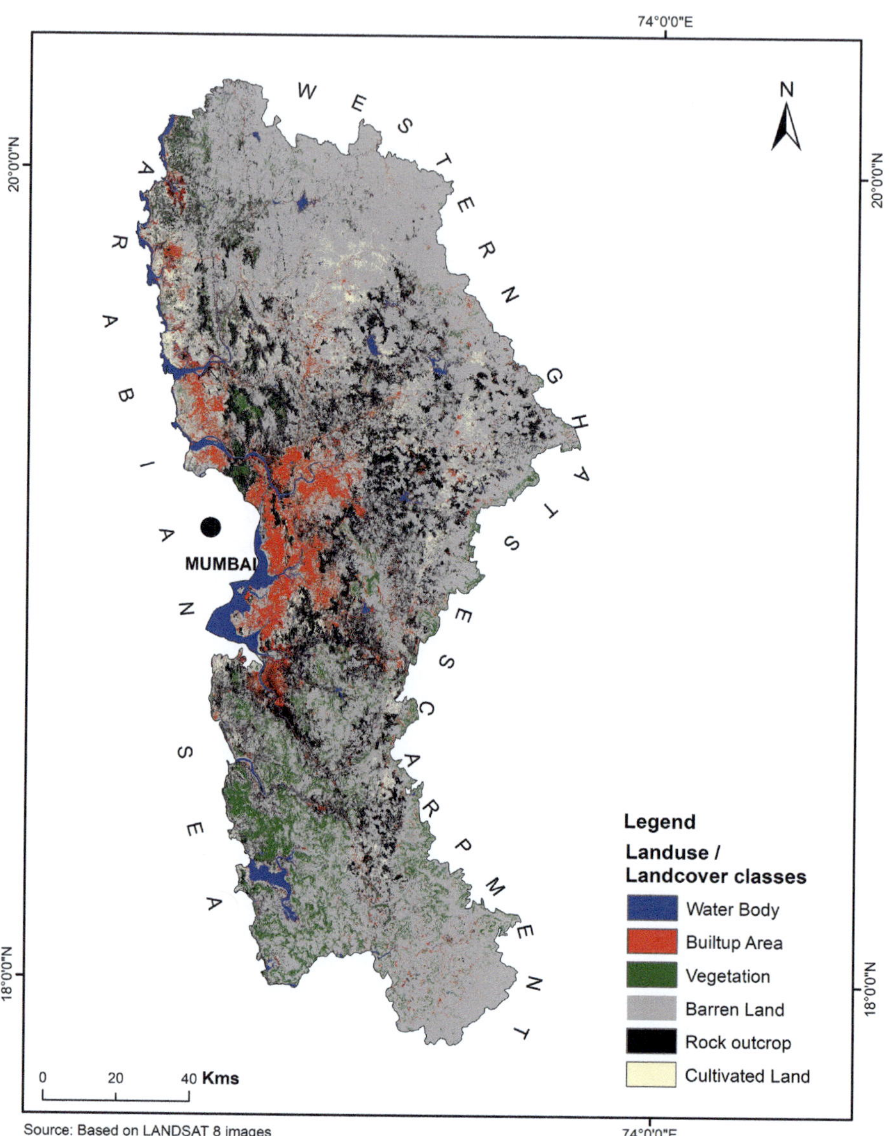

FIGURE 5.8   Land Use and Land Cover Map of North Konkan, Maharashtra.

TABLE 5.3
Land Use and Land Cover in North Konkan

Land Use/Land Cover Class	Area (km²)	Percentage Area
Barren land	3,345.09	20.467
Coastal sand	116.37	0.712
Rock outcrop	1,689.25	10.336
Built-up areas	3,225.69	19.737
Vegetation	1,857.56	11.368
Water body	148.18	0.906
Cultivated area	5,960.42	36.47
Unclassified area	0.99	0.004
Total	16,343.55	100

(*Source*: Based on LANDSAT 8 images)

5.2.3.8 Vegetation Cover

Although the general observation regarding the role of vegetation cover in slope stability is that vegetation cover promotes slope stability, it is difficult to determine the exact effect of vegetation on slope stability. This is mainly due to the view that the effect of moisture differences in different types of vegetation may develop cracks in the upslope direction and may lead to deep rotational landslides (Varnes, 1984). Therefore, the density of vegetation cover is an important parameter in the slope instability process. Dense and healthy vegetation helps to stabilise the slope, whereas slopes devoid of vegetation cover or poor vegetation are more susceptible to slope failure.

To delineate vegetation density classes in the North Konkan region, NDVI image has been generated from LANDSAT 8 images in Erdas Imagine using the following formula:

$$NDVI = (R - IR) / (R + IR)$$

$$NDVI = (Band\ 4 - Band\ 3) / (Band\ 4 + Band\ 3$$

The NDVI image has been classified into four vegetation density classes on the basis of NDVI values: viz., densely vegetated, moderately vegetated, sparsely vegetated, and barren land (Figure 5.10). Further, these classes have been assigned weights to determine landslide susceptibility zones.

5.2.3.9 Slope Aspect

Saturation of slope-forming material due to rainfall is the most important factor responsible for slope failures in India, particularly in the Western Ghat region.

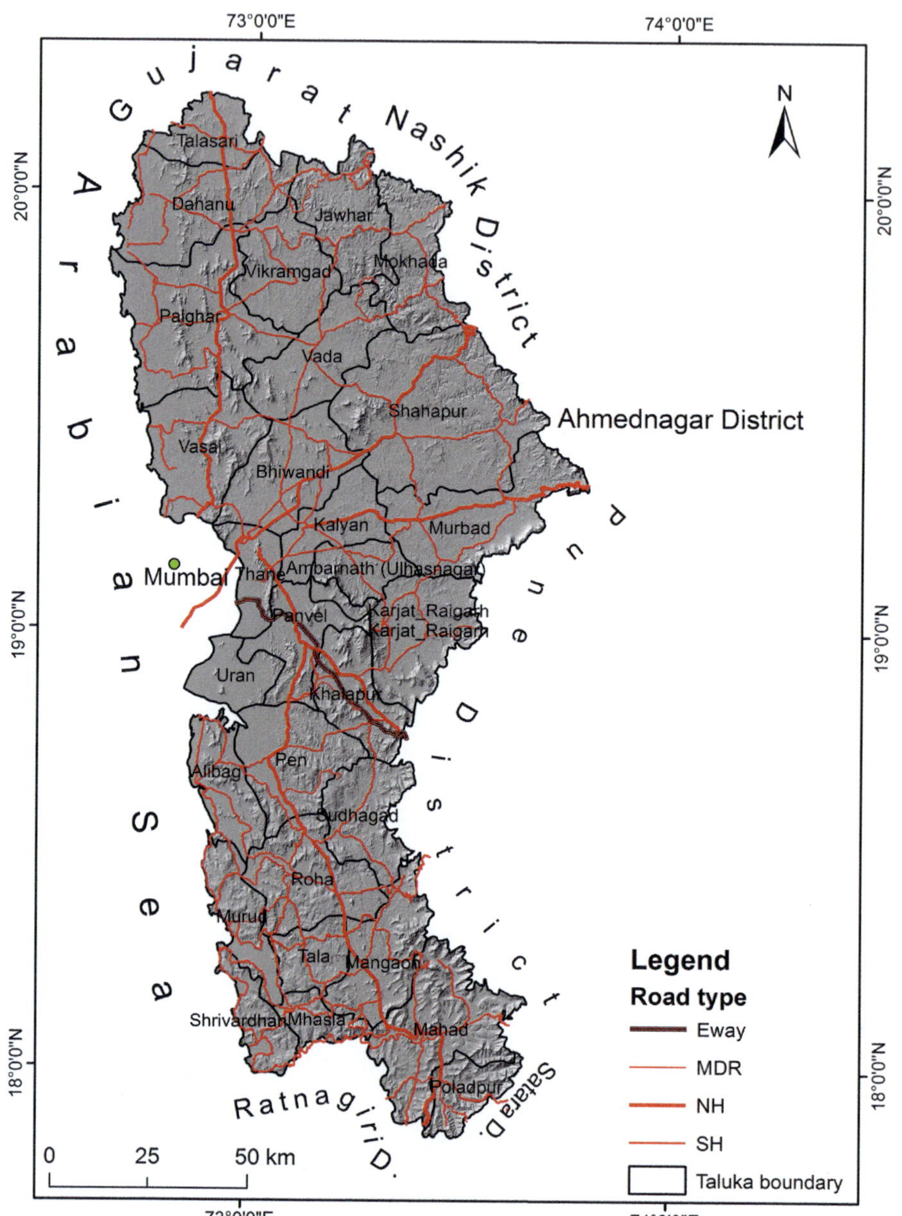

FIGURE 5.9   Road Map of North Konkan, Maharashtra.

Note: Eway, Expressway; NH, National highway; SH, State highway; MDR, Main district road.

The slopes facing rainstorms are more saturated and become unstable. Therefore, slope direction or aspect is considered to be another major factor controlling slope stability in this region. Nagarajan et al. (2000) claimed that the south, south-west, west, and north-west portions of the Konkan region receive maximum rainfall and hence are more susceptible to slope failures. Many other studies conducted in the Western Ghats and Konkan region considered the slope aspect in LSZ mapping.

The slope aspect map has been derived from SRTM DEM. Nine slope aspect classes have been identified and used for assigning weights in LSZ mapping (Figure 5.11).

5.2.3.10 Rainfall

In most cases, slope instability is caused by saturation of the slopes by rainwater. The slope-forming mass once saturated with rainwater becomes heavy which leads to downslope movement. Slope failures in India are induced by intense monsoon rainfall. Therefore, consideration of rainfall as a triggering factor in LSZ mapping is necessary. Ideally, the amount and intensity of rainfall causing slope failure can effectively be used as a data input layer in determining rainfall threshold for slope failures. However, it is very difficult to obtain real-time information about slope failures and their corresponding rainfall data.

To get generalised information about spatial patterns of rainfall, the average annual rainfall for 29 stations in the study area have been considered as data input for LSZ in North Konkan. This data have been interpolated using kriging operation in ArcGIS to derive the thematic layer of average annual rainfall. The data have been classified into five categories: viz., <2,500mm., 2,500–2,750mm., 2,750–3,000mm., 3,000–3,250mm. and >3,250mm. Numerical weights have been assigned to respective rainfall classes for LSZ mapping (Figure 5.12).

5.2.4 BIS (Bureau of Indian Standards) Based Landslide Susceptibility Zonation (LSZ)

The delineation of areas on the basis of their actual and potential susceptibility to slope failure is an important step in landslide susceptibility assessment. Varnes (1984) described landslide susceptibility zonation (LSZ) as 'division of land surface into areas and ranking of these areas according to the degree of actual or potential susceptibility from landslides on the slopes'. The basic principle of LSZ is to estimate potential areas susceptible to slope failures on the basis of past and present slope failures recorded in unique geo-environmental conditions.

Practically, LSZ involves the weighing and integration of landslide causative factors on the basis of their relative contribution in the process of slope instability. Therefore, ranking the landslide causative and triggering factors and assigning weights are critical steps in LSZ mapping. Selection of LSZ method is determined by the rating scheme to rank and weigh landslide causative

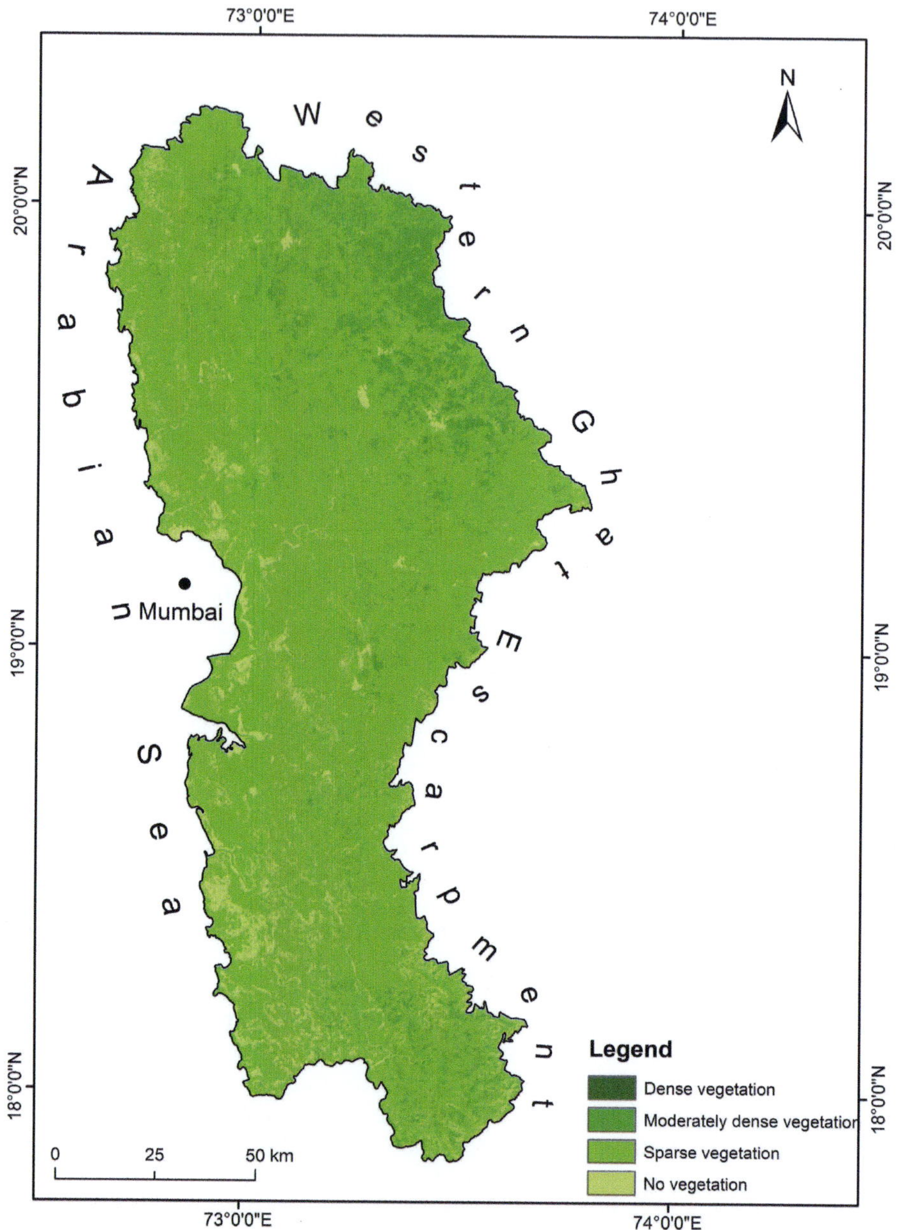

FIGURE 5.10 NDVI Map of North Konkan, Maharashtra.

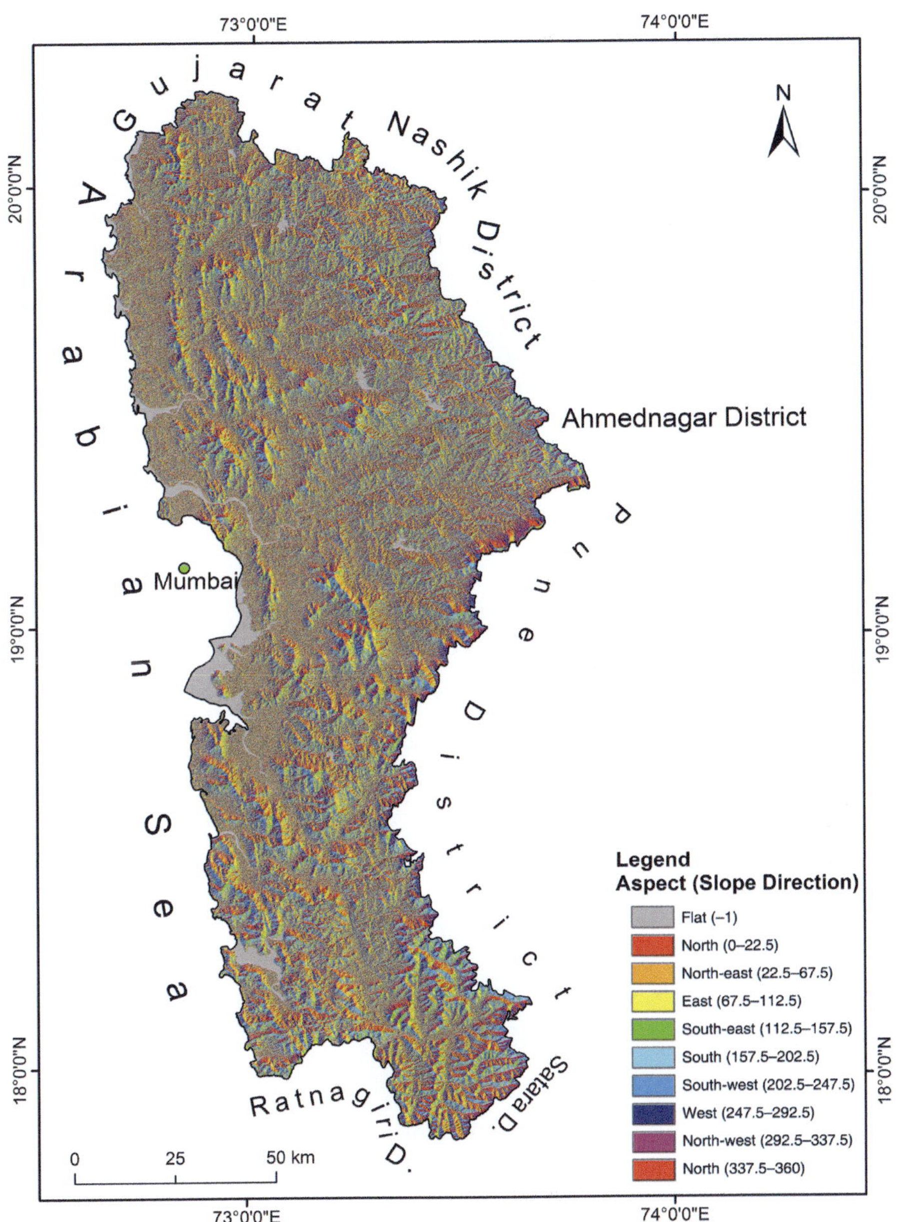

FIGURE 5.11 Slope Aspect Map of North Konkan, Maharashtra.

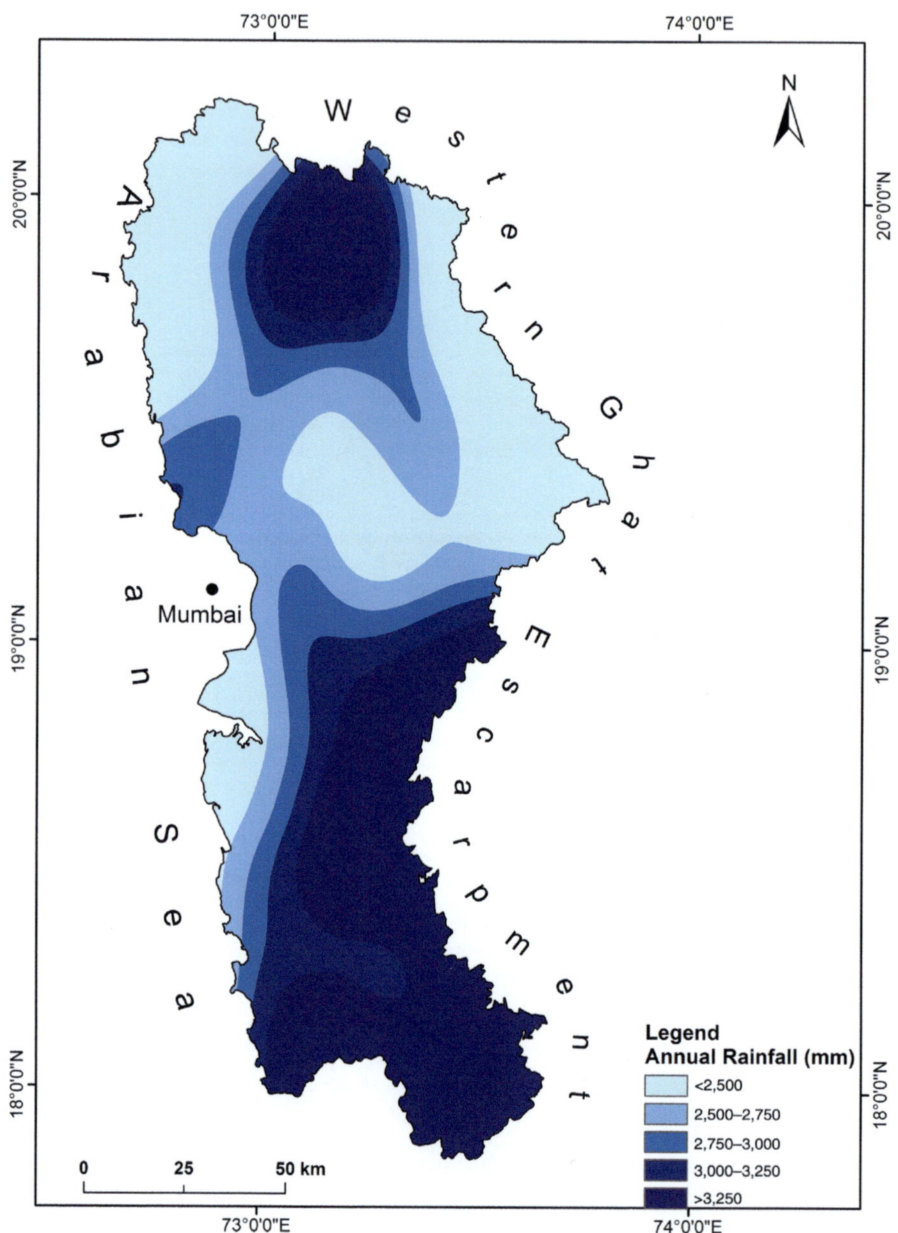

FIGURE 5.12   Rainfall Distribution Map of North Konkan, Maharashtra.

and triggering factors. From knowledge-based (heuristic) approaches to the multi-criteria decision-making approach, several methods to map landslide susceptibility have been proposed and applied in different geo-environmental settings. However, very less attention is given to assess which method is more applicable in different geo-environmental settings. This section describes the application of Bureau of Indian Standards (1998) based methodology for LSZ mapping of the study area.

5.2.5 BIS-Based LHEF (Landslide Hazard Evaluation Factors) Rating Scheme

In India, the Bureau of Indian Standards (1998) has developed a methodology to delineate landslide susceptibility zones at a macro level using terrain evaluation factors, which is given in the document IS 14496, part 2 of BIS (1998). Six primary geo-environmental factors, including slope morphometry, lithology, structure, relative relief, land use and land cover, and hydrological conditions of each slope facet, have been considered in LSZ.

Bureau of India Standards (1998) suggested a numerical rating system (LHEF) to assign numerical weights to each category of landslide causative factor (Table 5.4). The numerical weights assigned to all thematic layers of causative factors have been added to obtain the total estimated susceptibility (TEHD). The TEHD values have been classified using natural breaks method to determine five landslide susceptibility classes: viz. very high susceptibility, high susceptibility, moderate susceptibility, low susceptibility, and very low susceptibility.

The BIS-based LHEF rating scheme is a very simple and cost-effective method of landslide susceptibility mapping. However, subjectivity in the weight assignment procedure exists in this method, which can affect the level of accuracy of susceptibility zonation map. Moreover, this method does not consider landslide distribution, and therefore, it's very difficult to test its validity. Ghosh et al. (2009) evaluated the effectiveness of the existing BIS method in the Darjeeling Himalayas by adopting the WoE model. Ghosh et al. (2009) proposed a modified BIS model for the Himalayan region, which is based on relationships of landslide causative factors with landslide distribution and found it a more effective method for LSZ.

Although the LHEF rating scheme developed by BIS is prepared for macro-level LSZ in India, this method applies to the landslide susceptibility and geo-environmental conditions in the Himalayan region. However, the Western Ghats and rugged coastal plains exhibit different geological and physiographic conditions, where the applicability of BIS-based LHEF rating scheme is limited. Therefore, it was important to modify the BIS-based LHEF rating scheme, taking into consideration the lithological and geomorphic characteristics of the Western Ghat region and coastal plains. This section deals with the LSZ mapping in the North Konkan region of Maharashtra state using a modified form of BIS-based LHEF rating scheme.

TABLE 5.4

Bureau of Indian Standards-Based LHEF Rating Scheme

Landslide Causative Factor	Description		Category	Maximum LHEF Ratings
Lithology	i	Rock type	Type 1	
	ii	Soil type	Quartzite and limestone	0.2
			Granite and grabbo	0.3
			Gneiss	0.4
			Type 2	
			Well-cemented terrigenous sedimentary rocks, dominantly sandstone with minor beds of clay stones	1.0
			Poorly cemented terrigenous sedimentary rocks dominantly sand rock with minor clay shale beds	1.3
			Type 3	
			Slate and phyllite	1.2
			Schist	1.3
			Shale with interbedded clayey and non-clayey rocks	1.8
			Highly weathered shale, phyllite and schist	2.0
			Older, well-compacted alluvial fill material	0.8
			Clayey soil with naturally formed surface	1.0
			Sandy soil with naturally formed surface (alluvial)	1.4
			Debris comprising mostly rock pieces mixed with clayey, sandy soil (Colluvial)	
			– Older, well compacted	1.2
			– Younger, loose material	2.0
Structure		Relationship of structural discontinuities with slope	I > 30°	0.20
			II 21°–30°	0.25
	i	Relationship of parallelism between the slope and the discontinuity	III 11°–20°	0.30
			IV 6°–10°	0.40
			V < 5°	0.50
	ii	Relationship of dip of discontinuity and inclination of slope	I > 10°	0.3
			II 0°–10°	0.5
			III 0°	0.7
	iii	Dip of discontinuity	IV 0°–(−10°)	0.8
	iv	Depth of soil cover	V > (−10°)	1.0

TABLE 5.4 (*Continued*)
Bureau of Indian Standards-Based LHEF Rating Scheme

Landslide Causative Factor	Description	Category	Maximum LHEF Ratings
		I < 15°	0.2
		II 16°–25°	0.25
		III 26°–35°	0.30
		IV 36°–45°	0.40
		V > 45°	0.50
		< 5 m	0.65
		6–10 m	0.85
		11–15 m	1.30
		16–20 m	2.0
		> 20 m	1.20
Slope morphometry	i Escarpment/cliff	> 45°	2.0
	ii Steep slope	36°–45°	1.7
	iii Moderately steep slope	26°–35°	1.2
	iv Gentle slope	16°–25°	0.8
	v Very gentle slope	≤ 15°	0.5
Relative relief	i Low	< 100 m	0.3
	ii Medium	101 – 300 m	0.6
	iii High	> 300 m	1.0
Land use and land cover	i Agricultural land/populated flat land	–	0.6
			0.8
	ii Thickly vegetated forest area		1.2
	iii Moderately vegetated forest area		1.5
	iv Sparsely vegetated forest area		2.0
	v Barren land		
Hydrological conditions	i Flowing	–	1.0
	ii Dripping		0.8
	iii Wet		0.5
	iv Damp		0.2
	v Dry		0.0

(*Source*: Bureau of India Standards – IS 14496, part B, 1998)

TABLE 5.5

Landslide Susceptibility Zones on the Basis of TEHD Values

S. No.	Zone	Range of TEHD Value	Description
1	I	< 3.5	Very low susceptibility
2	II	3.5–5.0	Low susceptibility
3	III	5.1–6.0	Moderate susceptibility
4	IV	6.1–7.5	High susceptibility
5	V	> 7.5	Very high susceptibility

(*Source*: BIS 1998)

5.2.6 Modified LHEF (Landslide Hazard Evaluation Factors) Rating Scheme for LSZ

To map landslide susceptibility in the North Konkan region, the modified form of BIS (1998) based LHEF rating scheme has been adopted. Ten landslide causative factors, including slope morphometry, elevation, structure (lineaments, dykes, and faults), lithology, land use and land cover, density of vegetation cover, distance from drainage, distance from road, slope aspect, and rainfall, have been considered for determining LHEF ratings. Since the lithological characteristics of the North Konkan exhibit the characteristics of the Deccan Basaltic Province (DVP), the weights to the lithology are assigned on the basis of basaltic lava formations of different ages. Slope aspect is another major factor responsible for slope instability in the Western Ghats and Konkan region. Therefore, slope aspect is also considered for LSZ in the study area. Surface hydrological conditions also contribute to slope instability, especially in the Himalayas and Western Ghats. Proximity to drainage has been used to determine hydrological conditions because slopes closer to drainage lines remain wet and damp throughout the year, and dripping and water flow is common during the rainy season. To determine vegetation density classes, reclassified NDVI image has been used, and subsequently LHEF ratings are assigned. All the thematic layers of landslide causative factors have been reclassified using the modified LHEF rating scheme. The details of LHEF ratings adopted for LSZ in North Konkan are given in Tables 5.6 and 5.7.

5.2.7 BIS-Based Landslide Susceptibility Zone

The LSZ map (Figure 5.15) prepared for the North Konkan region using the modified LHEF rating scheme of the Bureau of Indian Standards (1998) depicts that 4,339.97 km² (26.55% of the total study area) in North Konkan falls in high to

TABLE 5.6
Modified BIS-Based LHEF (Landslide Hazard Evaluation Factors)
Scheme for LSZ

S. No.	Parameter	BIS (1998) Based LHEF Rating		Modified Rating Scheme for LSZ	
1	Lithology	Rock type: Quartzite	0.2	Karla	2.0
		Granite	0.3	Indrayani	1.8
		Gneiss	0.4	Megacryst flows	1.3
		Sandstone	1.0	Diveghat	0.8
		Slate/phyllite	1.2	Purandargarh	0.8
		Schist	1.3	Elephanta	0.4
		Shale	1.8	Laterite	0.3
		Soil type: Old alluvial	0.8	Salher	0.2
		Clayey	1.0	Alluvium	0.2
		Sandy	1.4	Borivali	0.2
		Debris	1.2	(on the basis of	
		Colluvial	2.0	formations)	
2	Structure	Parallelism between slope and structural discontinuities	1.0	Lineaments	1.0
				Dykes	1.0
				Faults	1.0
3	Slope gradient (degrees)	> 45	2.0	> 45	2.0
		36–45	1.7	35–45	1.7
		26–35	1.2	25–35	1.2
		16–25	0.8	15–25	0.8
		< 15	0.5	< 15	0.5
4	Elevation (m)	< 100	0.3	<100	0.3
		100–300	0.6	100–300	0.6
		> 300	1.0	> 300	1.0
5	Land use/land cover	Agriculture/built-up	0.6	Barren land	2.0
		Thickly vegetated	0.8	Rocky outcrop	1.5
		Moderately vegetated	1.2	Vegetation	1.2
		Sparsely vegetated	1.5	Agricultural land	0.6
		Barren land	2.0	Built-up areas	0.6
				Coastal sand	0.3
				Water body	0.3
6	Hydrological conditions	Flowing	1.0	Proximity to drainage	1.0
		Dripping	0.8		
		Wet	0.5		
		Damp	0.2		
		Dry	0		

(Continued)

TABLE 5.6 (*Continued*)
Modified BIS-Based LHEF (Landslide Hazard Evaluation Factors) Scheme for LSZ

S. No.	Parameter	BIS (1998) Based LHEF Rating		Modified Rating Scheme for LSZ	
7	Aspect	—	—	West	0.9
				South	0.5
				East	0.5
				North	0.2
				Flat	0.1
				South-west	0.9
				South-east	0.3
				North-west	0.1
				North-east	0.1
8	NDVI	–	–	Dense vegetation	1.5
				Moderately vegetated	1.2
				Sparsely vegetated	0.8
				Barren land	0.1
9	Distance from road (m)	–	–	< 100 m	1.0
				> 100 m	0
10	Mean annual rainfall (mm)	–	–	< 2,500	0.2
				2,500–2,750	0.4
				2,750–3,000	0.6
				3,000–3,250	0.8
				> 3,250	1.0

(*Source*: Bureau of Indian Standards 1998; Ghosh et.al 2009)

TABLE 5.7
Landslide Susceptibility Zones Using Modified BIS Method

S. No.	Zone	Range of TEHD value	Description
1	I	< 6.9	Very low susceptibility
2	II	6.9–8.3	Low susceptibility
3	III	8.3–9.7	Moderate susceptibility
4	IV	9.7–11.2	High susceptibility
5	V	> 11.2	Very high susceptibility

(*Source*: BIS, 1998; Ghosh et al., 2009)

TABLE 5.8

Landslide Susceptibility Zones Using Modified BIS Method

Zone	Description	Range of TEHD Value	Area (km²)	Percentage Area
I	Very low susceptibility	< 6.9	3,428.13	20.97
II	Low susceptibility	6.9–8.3	4,206.84	25.74
III	Moderate susceptibility	8.3–9.7	4,369.41	26.74
IV	High susceptibility	9.7–11.2	3,015.12	18.45
V	Very high susceptibility	> 11.2	1,324.05	8.10
Total			**16,343.55**	**100**

Source: (BIS, 1998)

very high susceptibility classes, whereas 4,369.41 (26.74% of the study area) falls in the moderate susceptibility class (Table 5.8).

The LSZ map based on the modified BIS method depicts that the potential landslide susceptibility zones in North Konkan with varying degrees of susceptibility are observed in concentrated pockets. Broadly, landslide hazard (LH) zones in the study area can be categorised into three classes:

1 High to very high susceptibility zone
2 Moderate susceptibility zone
3 Low susceptibility zone

The spatial pattern of LH zones in the study area is briefly discussed below.

5.2.7.1   High to Very High Susceptibility Zone

The distribution of areas with high to very high susceptibility in the North Konkan region shows that these areas are located in different isolated pockets throughout the study area. Six concentrated areas representing high to very high susceptibility zones are observed: viz., Jawhar plateau, coastal hill near Palghar, Malshej Ghat, Matheran Hills, Karnala Hills, and Mahad-Poladpur Circle (Figure 5.13).

The north-east corner of the study area falls in high to very high susceptibility zones. Slope failures in this area are associated with the plateau margins of the dissected Jawhar plateau at an elevation of 250–350 m. This area is dominated by Salher lava formations followed by megacryst flows of Ratangarh formations with moderately weathered surface. The region is intruded by numerous dykes aligned in the north-south direction and shows high structural control. Major communication routes passing through this area, including Sai-Vavar Road (SH 28), Wada-Khodala Road (SH 34), and Jawhar-Mokhada Road (SH 36), are affected by slope failures frequently during the monsoon season.

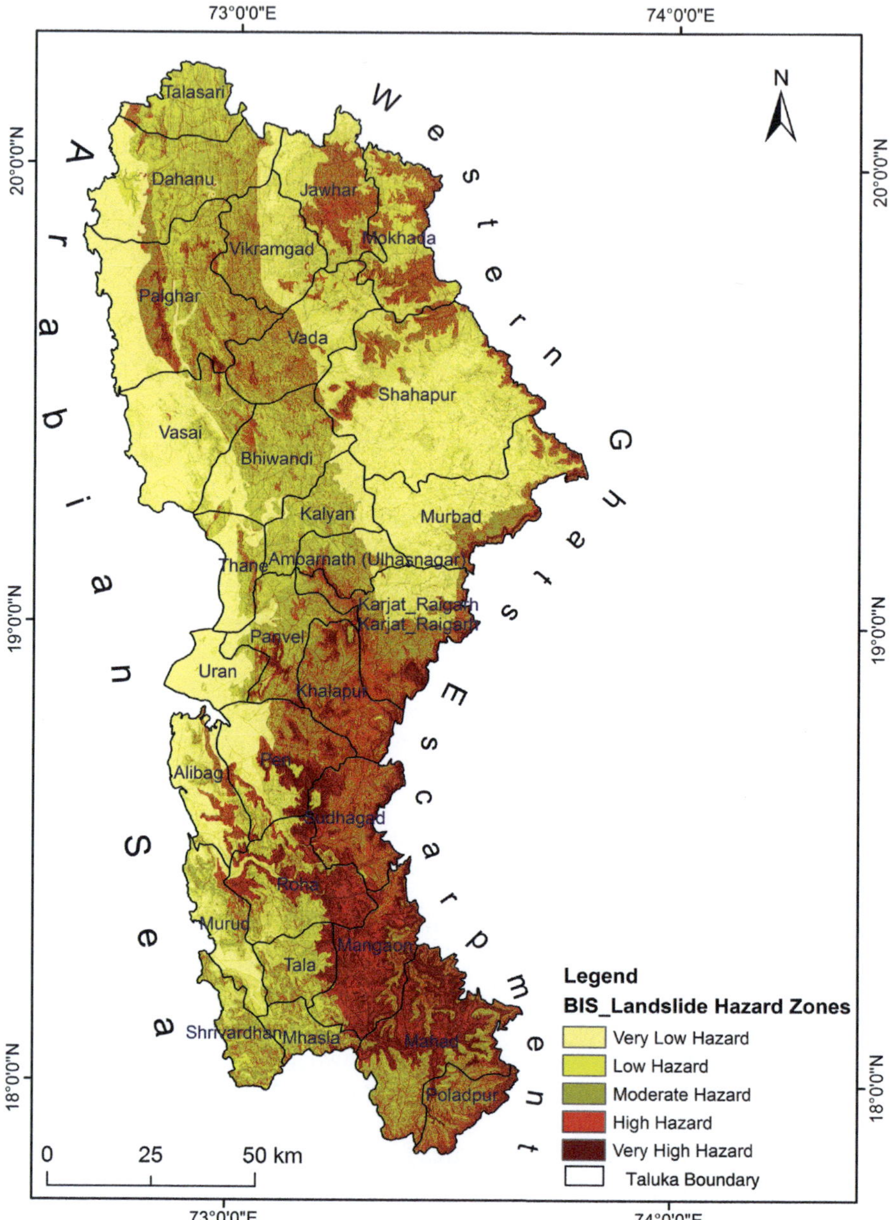

FIGURE 5.13 Landslide Susceptibility Zonation Map Using Modified BIS Method.

The first zone falling in high to very high susceptibility classes is located near the coastal hill range to the east of Palghar town. The total extent of this hill is about 26 km in the north-south direction. Steep natural slope (35° and above), high structural discontinuities, westwards slope aspect (facing the south-west monsoon winds), and heavy annual rainfall are major factors contributing to slope failures in this area. Lithologically, the area comprises Karla formations, which is a relatively younger formation and more susceptible to slope failures. None of the reported landslides are located in this area, possibly due to lesser human interference. However, development activities, such as construction of roads, tunnels, and railway lines, may cause slope failure in this area and may also pose a serious threat to the settlements located in the foothills.

The Malshej Ghat area located at the eastern border of the study area is the second zone of high to very high susceptibility classes. About a 10 km stretch along NH 222 passing through this section is affected by slope failures every year during the monsoon season. Steep to near-vertical slopes (35°–73° at places), north-west to south-east trending lineaments, westwards slope aspect, and heavy monsoon rainfall are important landslide causative factors in this area.

The third landslide susceptibility zone of the study area falling in the high susceptibility class is Matheran Hill, an isolated hill complex located at the northern border of Raigad district near Neral town. The area is located at 250 m ASL. The LSZ of this hill complex falls in high to very high susceptibility zones. The Matheran hill complex is characterised by steep slopes (30°–35°) and a predominance of loose, weathered, and coarse slope-forming material. The lithology of this area is composed of lava flows, classified under Indrayani and Ratangarh formations, capped with low-level laterites at the hilltop. Besides natural causative factors, modification of slopes by anthropogenic activities, such as the construction of roads and narrow-gauge railway line, also results in slope instability along communication routes.

The Karnala Hill complex, located about 20 km south of Panvel, is another major concentration of high to very high susceptibility zones. This area is characterised by steep hill slopes with angles ranging from 20° to 25°), north-south trending lineaments and faults, and moderate to dense vegetation cover. Minor slope failures are quite common along the Mumbai-Goa Highway (NH 17) passing through this stretch. However, no major slope events have been reported in the interior parts or off the road. However, it can be inferred that even minor human intervention in this hill complex may cause slope failures in this area.

The Mahad-Poladpur Circle is located at the south-east corner of Raigad district. This area is characterised by highly rugged topography with a slope angle between 25° and 35° and dominated by west and south-west slope aspect. Heavy rainfall and a dense network of north-west to south-east trending lineaments, combined with steep slopes, cause slope failures in this area. The traffic along major communication routes, including Poladpur-Mahabaleshwar Road (SH 72), Mumbai-Goa Highway (NH 17), and Mahad-Bhor Road (SH 70), is affected by slope failures in this area.

5.2.7.2 Moderate Susceptibility Zone

About 26.74% of the total geographical area of North Konkan falls in the moderate landslide susceptibility zone. The spatial distribution of this zone is associated with the western foothills of the Western Ghat region of the study area and central plains. Major parts of Thane and Palghar districts falls in this category. Besides, small isolated hills with elevation less than 300 m. ASL, located in the south Raigad district, also fall in the moderate susceptibility zone.

5.2.7.3 Low and Very Low Susceptibility Zone

Almost half (46.72%) of the total geographical area of North Konkan falls in low to very low susceptibility classes. The distribution of this zone is associated with coastal plains throughout the study area.

5.2.8 Validation

To validate the results of the LSZ in the North Konkan region, the modified BIS-based LSZ map has been compared with the actual distribution of the slope failure events in the study area (Figure 5.14).

The LH zones are compared with the actual distribution of slope failure sites recorded during the field survey to evaluate the validity of modified BIS method. The result shows that 75.69% of the recorded slope failure events fall in high to very high susceptibility classes, which confirms the reliability of the modified BIS method.

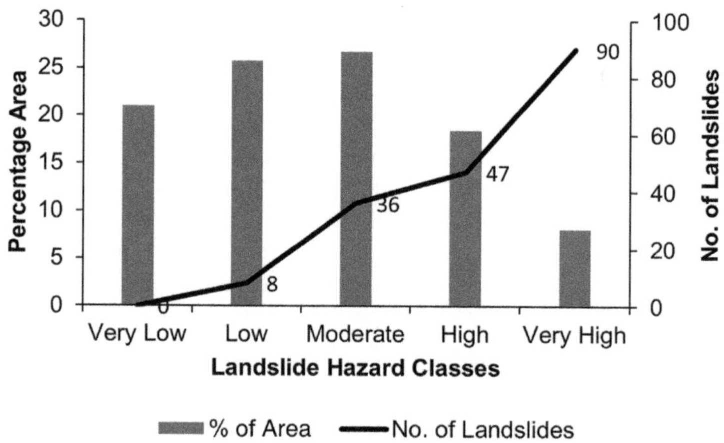

FIGURE 5.14 Landslide Susceptibility Zones Based on Modified BIS Method.

5.3 LANDSLIDE SUSCEPTIBILITY ZONATION USING MULTI-CRITERIA DECISION-MAKING APPROACH: CASE STUDY 2 (WESTERN GHATS AND WESTERN COASTAL AREAS OF INDIA)

5.3.1 INTRODUCTION

Decision-making in situations controlled by many factors is a complex process. The multi-criteria decision-making approach involves prioritisation of a set of elements for decision-making in a complex phenomenon. The analytic hierarchy process, developed by Saaty (1980), is the most widely used MCDA tool.

AHP is a theory of measurement through pairwise comparisons and relies on the judgements of experts to derive priority scales (Saaty, 2008). AHP involves defining the problem to be sought; defining criteria, sub-criteria, and alternatives; and prioritising them on the basis of their relative significance to derive the most suitable alternatives.

In AHP, the prioritisation of the alternatives is done through the following steps.

1 Definition of the problem for which decision is to be taken.
2 Structuring the decision hierarchy with goal, criteria, sub-criteria, and alternatives.
3 Preparation of pairwise comparison matrices using scale of absolute numbers (Table 5.9), where each element is compared with other elements to derive priorities. Such matrices are constructed for both inter-parameter and intra-parameter comparisons.
4 The priorities obtained from pairwise comparisons are used to assign weights to all the parameters.
5 Calculation of consistency index (CI) and consistency ratio (CR) to evaluate the consistency of judgements using standard values from the index of consistency (Table 5.10). The most common method adopted for calculation of consistency is the eigen vector approach.

The consistency index is expressed as

$$CI = (\lambda_{max} - n)/(n - 1),$$

where λ_{max} is the principal eigen value, and n is the number of elements or order of the matrix.

The consistency ratio is calculated as

$$CR = CI/RI,$$

where CI is the consistency index, and RI is the average of consistency index based on Saaty's (1980) order of matrix (i.e., standard value for index of constancy on the basis of order of matrix). The judgements used for pairwise comparison are considered to be consistent if the CR value is less than 0.1.

TABLE 5.9
Fundamental Scale of Absolute Numbers

Intensity of Importance	Definition	Explanation
1	Equal importance	Two activities equally contribute to the object.
2	Weak or slight importance	Two activities almost equally contribute to the object.
3	Moderate importance	Experience and judgements slightly favour one activity over another.
4	Moderate plus	Experience and judgements slightly favour one activity over another.
5	Strong importance	Experience and judgements strongly favour one activity over another.
6	Strong plus	Experience and judgements strongly favour one activity over another.
7	Very strong or demonstrated importance	Experience and judgements very strongly favour one activity over another; its dominance is demonstrated in practice.
8	Very, very strong importance	Experience and judgements very strongly favour one activity over another; its dominance is demonstrated in practice.
9	Extreme importance	The evidence favouring one activity over another is of the highest possible order of affirmation.
Reciprocals of above	If activity i has one of the above nonzero numbers assigned to it when compared with activity j, then j has the reciprocal value when compared with i	A reasonable assumption.
1.1–1.9	If the activities are very close	May be difficult to assign the best value, but when compared with other contrasting activities, the size of the small numbers would not be too noticeable, yet they can still indicate the relative importance of the activities.

(*Source*: Saaty 2008)

TABLE 5.10
Random Index (RI)

Order of Matrix/ No. of Parameters	1	2	3	4	5	6	7	8	9	10	11	12	13	14	15
Random Inconsistency (RI)	0	0	0.52	0.89	1.11	1.25	1.35	1.40	1.45	1.49	1.52	1.54	1.56	1.58	1.59

(*Source*: Saaty 2008)

The most important advantage of AHP is its ability to rank/prioritise alternatives on the basis of their relative importance over the other. Therefore, AHP is a proven technique to discriminate among the alternatives for more consistent and reliable judgements. Another major advantage of AHP is its ability to detect inconsistent judgements. AHP can effectively be used to synthesise intangible information more than any other tools.

5.3.2 Application of AHP in Landslide Susceptibility Zonation

AHP is being used in a wide range of applications the world over. A review carried out by Vaidya and Kumar (2006) suggests that a recent trend is to apply AHP in combination with other techniques to refine results. Since the last decade, the application of AHP in landslide susceptibility assessment, particularly in LSZ, is gaining more importance.

AHP is proven to be a reliable method to map landslide susceptibility in many studies carried out in Malaysia (Othman et al., 2012; Saadatkhah et al., 2014), Iran (Kornejady et al., 2014; Habibi, 2014; Boroumandi et al., 2015), Nepal (Bhatt et al., 2013), and India (Phukon et al., 2012; Saaty, 2013). This section describes the application of AHP for LSZ in the North Konkan region of Maharashtra.

5.3.3 Methodology

This section deals with LSZ mapping in the North Konkan region of Maharashtra using AHP. The thematic data layers of landslide causative factors, including slope angle, slope aspect, elevation, drainage, land use and land cover, vegetation cover, lithology, structure, roads, and rainfall, have been used as input data layers in a GIS environment. Data preprocessing and rasterisation of these thematic data layers is done in ArcGIS.

To identify landslide susceptibility zones, geo-environmental parameters (criteria) and their subclasses (sub-criteria) have been considered to determine the priorities. To derive priorities, comparison between all landslide causative factors has been carried out. Using Saaty's (1980) scale of absolute numbers, pairwise comparison matrices have been constructed for inter-parameter and intra-parameter comparisons. To derive priorities, numbers are assigned on the basis of relative importance of each parameter in the process of slope failure. To evaluate the consistency of judgements, principal eigen vector approach has been adopted by calculating CI and CR.

After the finalisation of priorities, all the thematic data layers have been reclassified using final weights derived from pairwise comparisons. Reclassified thematic data layers (Figures 5.14–5.23) are integrated in a GIS environment using 'weighted sum' operation. The output map has been classified using natural break methods into five landslide susceptibility classes: viz., very high susceptibility, high susceptibility, moderate susceptibility, low susceptibility, and very low susceptibility.

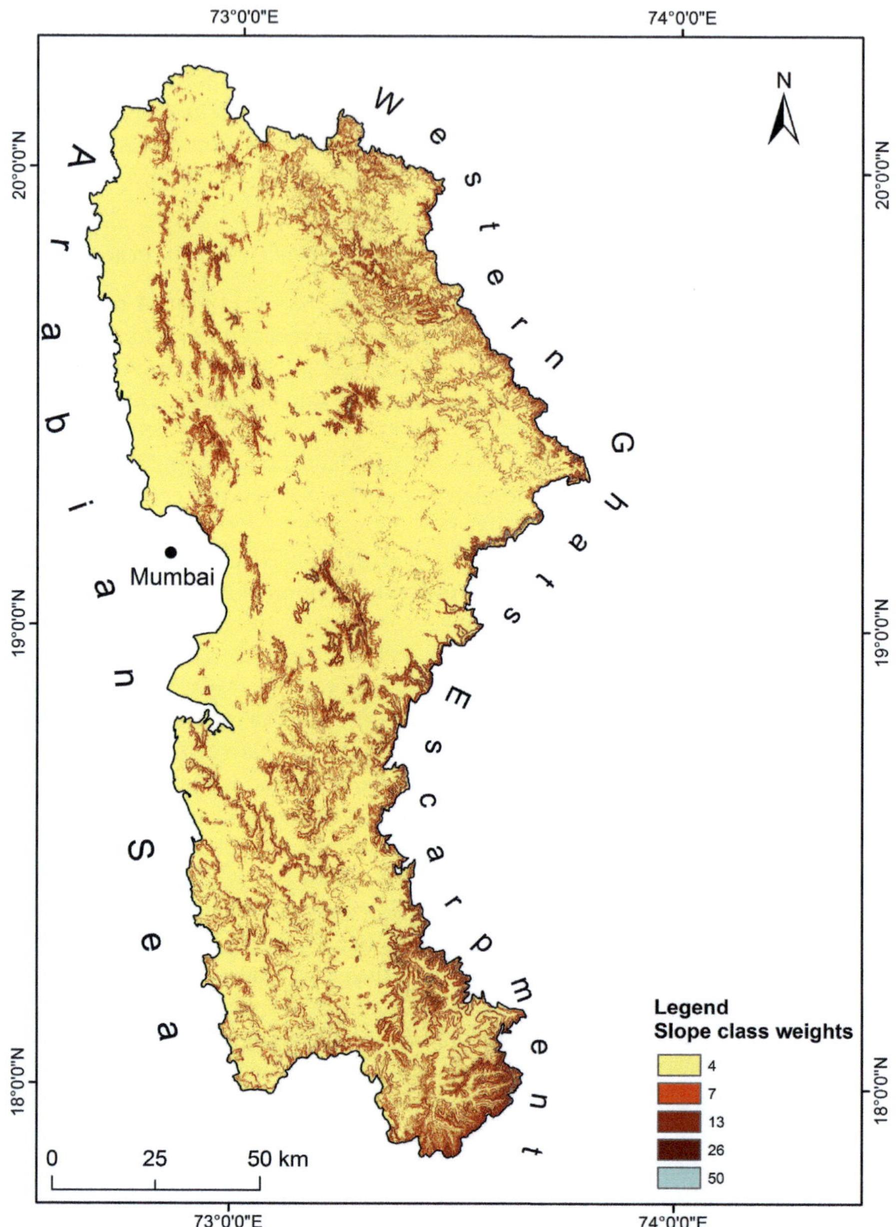

FIGURE 5.15   Reclassified Thematic Data Layer of Slope Map Using AHP Method.

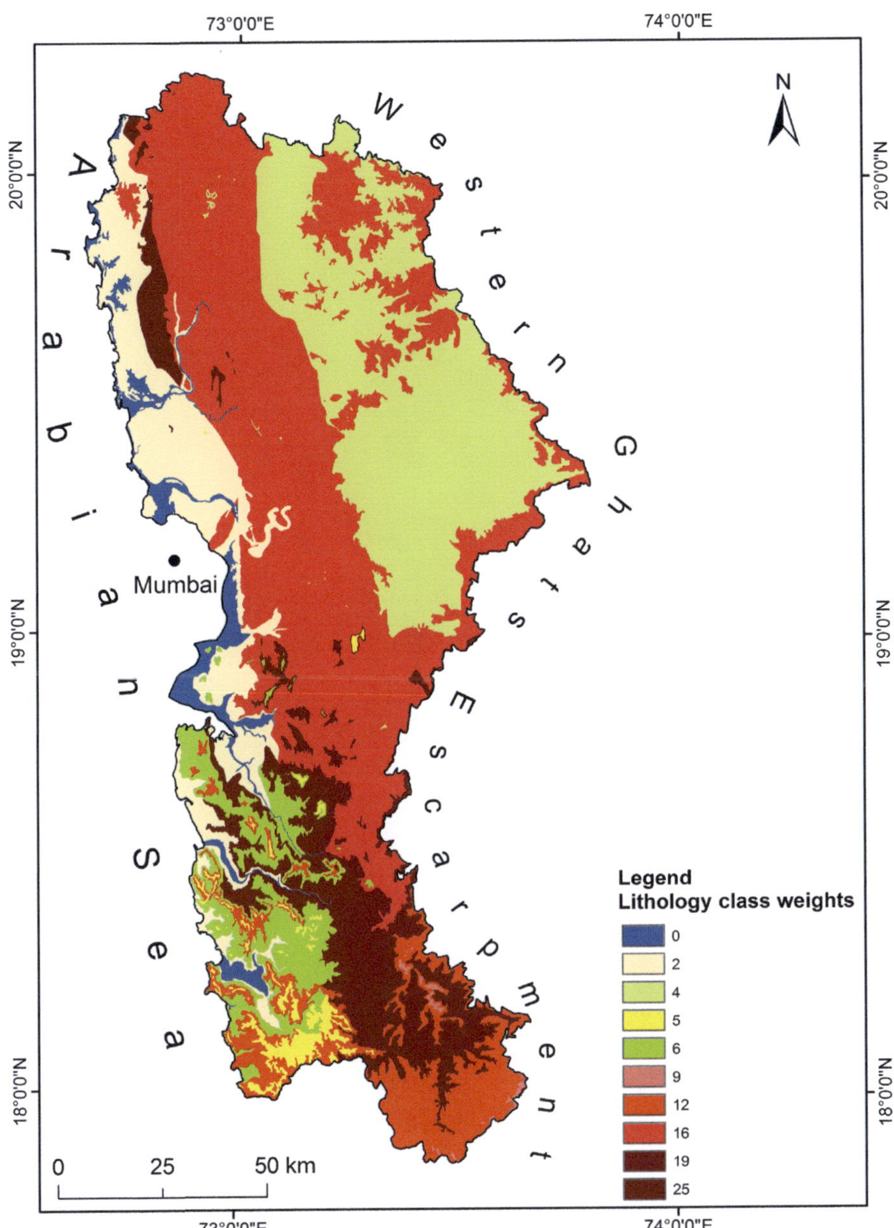

FIGURE 5.16 Reclassified Thematic Data Layer of Lithology Map Using AHP Method.

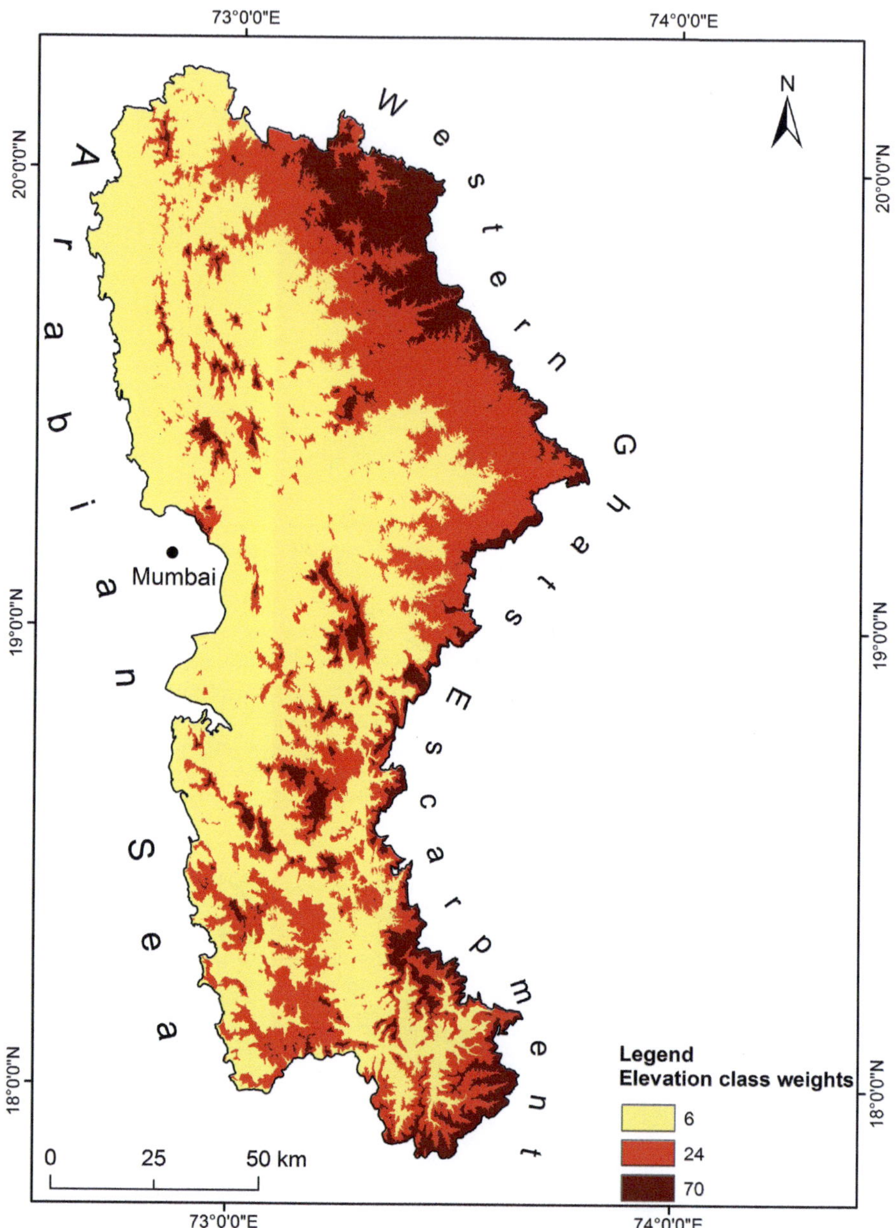

FIGURE 5.17 Reclassified Thematic Data Layer of Relief Map Using AHP Method.

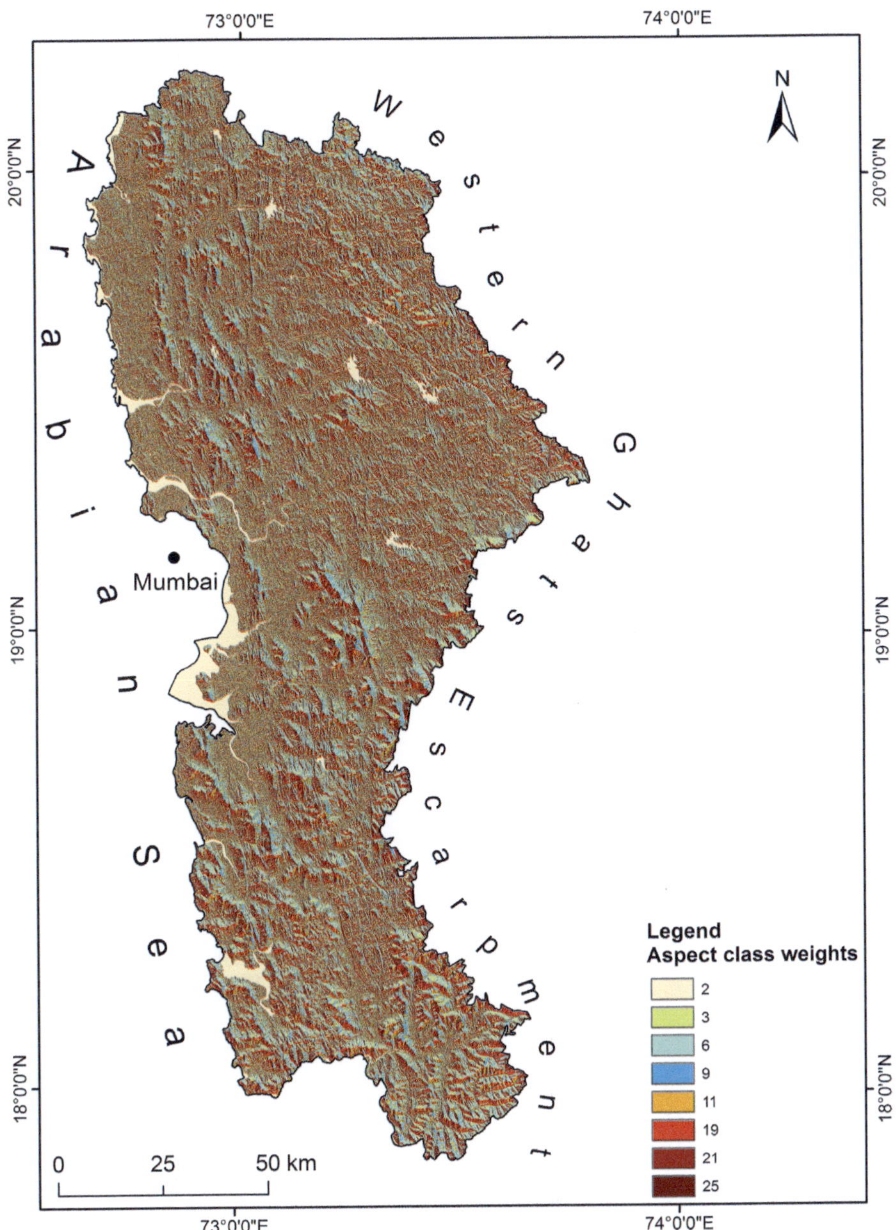

FIGURE 5.18 Reclassified Thematic Data Layer of slope Aspect Map Using AHP Method.

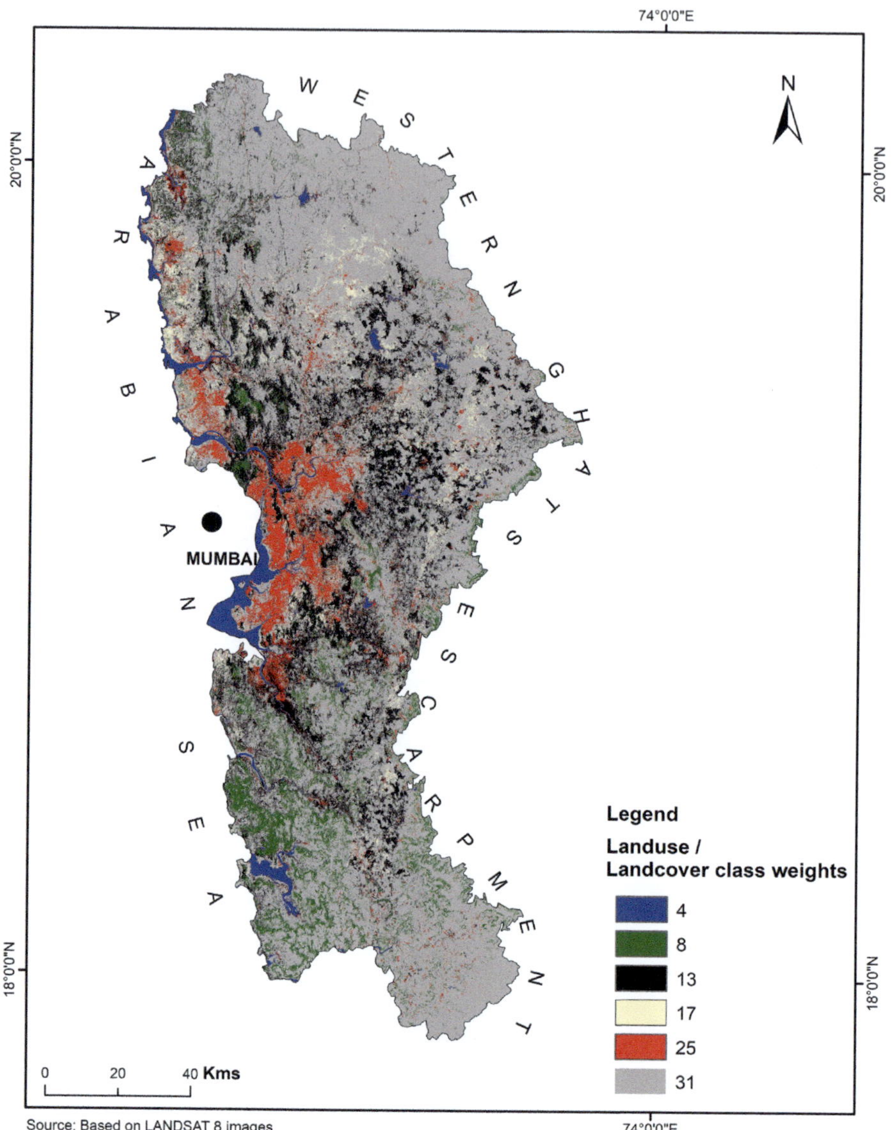

FIGURE 5.19 Reclassified Thematic Data Layer of Land Use and Land Cover Map Using AHP Method.

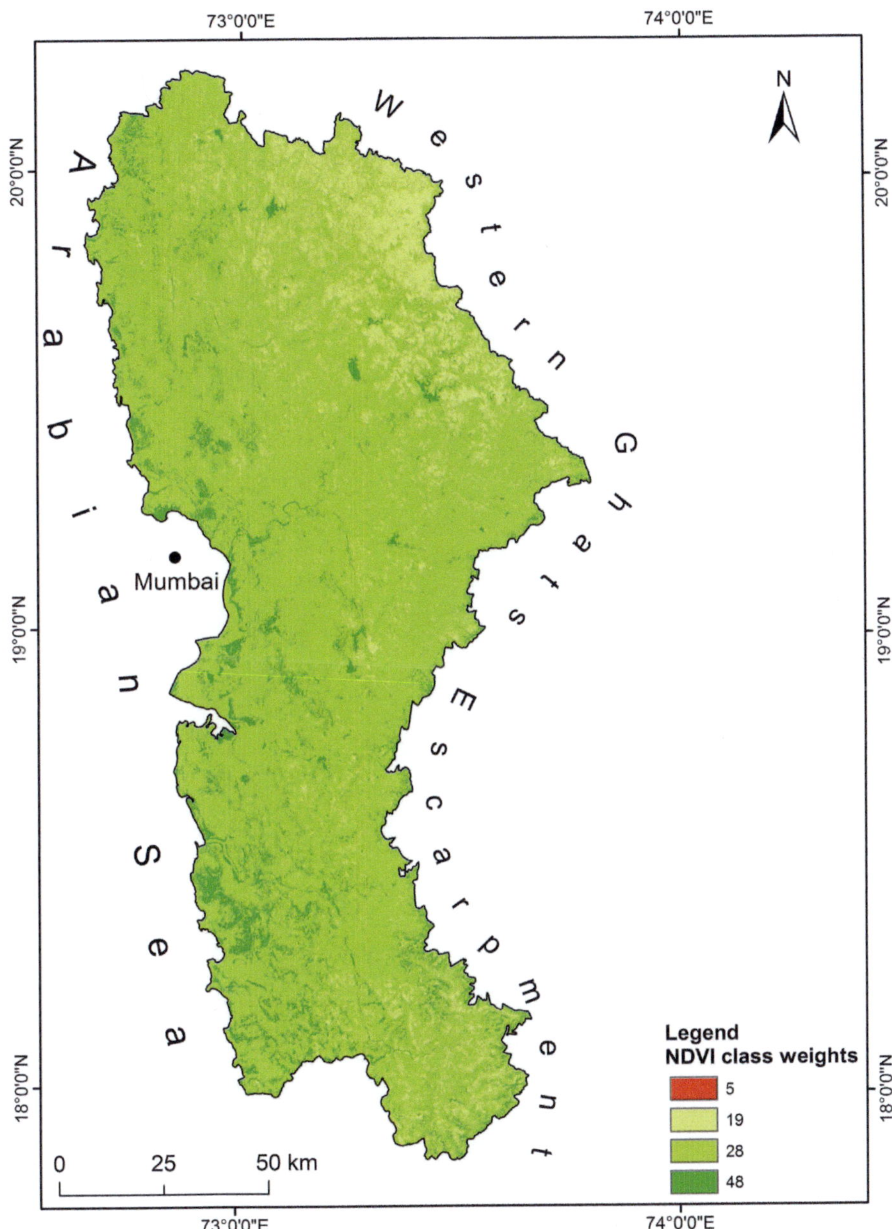

FIGURE 5.20 Reclassified Thematic Data Layer of NDVI Map Using AHP Method.

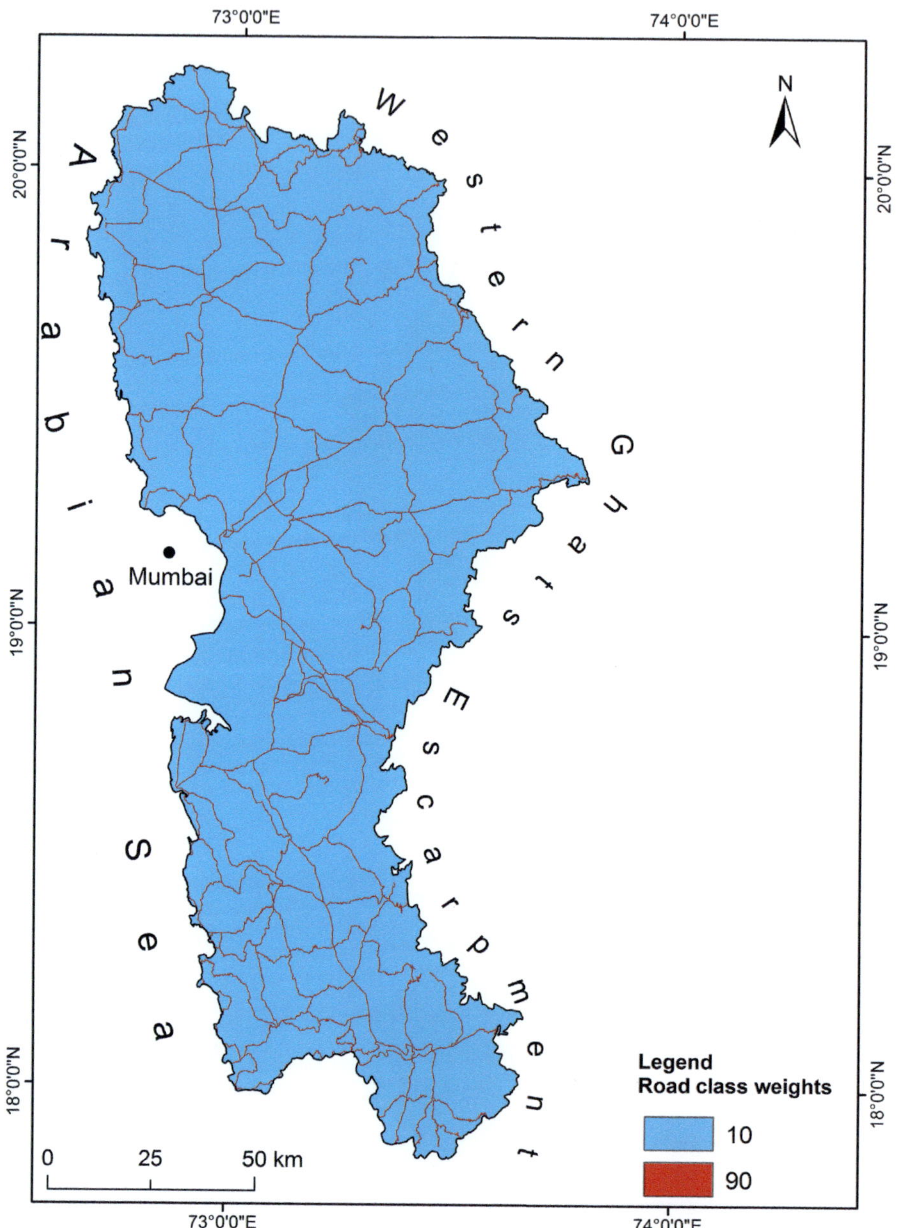

FIGURE 5.21   Reclassified Thematic Data Layer of Road Map Using AHP Method.

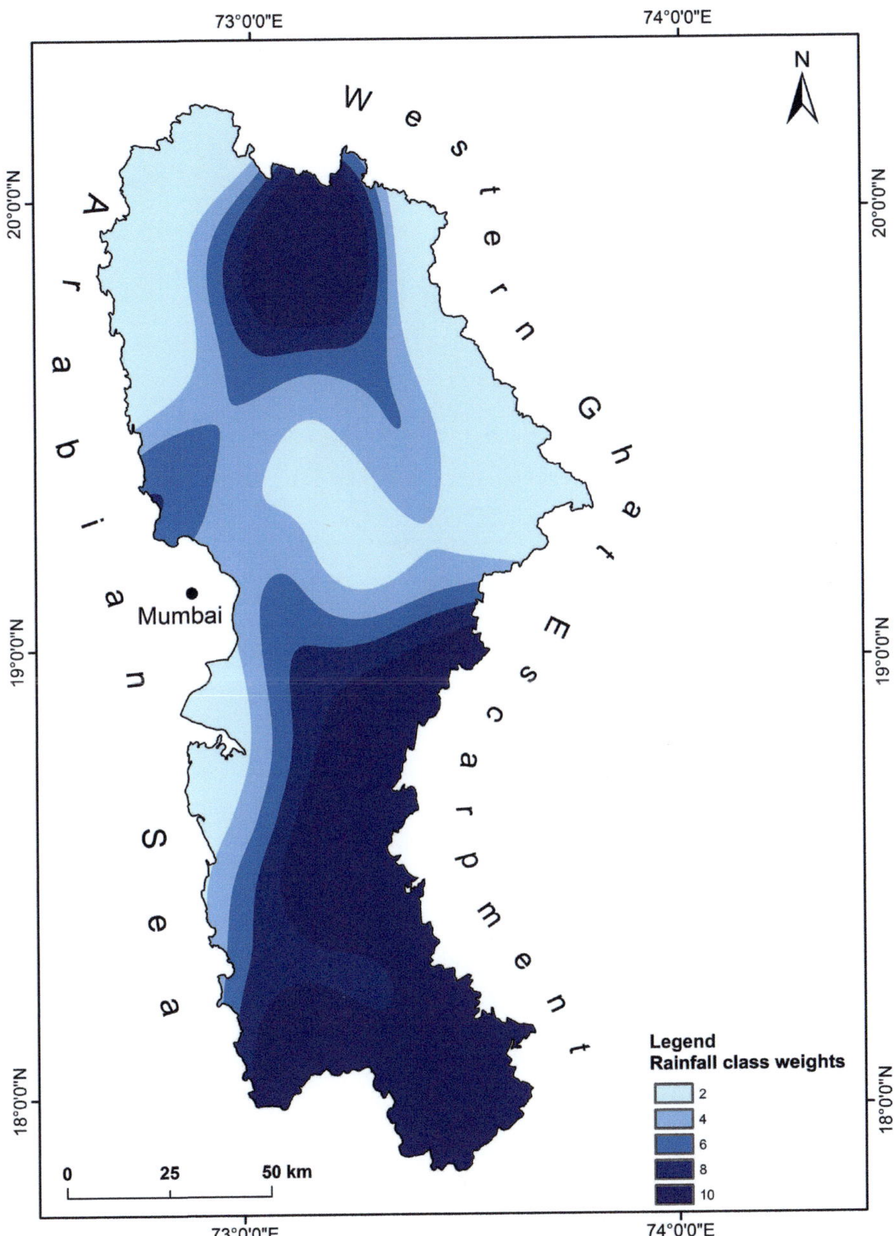

FIGURE 5.22   Reclassified Thematic Data Layer of Rainfall Map Using AHP Method.

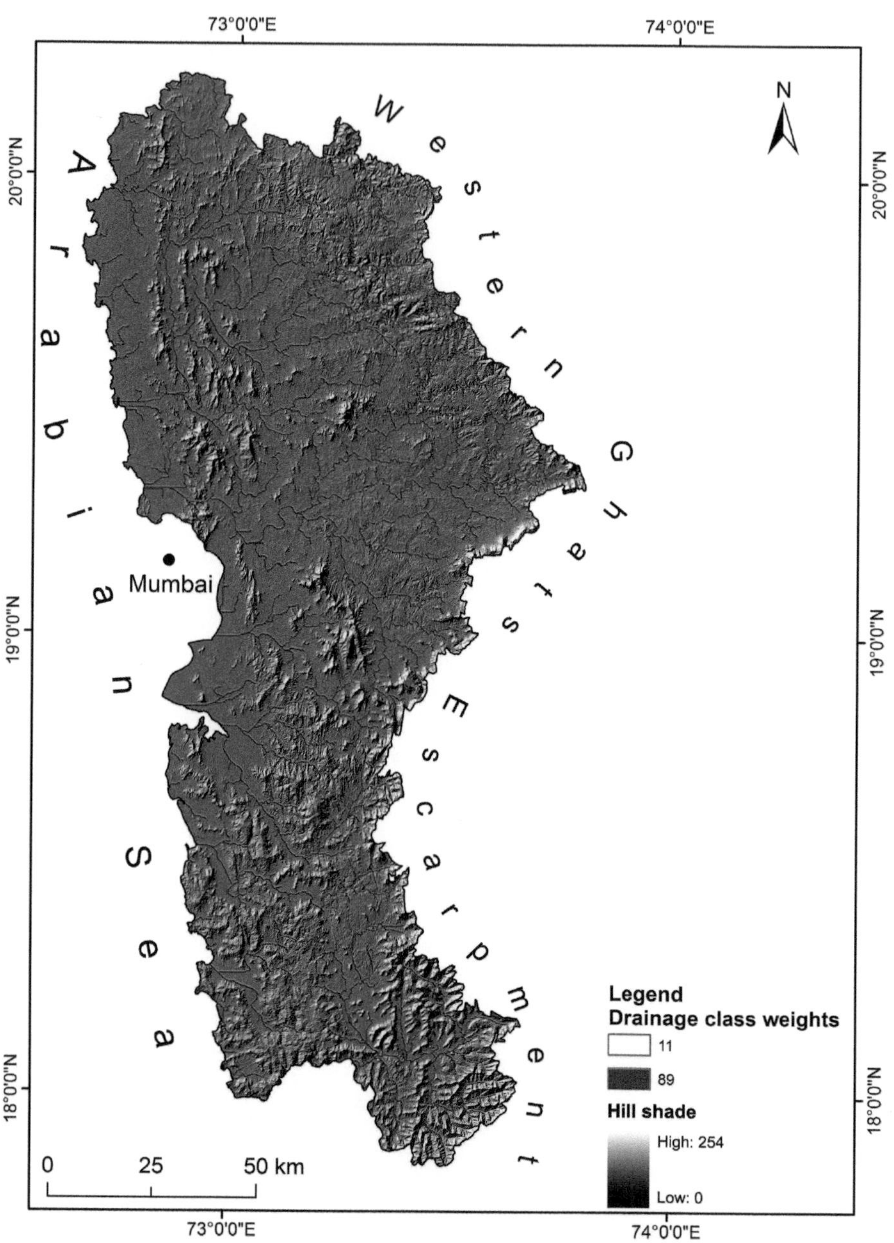

FIGURE 5.23   Reclassified Thematic Data Layer of Drainage Map Using AHP Method.

5.3.4 Pairwise Comparison of Landslide Causative and Triggering Factors

Assigning weights for landslide causative factors is the most important step in LSZ mapping. It is therefore essential to determine the relative importance of each landslide causative factor in initiating slope failure. AHP provides a framework to derive priorities based on pairwise comparisons, where each parameter is compared with the remaining parameters to determine the weights for all causative factors/parameters. Eleven pairwise comparison matrices have been constructed, for both inter-parameter comparisons and intra-parameter comparisons, and weights for each parameter have been derived. To evaluate the consistency of judgements used in pairwise comparisons, CI and CR have been computed. The details of the paired comparisons are given in Tables 5.11–5.22.

5.3.5 AHP-Based Landslide Susceptibility Zones

The AHP-based LSZ map prepared for the North Konkan region shows different concentrations of landslide susceptibility zones on the basis of zonation analysis (Table 5.23).

The LSZ map produced using the AHP method shows that 2,507.74 km^2 (15.34%) of the total geographical area of North Konkan falls in high to very high susceptibility classes, whereas 26.77% and 57.89% of the area fall in moderate and very low to low susceptibility zones, respectively. The spatial distribution of landslide susceptibility zones is discussed as follows (Figure 5.25).

5.3.5.1 High to Very High Susceptibility Zones

Only 15.34% of the total area falls in high to very high susceptibility classes. However, the areas showing high landslide susceptibility are concentrated in isolated pockets. The major concentration of these zones is observed along the western slopes of the Western Ghat section throughout the study area. Another major area with high to very high susceptibility is observed along the margins of the highly dissected Jawhar plateau, which is intruded by a dense network of structural discontinuities. In this zone, the main road routes, including Wada-Khodala Road (SH 34) and Sai-Vavar Road (SH 28), have been affected by several slope failure events, per the records available from public works departments.

Matheran Hill, located on the borders of Thane and Raigad districts, is a denudational hill complex that also falls in high to very high susceptibility classes. Steep natural slopes and moderate to highly weathered surfaces, combined with heavy rainfall, cause slope instability in this area. Besides these, the Karnala Hill areas and rugged topography in the south-east part of Raigad district fall in high to very high susceptibility classes (Figure 5.25).

TABLE 5.11

Pairwise Comparison Matrix for Landslide Causative Parameters

Parameters	Slope	Lithology	Structure	Drainage	Rainfall	Land Use and Land Cover	Road	Slope Aspect	Vegetation	Relief	Weights
Slope	1	2	2	3	4	4	5	5	3	4	**0.235253**
Lithology	0.5	1	2	4	4	3	4	4	3	2	**0.178904**
Structure	0.5	0.5	1	2	3	3	4	5	2	0.3333	**0.124146**
Drainage	0.3333	0.25	0.5	1	2	0.5	3	3	0.3333	0.5	**0.065030**
Rainfall	0.25	0.25	0.3333	0.5	1	2	2	3	0.5	0.5	**0.058740**
Land Use and Land Cover	0.25	0.3333	0.3333	2	0.5	1	2	2	0.3333	0.5	**0.055899**
Road	0.2	0.25	0.25	0.3333	0.5	0.5	1	2	0.3333	0.25	**0.034686**
Slope Aspect	0.2	0.25	0.2	0.3333	0.3333	0.5	0.5	1	0.5	1	**0.036372**
Vegetation	0.3333	0.3333	0.5	3	2	3	3	2	1	0.5	**0.091887**
Relief	0.25	0.5	3	2	2	2	4	1	2	1	**0.119078**

Consistency Index (CI) = 0.11237 Consistency Ratio (CR) = 0.07542

TABLE 5.12

Pairwise Comparison Matrix for Slope Gradient (Degrees)

Slope (°)	> 45	35–45	25–35	15–25	< 15	Weights
> 45	1	3	5	7	9	**0.502819**
35–45	0.3333	1	3	5	7	**0.260232**
25–35	0.2	0.3333	1	3	5	**0.13435**
15–25	0.1428	0.2	0.3333	1	3	**0.067778**
< 15	0.1111	0.1428	0.2	0.3333	1	**0.034821**

Consistency Index (CI) = 0.060652 Consistency Ratio (CR) = 0.054153

TABLE 5.13

Pairwise Comparison Matrix for Lithology Classes

Lithology Type	Karla	Indrayani	Megacryst	Diveghat	Purandargarh	Elephanta	Laterite	Salher	Borivali	Alluvium	Weights
Karla	1	2	3	3	4	4	4	5	7	8	**0.249484**
Indrayani	0.5	1	2	2	3	4	5	5	6	7	**0.185546**
Megacryst	0.3333	0.5	1	2	3	3	5	6	7	7	**0.158332**
Dive ghat	0.3333	0.5	0.5	1	2	3	4	5	5	6	**0.120948**
Purandargarh	0.25	0.3333	0.3333	0.5	1	2	3	4	5	6	**0.08851**
Elephanta	0.25	0.25	0.3333	0.3333	0.5	1	2	3	4	5	**0.064915**
Laterite	0.25	0.2	0.2	0.25	0.3333	0.5	1	3	4	6	**0.055355**
Salher	0.2	0.2	0.1666	0.2	0.25	0.3333	0.3333	1	3	5	**0.038362**
Borivali	0.1428	0.1666	0.1428	0.2	0.2	0.25	0.25	0.3333	1	2	**0.022185**
Alluvium	0.125	0.1428	0.1428	0.1666	0.1666	0.1666	0.1666	0.2	0.5	1	**0.016365**

Consistency Index (CI) = 0.0691   Consistency Ratio (CR) = 0.0464

TABLE 5.14

Pairwise Comparison Matrix for Distance from Lineaments (m)

Distance from Lineaments (m)	< 50	> 50	Weights
< 50	1	9	**0.9**
> 50	0.1111	1	**0.1**

Consistency Index (CI) = −1 Consistency Ratio (CR) = 0

TABLE 5.15

Pairwise Comparison Matrix for Elevation (m)

Elevation (m)	< 100	100–300	> 300	Weights
< 100	1	4	8	**0.693063**
100–300	0.25	1	5	**0.244694**
> 300	0.125	0.125	1	**0.062243**

Consistency Index (CI) = −0.00101 Consistency Ratio (CR) = 0.00175

TABLE 5.16

Pairwise Comparison Matrix for Distance from Drainage Line (m)

Distance from Stream (m)	< 100	> 100	Weights
< 100	0.888889	0.888889	**0.888889**
> 100	0.111111	0.111111	**0.111111**

Consistency Index (CI) = 0 Consistency Ratio (CR) = 0

TABLE 5.17

Pairwise Comparison Matrix for Land Use and Land Cover

Land Use Type	Barren Land	Rocky Outcrop	Cultivated Land	Built-up Area	Vegetation Cover	Coastal Sand	Water Body	Weights
Barren Land	1	1	3	4	5	7	9	**0.311342**
Rocky Outcrop	1	1	2	3	4	5	7	**0.252528**
Cultivated Land	0.3333	0.5	1	2	3	6	7	**0.163006**
Built-up Area	0.25	0.3333	0.5	1	3	6	8	0.132177
Vegetation Cover	0.2	0.25	0.3333	0.3333	1	4	7	0.083455
Coastal Sand	0.1428	0.2	0.1666	0.1666	0.25	1	2	0.034622
Water Body	0.1111	0.1428	0.1428	0.125	0.1428	0.5	1	0.022871

Consistency Index (CI) = 0.0849856 Consistency Ratio (CR) = 0.064383

TABLE 5.18

Pairwise Comparison Matrix for Distance from Road (m)

Distance from Road (m)	< 100	> 100	Weights
< 100	1	9	**0.9**
> 100	0.1111	1	**0.1**

Consistency Index (CI) = 0 Consistency Ratio (CR) = 0

TABLE 5.19

Pairwise Comparison Matrix for NDVI

	Dense	Moderate	Sparse	Barren	Weights
Dense	1	2	3	7	**0.479831**
Moderate	0.5	1	2	5	**0.28091**
Sparse	0.333333	0.5	1	5	**0.185719**
Barren	0.142857	0.2	0.2	1	**0.053539**

Consistency Index (CI) = 0.02706 Consistency Ratio (CR) = 0.030066

5.3.5.2 Moderate Susceptibility Zones

More than one-fourth (26.77%) of the total study area falls in moderate susceptibility zone. It covers almost half the area of Raigad district. This area is characterised by small isolated hills with an elevation between 200 and 300 m, few of
which are covered with laterites at places. Major road corridors passing through
this area are frequently affected by slope failures during the peak rainy season.

5.3.5.3 Low and Very Low Susceptibility Zones

More than half the total geographical area (57.88%) of North Konkan falls in low
to very low susceptibility zones. The major part of Thane district and coastal plains
throughout the study area are safe with regard to landslide susceptibility zonation.

5.3.6 Validation of AHP-Based Landslide Susceptibility Zonation Map

To validate the results obtained from AHP-based LSZ maps, LH zones are
compared with the distribution of slope failures recorded during field surveys
(Figure 5.26).

The result shows that 73.48% of the recorded slope failure events fall in high to
very high susceptibility classes. It reveals that the AHP-based LSZ model can also be
used effectively to identify potential landslide zones with a desired level of accuracy.

TABLE 5.20

Pairwise Comparison Matrix for Slope Aspect

	West	South-West	South	South-East	East	North-East	North-West	North	Flat	Weights
West	1	2	2	3	4	5	5	6	7	**0.254987**
South-West	0.5	1	2	3	3	4	5	6	8	**0.205098**
South	0.5	0.5	1	3	4	4	6	6	7	**0.185241**
South-East	0.3333	0.3333	0.3333	1	2	3	4	5	6	**0.111018**
East	0.25	0.3333	0.25	0.5	1	3	4	5	6	**0.094454**
North-East	0.2	0.25	0.25	0.3333	0.3333	1	2	4	6	**0.062547**
North-West	0.2	0.2	0.1666	0.25	0.25	0.5	1	2	4	**0.040871**
North	0.1666	0.1666	0.1666	0.2	0.2	0.25	0.5	1	2	**0.026947**
Flat	0.1428	0.125	0.1428	0.1666	0.1666	0.1666	0.25	0.5	1	**0.018837**

Consistency Index (CI) = 0.089777 Consistency Ratio (CR) = 0.061984

TABLE 5.21

Pairwise Comparison Matrix for Rainfall (mm)

Annual Rainfall (mm)	> 3,250	3,000–3,250	2,750–3,000	2,500–2,750	< 2,500	Weights
> 3,250	1	2	4	6	8	**0.451482**
3,000–3,250	0.5	1	3	5	7	**0.302592**
2,750–3,000	0.25	0.3333	1	2	4	**0.128171**
2,500–2,750	0.1666	0.2	0.5	1	3	**0.078676**
< 2,500	0.125	0.1428	0.25	0.3333	1	**0.039078**

Consistency Index (CI) = 0.03106479 Consistency Ratio (CR) = 0.02773642

TABLE 5.22

Final Priorities for Landslide Causative Factors

Parameters	Weights of Sub-parameters										Overall Weights
	1	2	3	4	5	6	7	8	9	10	
Slope	0.502819	0.260232	0.13435	0.067778	0.034821	0.064915	–	–	–	–	0.235253787
Lithology	0.249484	0.185546	0.158332	0.120948	0.08851	0.064915	0.055355	0.038362	0.022185	0.016365	0.178904835
Structure	0.9	0.1	–	–	–	–	–	–	–	–	0.124146128
Elevation	0.693063	0.244694	0.062243	–	–	–	–	–	–	–	0.065030359
Drainage	0.888889	0.111111	–	–	–	–	–	–	–	–	0.058740386
Land Use and Land Cover	0.311342	0.252528	0.163006	0.132177	0.083455	0.034622	0.022871	–	–	–	0.055899964
Roads	0.9	0.1	–	–	–	–	–	–	–	–	0.034686217
NDVI	0.479831	0.28091	0.185719	0.053539	–	–	–	–	–	–	0.036372765
Aspect	0.254987	0.205098	0.185241	0.111018	0.094454	0.062547	0.040871	0.026947	0.018837	–	0.091887411
Rainfall	0.451482	0.302592	0.128171	0.078676	0.039078	–	–	–	–	–	0.119078147

(*Source*: Based on pairwise comparison using AHP method)

Note: Overall weights are increased up to 2 digits multiplied by 100 to overcome the problem of calculation.

TABLE 5.23

AHP-Based Landslide Susceptibility Zones

Susceptibility Zone	Description	Area (km²)	Percentage Area
I	Very low susceptibility	3,015.83	18.45
II	Low susceptibility	6,445.56	39.44
III	Moderate susceptibility	4,374.42	26.77
IV	High susceptibility	2,063.01	**12.62**
V	Very high susceptibility	444.73	**2.72**
Total		16,343.55	100

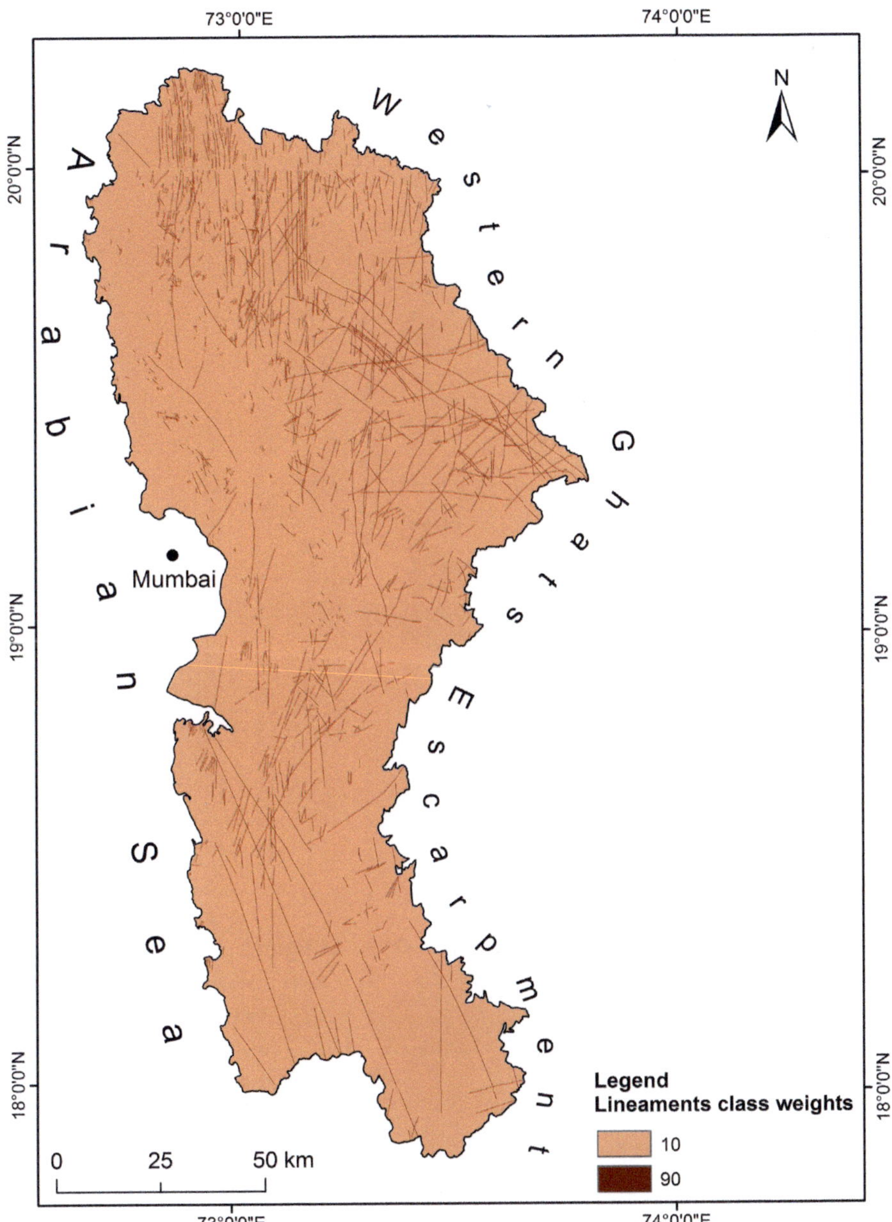

FIGURE 5.24 Reclassified Thematic Data Layer of Lineaments Map Using AHP Method.

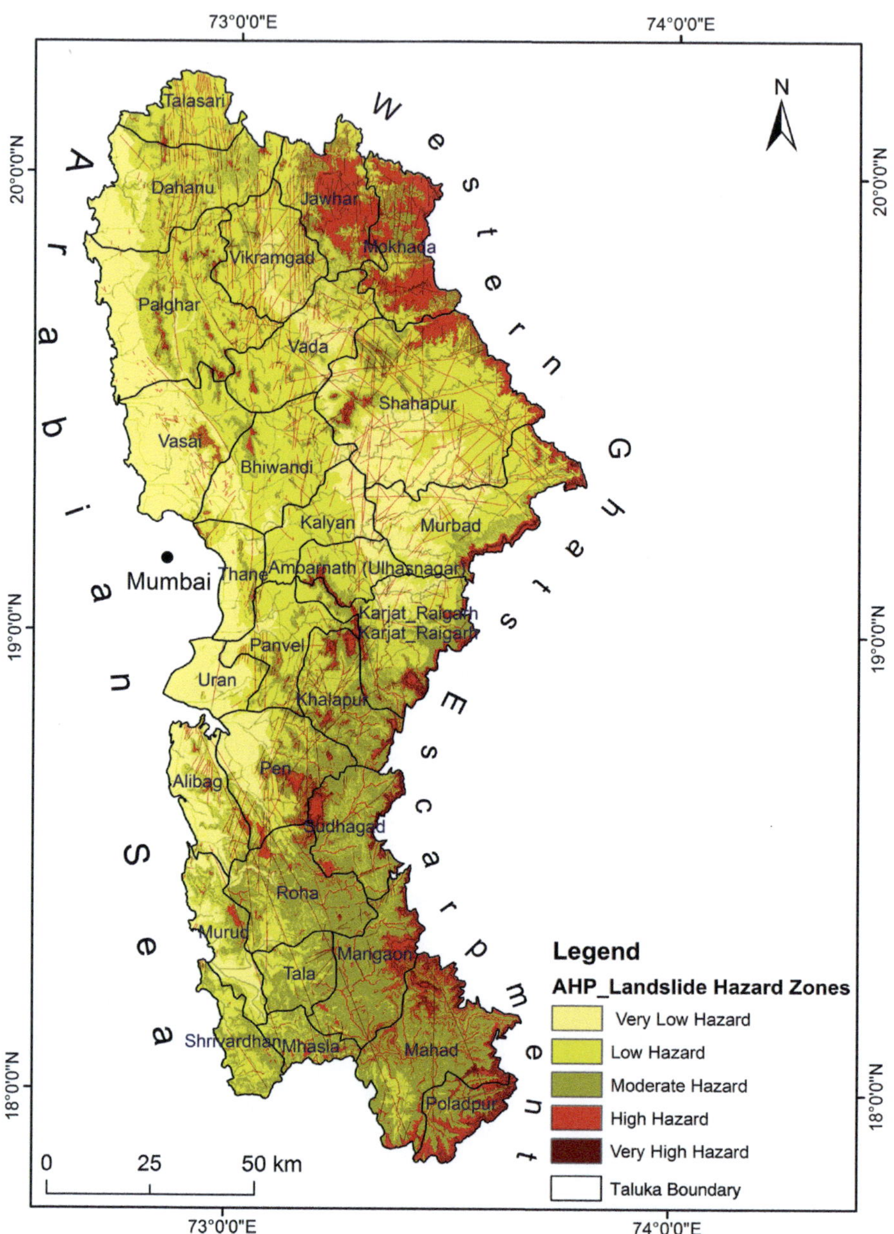

FIGURE 5.25 Landslide Susceptibility Zonation Map Using AHP Method.

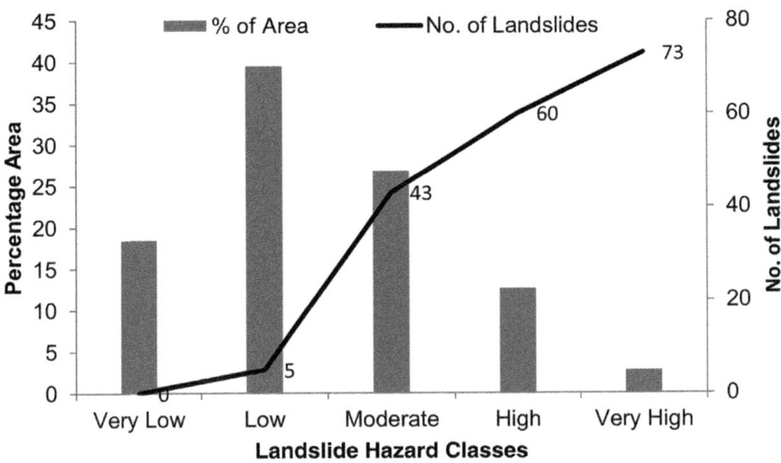

FIGURE 5.26   Landslide Susceptibility Zones Using AHP Method.

5.4   LANDSLIDE SUSCEPTIBILITY ZONATION USING MULTI-CRITERIA DECISION-MAKING APPROACH: CASE STUDY 3 (A CASE STUDY FROM KENTUCKY, USA)

5.4.1   THE STUDY AREA

Kentucky is one of the states of the United States of America, located in the south-eastern part of the country. Illinois, Indiana, and Ohio are north of Kentucky, whereas West Virginia (north-east) and Virginia (east) are the neighbouring regions of Kentucky. Kentucky lies between 36° 30′ N to 39° 09′ N latitude and 81° 58′ W and 89° 34′ W longitude (Figure 5.27). The total geographical area of this region is 104,656 km². The total population of Kentucky is 45 million.

The average relief of the region is 230 m, with the highest elevation of 1,265 m at Black Mountain. A significant portion of Kentucky lies in the Appalachian mountainous region. Physiographically, Kentucky is divided into five major areas: from west to east, they are the Mississippi embayment, Mississippian plateau, western coal fields, bluegrass, and eastern coal fields. The region is characterised by a humid subtropical climate. The average annual precipitation in this region is 1,200 mm. The Mississippi (west), Ohio (north), Big Sandy, and Tug Fork (east) are the important rivers bordering the region. Kentucky is mainly drained by the Kentucky, Tennessee, Cumberland, Green, and Licking Rivers.

Geologically, Kentucky is dominated by Mississippian, Ordovician, and Pennsylvanian groups. The western region is dominated by quaternary sediments, particularly those along major drainage lines. Almost the entire region is intruded by numerous faults. Although fault directions in the study area are not uniform everywhere, the general orientation of the majority of faults are in the north-east to south-west direction.

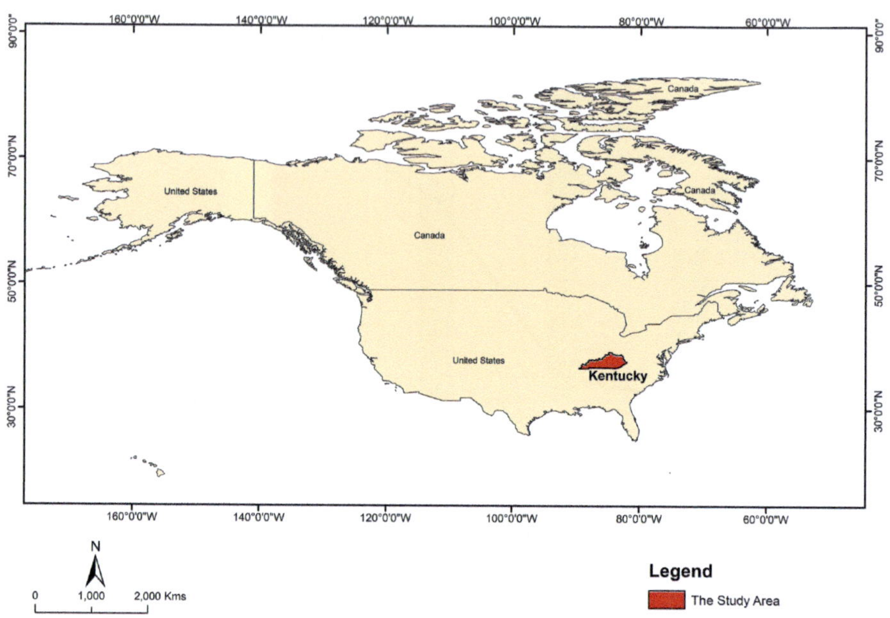

FIGURE 5.27   Location Map of Kentucky, USA.

5.4.2 OBJECTIVE OF THE STUDY

The present study focuses on delineating potential landslide-prone zones using multi-criteria decision-making approach in Kentucky.

5.4.3 MATERIALS AND METHODS

To delineate potential landslide susceptibility zones, five preparatory and triggering parameters, including slope, elevation, land use and land cover, rainfall, and roads, have been taken as input data layers. The AHP approach has been employed for landslide susceptibility zonation for the study area. Pairwise comparison matrices have been constructed for inter-parameter (level 1) as well as intra-parameter (level 2) comparisons. The scale of absolute numbers has been used to construct pairwise comparison matrices, and weights for each parameter have been derived. The eigen vector approach has been adopted to evaluate the consistency of judgements by calculating the consistency index and consistency ratio. The final weights derived from pairwise comparison are then used to assign numerical weightage to landslide preparatory and triggering parameters, and all data layers are integrated in a GIS environment using weighted overlay operation to generate the landslide susceptibility map of Kentucky.

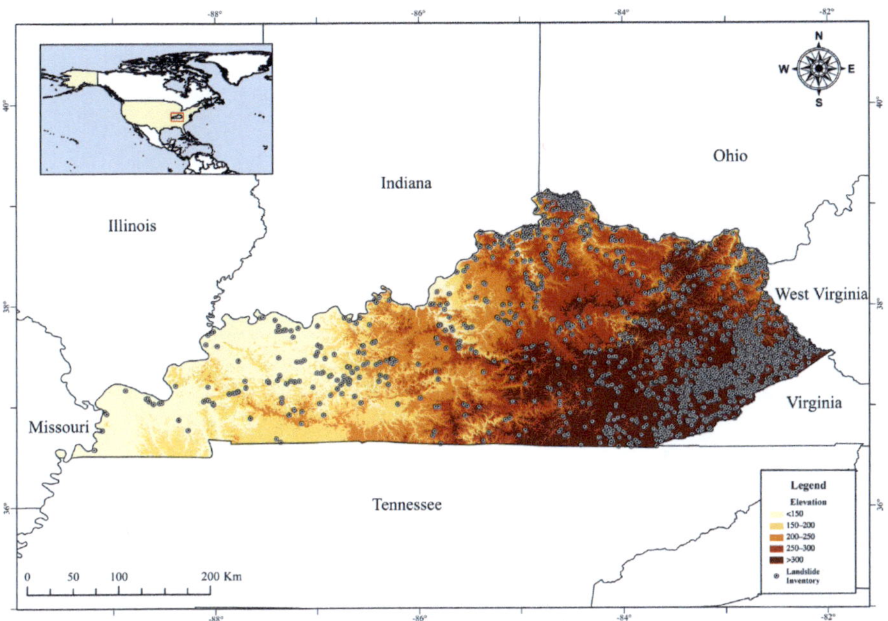

FIGURE 5.28 Distribution of Landslides in Kentucky, USA.

5.4.4 RESULTS

Based on past landslide records, the landslide distribution map of Kentucky has been produced. Landslide distribution in the study area is given in Figure 5.28.

The maximum number of landslides in Kentucky are concentrated in the eastern parts of the province, particularly in the eastern coalfields (the Appalachian region). The high concentration of landslides here is associated with high elevation, steeply sloping areas, heavy rainfall, etc., in the Appalachian region of Kentucky. However, landslides also occur in the western parts of the region (Mississippi embayment and plateau). Rainfall is observed to be an important landslide triggering factor in the study area. The average annual precipitation of the region progressively increases from west to east of the study area. The highest concentration of observed landslides is found in the eastern parts of the region.

Following are the details of thematic data layers of parameters used for landslide susceptibility zonation of Kentucky (Figures 5.29–5.33).

Final weights after pairwise comparison are given in Table 5.24.

Based on the final weights derived for each of the landslide parameters considered for landslide susceptibility mapping, the landslide susceptibility map has been produced. The landslide susceptibility map for Kentucky is presented in Figure 5.34.

Results

The LSZ map using multi-criteria decision-making approach for Kentucky has been carried out. The results are presented in Table 5.25.

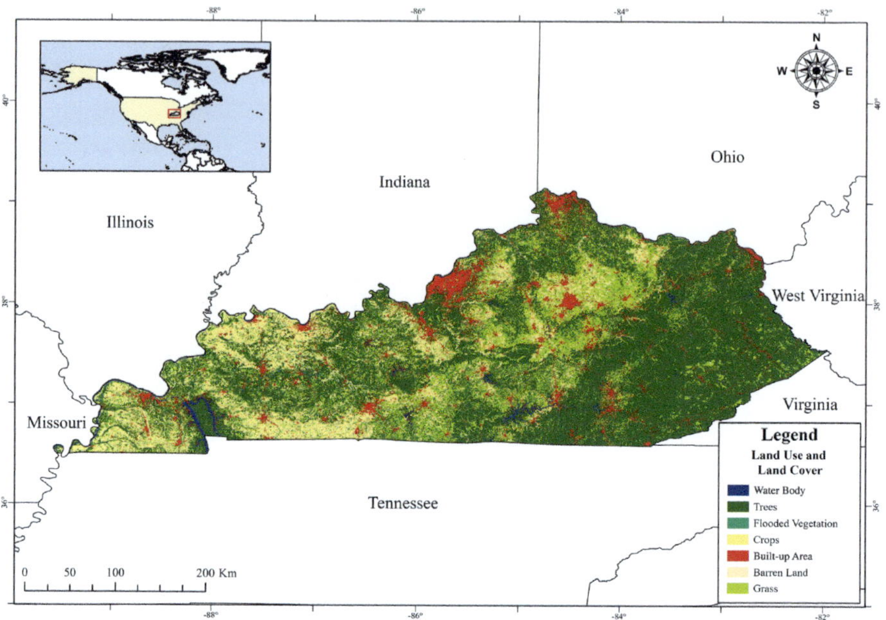

FIGURE 5.29   Land Use and Land Cover Map of Kentucky, USA.

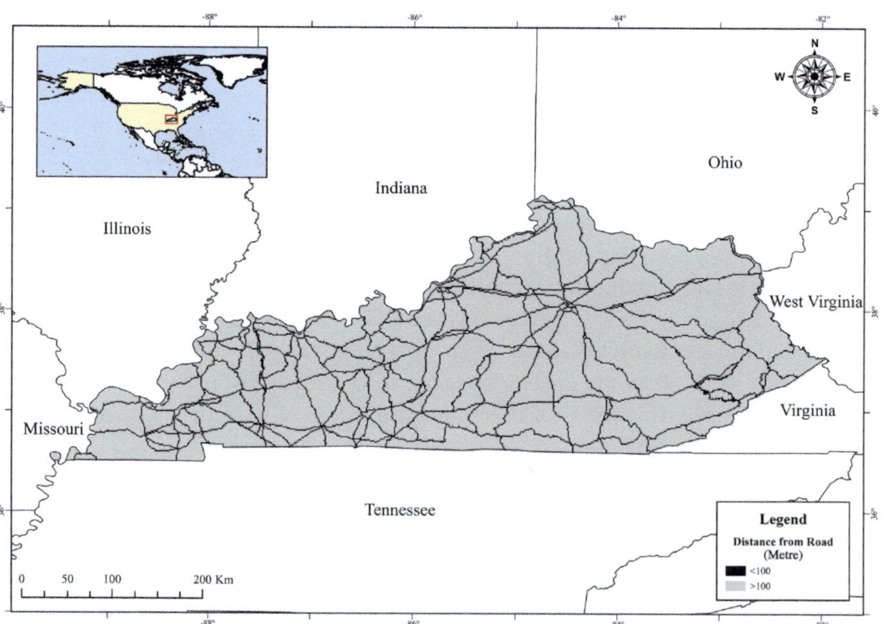

FIGURE 5.30   Road Map of Kentucky, USA.

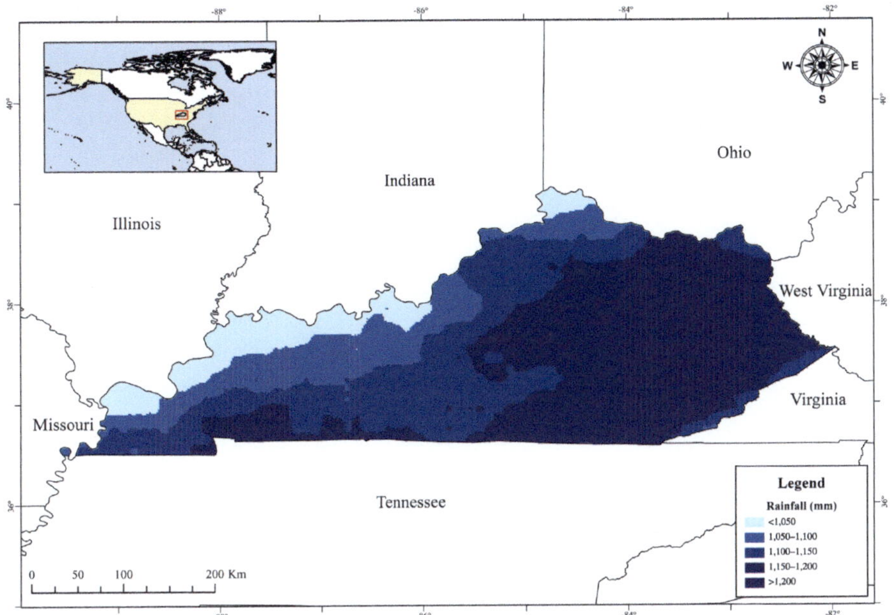

FIGURE 5.31 Rainfall Distribution Map of Kentucky, USA.

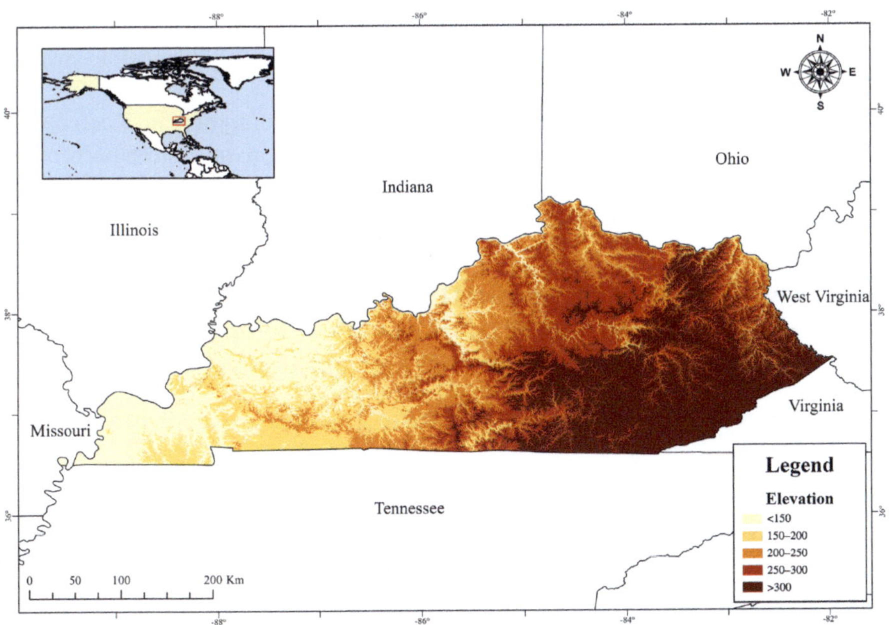

FIGURE 5.32 Elevation Map of Kentucky, USA.

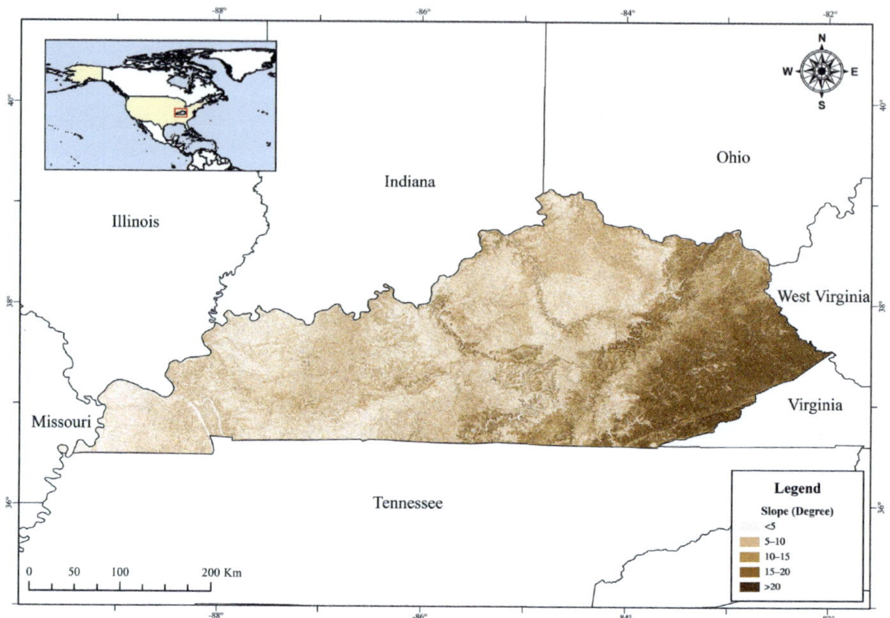

FIGURE 5.33 Slope Map of Kentucky, USA.

Table 5.25 clearly indicates that the highest number of observed landslides are classified as high to very high landslide susceptibility zones. Of the total geographical area of Kentucky, 36.38% is under high to very high landslide susceptibility zones.

It is evident from these results that landslide concentration in Kentucky is attributed to the Appalachian mountainous area, located in the eastern parts of the region. Landslide density is observed to be low in the western parts (Mississippi embayment) of the study area.

5.5 LANDSLIDE SUSCEPTIBILITY ZONATION USING MULTI-CRITERIA DECISION-MAKING APPROACH: CASE STUDY 4 (A CASE STUDY FROM VERMONT, USA)

This case study is an attempt to delineate landslide susceptibility zones in Vermont, USA.

About the Study Area

The US state of Vermont is located in the New England region of the north-eastern United States, covering an area of 24,900 km². Green Mountains in Vermont is a major mountainous region spread roughly in the north-south direction, covering most of the length of the region (Figure 5.35). The Green Mountain region is a tectonically active region, where several geomorphic processes, including mass movement, operate continuously. The Green Mountains

TABLE 5.24
Final Weights of AHP Pairwise Comparison Matrices

Parameter	Final Weights for Subclasses								Overall Weights
Land Use and Land Cover	Baren Ground	Built-up Area	Crops	Grass	Flooded Vegetation	Trees	Water Body	CR	
Weights	0.0262	0.0364	0.0558	0.0937	0.1233	0.2537	0.4111	0.0400	0.0368
Road	>400	301–400	201–300	101–200	1–100	–	–	–	–
Weights	0.0487	0.0872	0.1499	0.2569	0.4572	–	–	0.0030	0.0596
Rainfall	955–1,050	1,050–1,100	1,100–1,150	1,150–1,200	>1,200	–	–	–	–
Weights	0.0451	0.0856	0.1476	0.2713	0.4504	–	–	0.0050	0.1261
Elevation	0–150	150–200	200–250	250–300	>300	–	–	–	–
Weights	0.0451	0.0856	0.1476	0.2713	0.4504	–	–	0.0050	0.2523
Slope	0–5	5–10	10–15	15–20	>20	–	–	–	–
Weights	0.0451	0.0856	0.1476	0.2713	0.4504	–	–	0.0050	0.5251

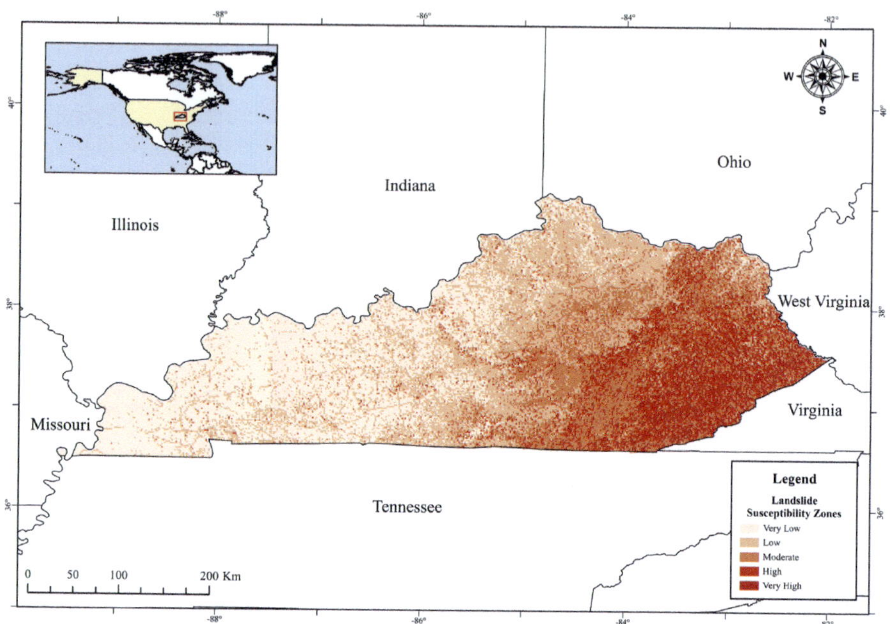

FIGURE 5.34 Landslide Susceptibility Map of Kentucky, USA.

TABLE 5.25

Landslide Susceptibility Zones in Kentucky, USA

S. No.	Landslide Susceptibility Class	Area (Sq. Km)	Percentage Area
1	Very low	35,870	34.27
2	Low	11,594	11.08
3	Moderate	19,126	18.27
4	High	20,391	19.48
5	Very high	17,701	16.90
Total		104,682	100

range is covered with Grenville basement rock. The upliftment of the Green Mountains is associated with tectonic activity, resulting in the formation of metamorphic rocks.

Freeze-thaw cycles and moist storms are responsible for the slope instability in this region (Figure 5.36).

There are five physiographic regions in Vermont: North-East Highlands, Green Mountains, Taconic Mountains, Champlain Lowlands, and Vermont Piedmont. Vermont is surrounded by the states of Massachusetts to its south,

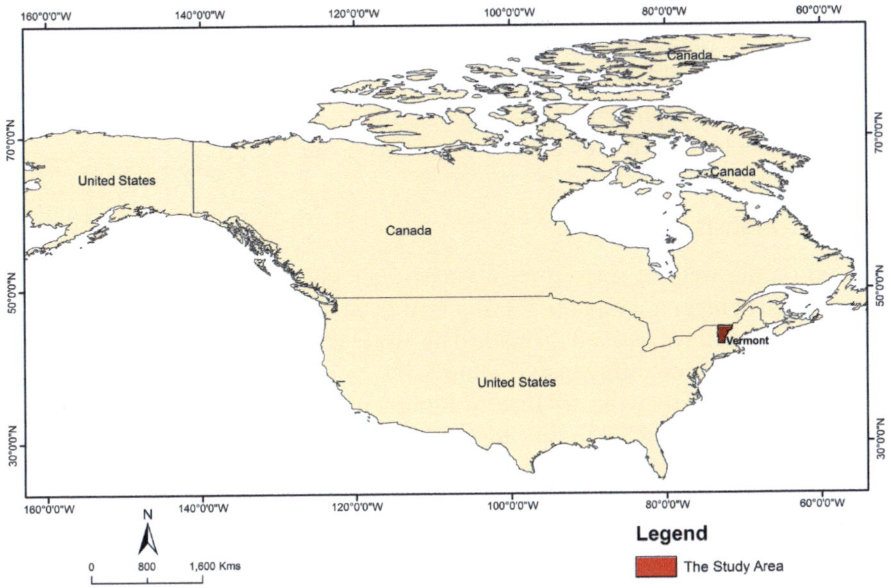

FIGURE 5.35 Location Map of Vermont, USA.

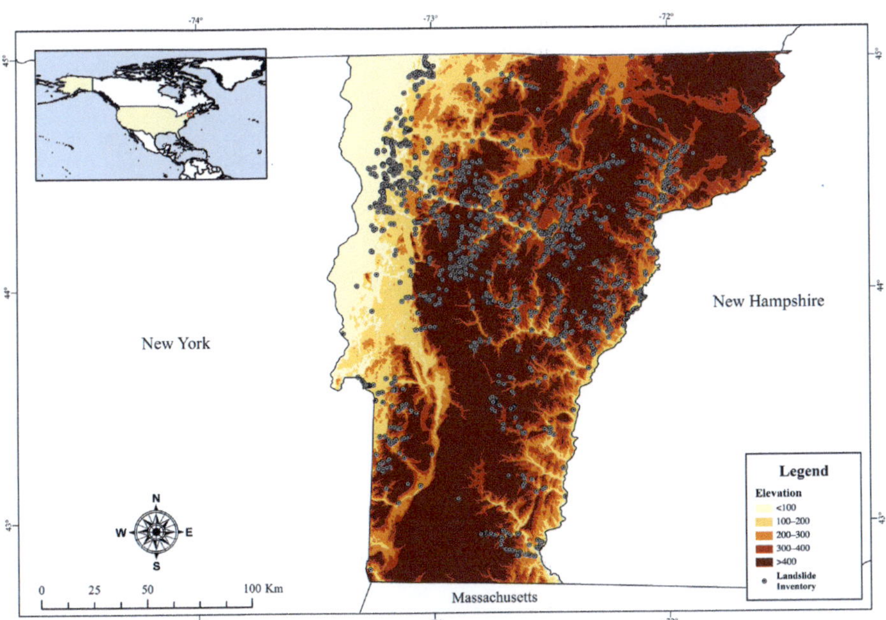

FIGURE 5.36 Landslide Inventory of Vermont, USA.

New Hampshire in the east, New York to the west, and the Canadian province of Quebec to the north. The parameters used as input data layer for LSZ in this case study include slope, rainfall, elevation, and land use and land cover (Figures 5.37–5.40). The AHP approach has been adopted for LSZ in Vermont. The details of this methodology have been already discussed in Section 5.4.

5.5.1 Results

Based on the weights derived from pairwise comparisons, thematic data layers of landslide preparatory and triggering factors are reclassified using final weights (Table 5.26). All reclassified thematic data layers are finally integrated to generate the landslide susceptibility map (Figure 5.41).

Table 5.27 clearly indicates that the highest number of observed landslides are in areas classified as high to very high landslide susceptibility zones. Of the total geographical area of Vermont, 12.92% comes under high to very high landslide susceptibility zones.

It is evident from these results that landslide concentration in Vermont is attributed to the Appalachian mountainous area, located in the eastern parts of the region. Landslide density is observed to be high in the eastern parts of the study area.

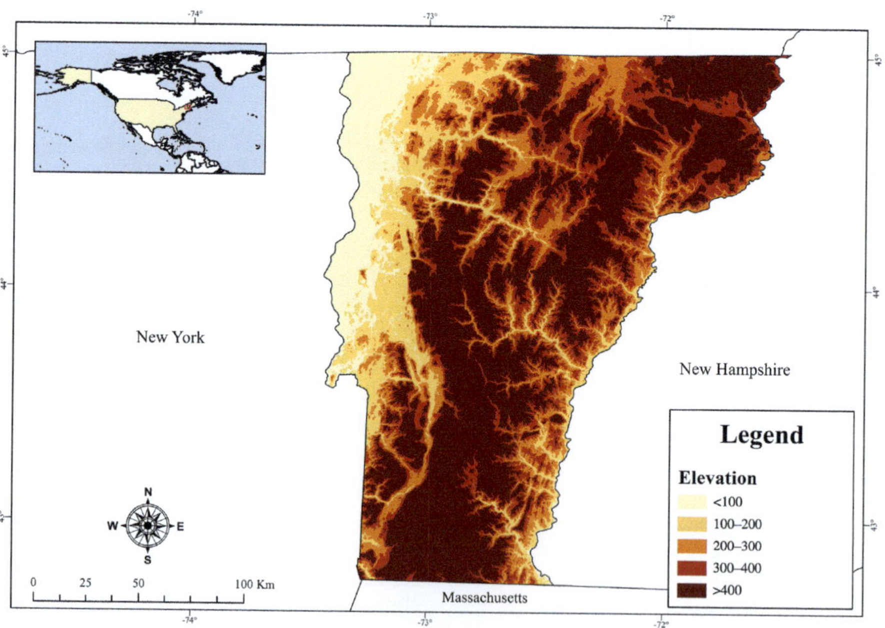

FIGURE 5.37   Elevation Map of Vermont, USA.

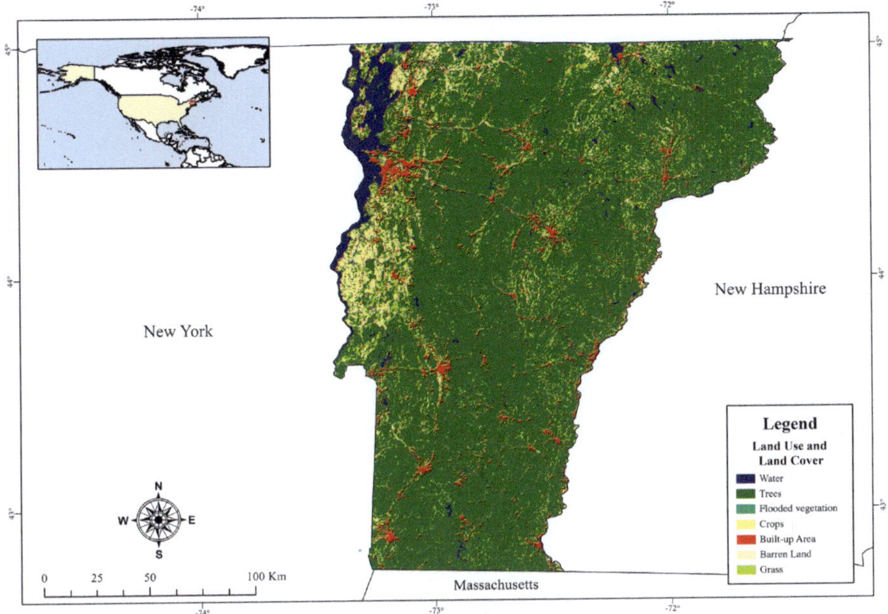

FIGURE 5.38 Land Use and Land Cover Map of Vermont, USA.

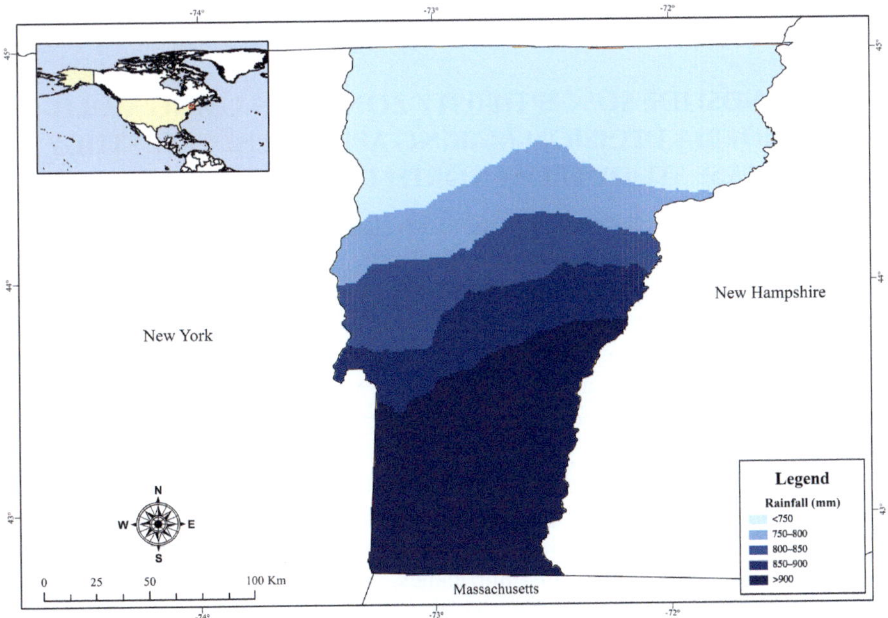

FIGURE 5.39 Rainfall Distribution Map of Vermont, USA.

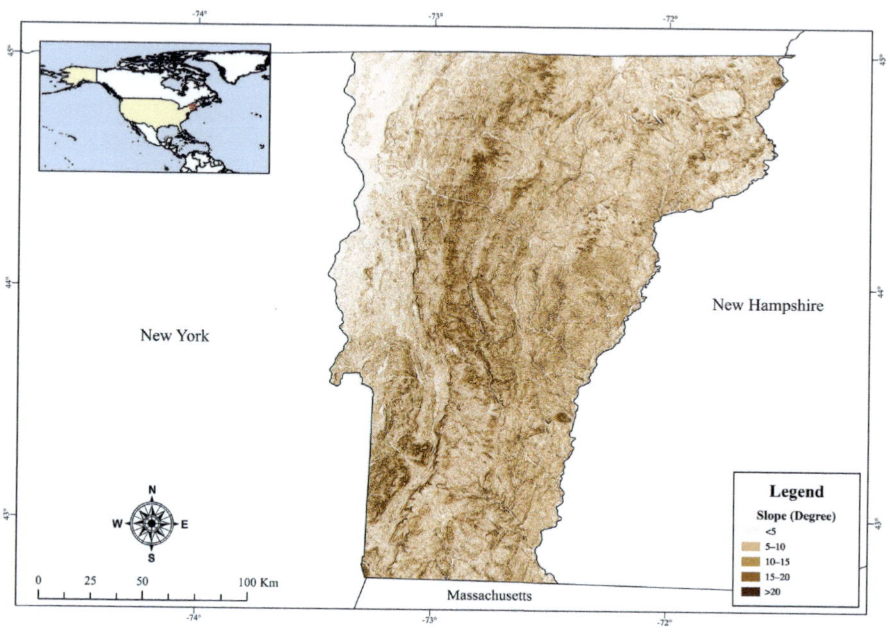

FIGURE 5.40   Slope Map of Vermont, USA.

5.6   LANDSLIDE SUSCEPTIBILITY ZONATION USING MULTI-CRITERIA DECISION-MAKING APPROACH: CASE STUDY 5 (A CASE STUDY FROM NORTH CAROLINA, USA)

North Carolina is a state located in the south-eastern region of the United States. It is bordered by Virginia in the north, Atlantic Ocean to the east, Georgia and South Carolina to the south, and Tennessee to the west. It is the ninth-most populated region of the United States with a population of 10.439 million (Figure 5.42). North Carolina consists of three main geographic regions: the Atlantic coastal plain, the central piedmont region, and the Appalachian Mountain region in the west. The coastal plain consists of sandy, narrow barrier islands separated from the mainland by inlets.

Landslides are most common in the mountain region of North Carolina because of steep slopes. The piedmont and coastal plain regions also have land-slides that are commonly related to human activities, such as making a road cut too steep. Large rainstorms, hurricanes, freeze-thaw processes, and human activities all can trigger landslides. There are many types of landslides, made of different types of material that travel at different speeds. Some landslides only consist of soil, called an earthslide. Some are a mixture of soil, rock, trees, and

TABLE 5.26

Final Weights of AHP Pairwise Comparison Matrices

Parameter	Final Weights for Subclasses								Overall Weights
Land Use and Land Cover	Baren Ground	Built-up Area	Crops	Grass	Flooded Vegetation	Trees	Water Body	CR	
Weights	0.0262	0.0364	0.0557	0.0937	0.1233	0.2536	0.4111	0.041	0.0368
Distance from Road (m)	>400	301–400	201–300	101–200	1–100	–	–	–	–
Weights	0.0487	0.0872	0.1499	0.2569	0.4572	–	–	0.04	0.0596
Rainfall (mm)	611–750	750–800	800–850	850–900	>900	–	–	–	–
Weights	0.0451	0.0856	0.1476	0.2713	0.4504	–	–	0.005	0.1261
Elevation (m ASL)	0–100	100–200	200–300	300–400	>400	–	–	–	–
Weights	0.0451	0.0856	0.1476	0.2713	0.4504	–	–	0.005	0.2522
Slope (degrees)	0–5	5–10	10–15	15–20	>20	–	–	–	–
Weights	0.0451	0.0856	0.1476	0.2713	0.4504	–	–	0.005	0.5251

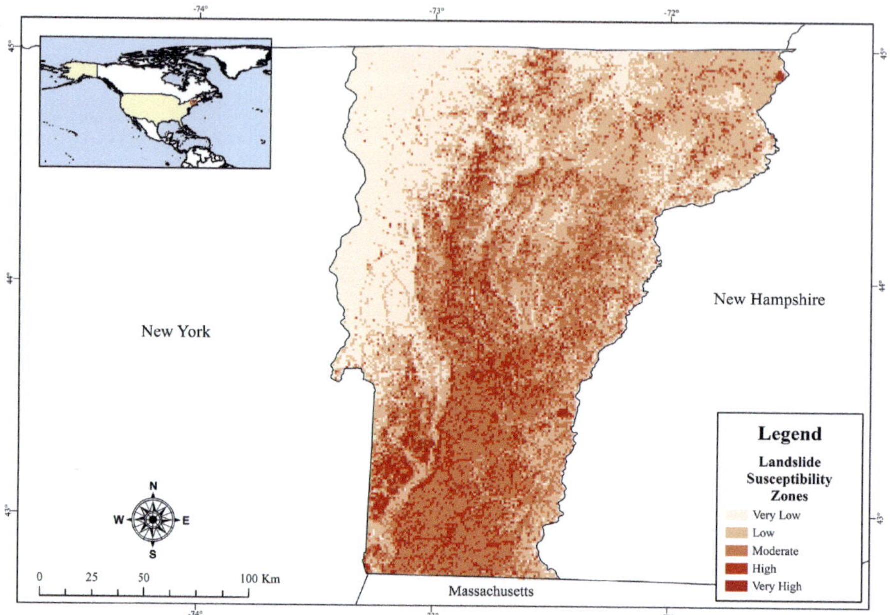

FIGURE 5.41 Landslide Susceptibility Zonation Map of Vermont, USA.

TABLE 5.27
Landslide Susceptibility Zones in Vermont, USA

S. No.	Landslide Susceptibility Class	Area (Sq. Km)	Percentage Area
1	Very low	6,263	25.24
2	Low	8,350	33.65
3	Moderate	6,994	28.19
4	High	2,447	9.86
5	Very high	757	3.06
Total		24,811	100

mud, called a debris flow. Other landslides contain only rock, called a rockfall or rockslide (Figure 5.43).

LSZ mapping has been carried out in North Carolina using the AHP model. The parameters used as input data layer for LSZ in this case study include slope, rainfall, elevation, and land use and land cover (Figures 5.44–5.48). The AHP approach has been adopted for LSZ in North Carolina. The details of this methodology have been already discussed in Section 5.4.

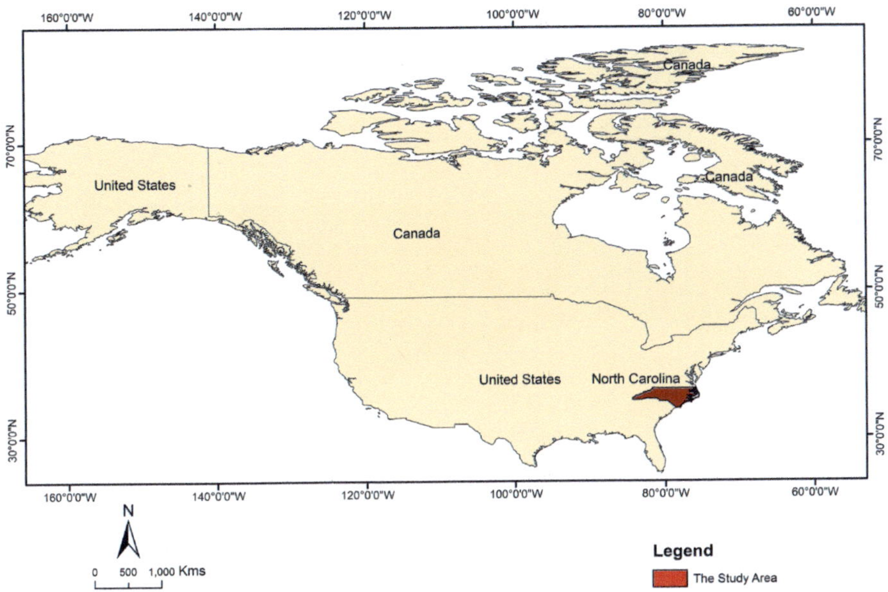

FIGURE 5.42 Location Map of North Carolina, USA.

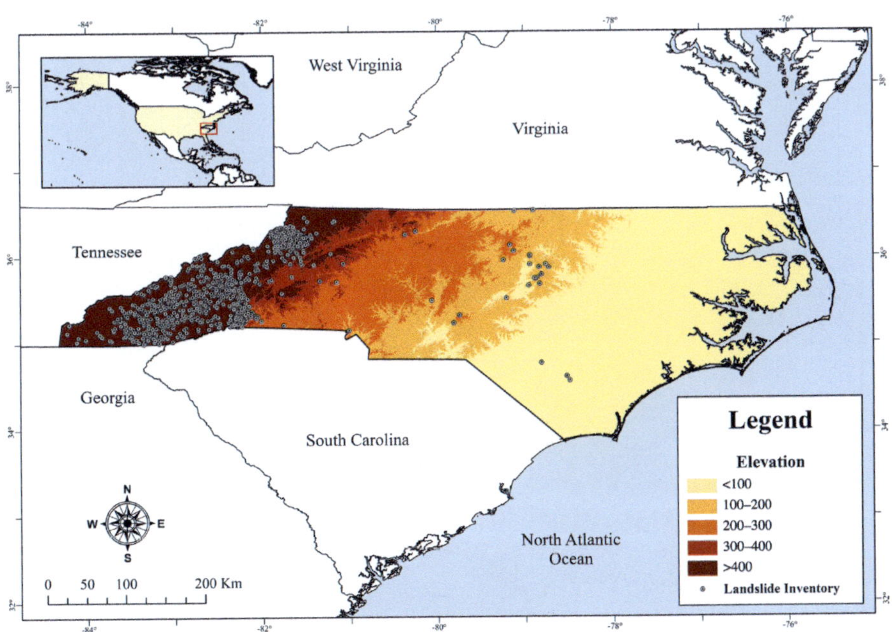

FIGURE 5.43 Distribution of Landslides in North Carolina, USA.

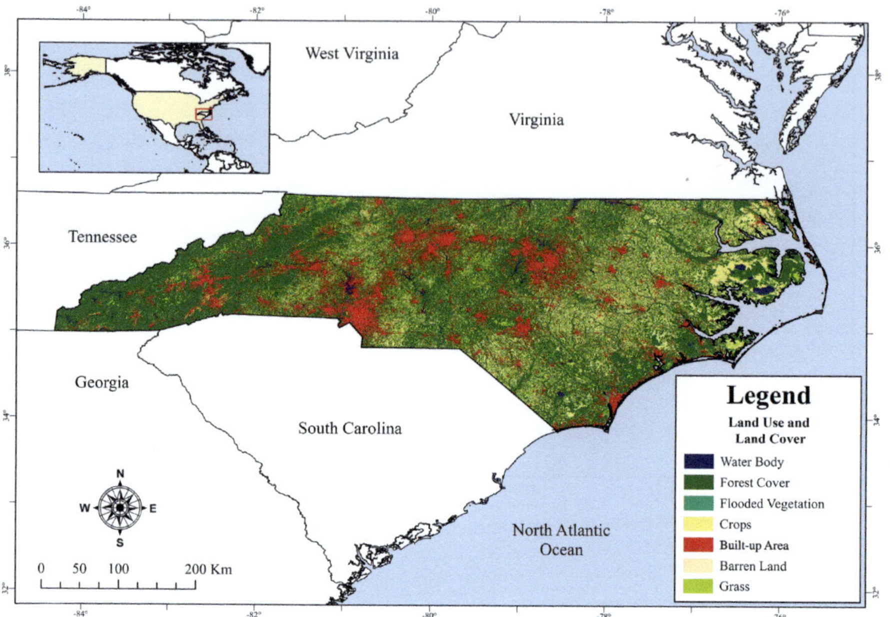

FIGURE 5.44 Land Use and Land Cover Map of North Carolina, USA.

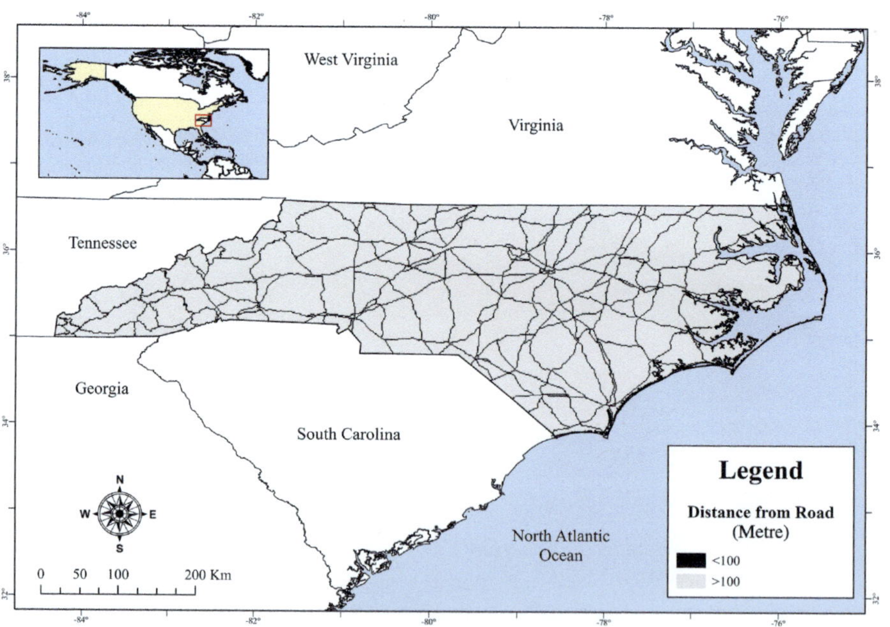

FIGURE 5.45 Road Map of North Carolina, USA.

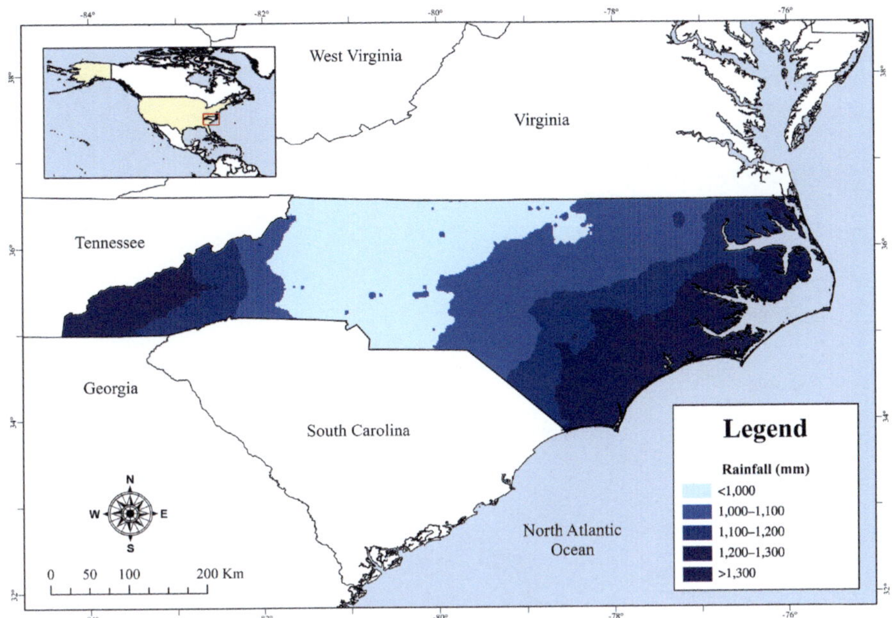

FIGURE 5.46 Rainfall Map of North Carolina, USA.

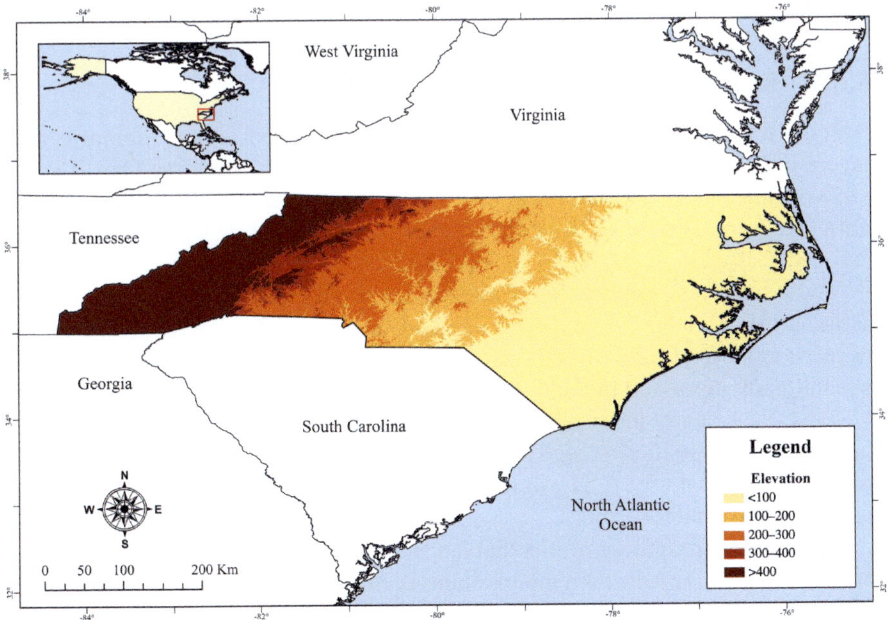

FIGURE 5.47 Elevation Map of North Carolina, USA.

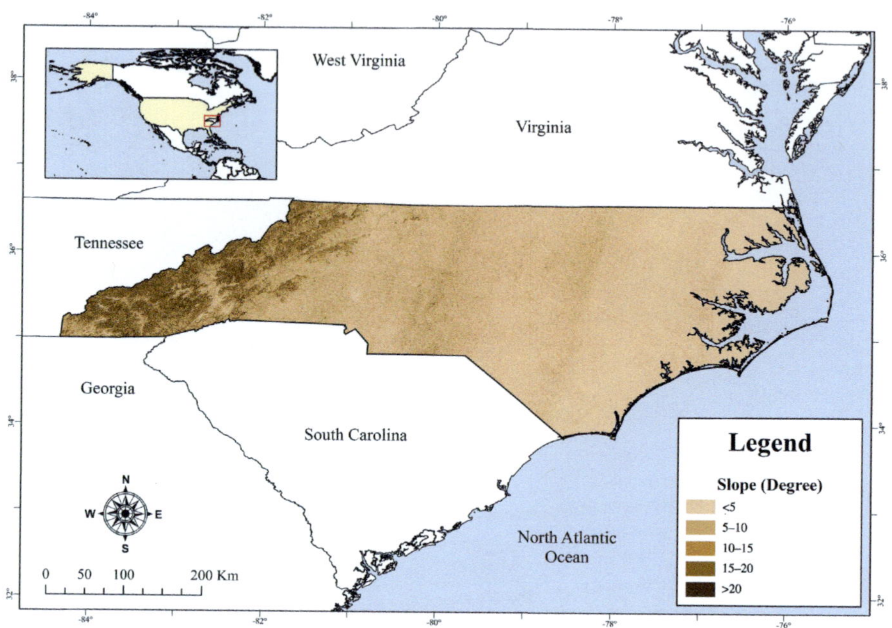

FIGURE 5.48   Slope Map of North Carolina, USA.

5.6.1 RESULTS

Based on the weights derived from pairwise comparisons, thematic data layers of landslide preparatory and triggering factors for North Carolina are reclassified using the final weights (Table 5.28). All reclassified thematic data layers are finally integrated to generate the landslide susceptibility map (Figure 5.49).

The parameters used as input data layer for LSZ in this case study include slope, rainfall, elevation, and land use and land cover. The AHP approach has been adopted for LSZ in North Carolina. The details of this methodology have been already discussed in Section 5.4.

Table 5.29 clearly indicates that the areas with the highest number of observed landslides are classified as high to very high landslide susceptibility zones. Of the total geographical area of North Carolina, 8.48% comes under high to very high landslide susceptibility zones.

It is evident from these results that landslide concentration in North Carolina is attributed to the Appalachian mountainous area, located in the eastern parts of the region. Landslide density is observed to be high in the eastern parts of the study area.

TABLE 5.28
Final Weights of AHP Pairwise Comparison Matrices

Parameter	Final Weights								Overall Weights
Land Use and Land Cover	Crops	Baren Ground	Grass	Flooded Vegetation	Water Body	Built-up Area	Trees	CR	
Weights	0.0262	0.0364	0.0558	0.0937	0.1233	0.2537	0.4111	0.0400	0.0368
Road	(>400)	301–400	201–300	101–200	1–100				
Weights	0.0487	0.0872	0.1499	0.2569	0.4572			0.0030	0.0596
Rainfall	876–1000	1001–1100	1101–1200	1201–1300	1301–1777 (>1300)				
Weights	0.0451	0.0856	0.1476	0.2713	0.4504			0.0050	0.1261
Elevation	0–100	100–200	200–300	300–400	>400				
Weights	0.0451	0.0856	0.1476	0.2713	0.4504			0.0050	0.2523
Slope	0–5	5–10	10–15	15–20	20–69 (>20)				
Weights	0.0451	0.0856	0.1476	0.2713	0.4504			0.0050	0.5251

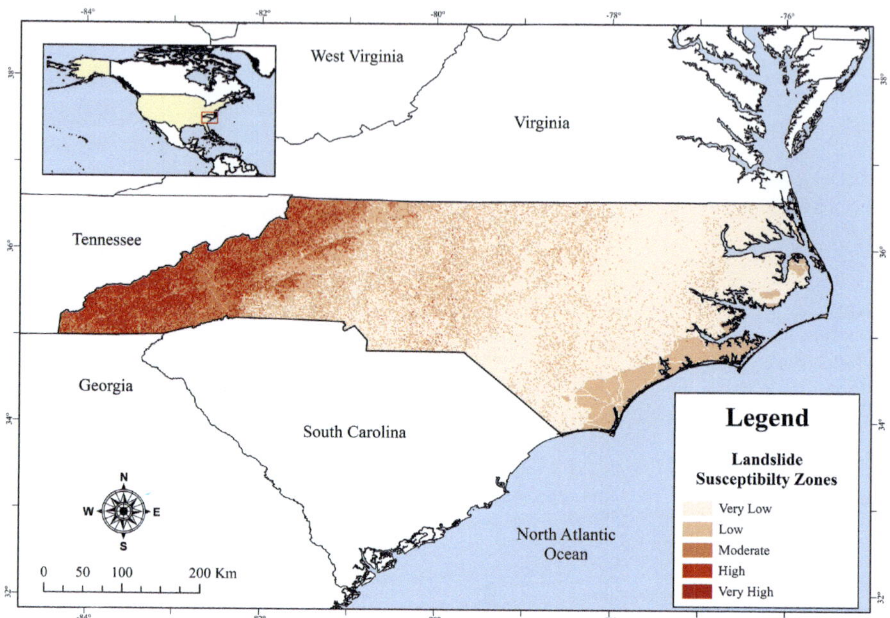

FIGURE 5.49   Landslide Susceptibility Map of North Carolina, USA.

TABLE 5.29

Landslide Susceptibility Zones in North Carolina, USA

S. No.	Landslide Susceptibility Class	Area (Sq. Km)	Percentage Area
1	Very low	71,514	56.73
2	Low	32,023	25.4
3	Moderate	11,835	9.39
4	High	5,731	4.54
5	Very high	4,966	3.94
	Total	126,069	100

5.7   LANDSLIDE SUSCEPTIBILITY ZONATION USING MULTI-CRITERIA DECISION-MAKING APPROACH: CASE STUDY 6 (A CASE STUDY FROM TOSCANA, ITALY)

The Toscana region is located in west-central Italy. It lies along the Tyrrhenian and Ligurian seas. It comprises the provinces of Massa-Carrara, Lucca, Pistoia, Prato, Firenze, Livorno, Pisa, Arezzo, Siena, and Grosseto. The Toscana region is bordered by Liguria to its north-west, Emillia-Romagna to the north, Marche to the east, Umbria to the south-east, and Lazio to its south.

The characteristic landscape in Toscana is gently rolling hills leading to sharply peaked mountains that pose a barrier between Tuscany and regions to the south (Figure 5.50). It is bordered in the north and north-east by the Tuscan-Emilian Apennines and the Apuan Alps, these being separated by a series of long valleys from the sub-Apennine hills of Mount Albano, Mount Pratomagno, and others. South of Siena, the surface rises to less fertile mountains and plateaus, such as the Metallifere Mountains, Mount Amiata, and Mount Argentario on the coast. The lowlands of the region are drained by the Arno River. The total area of the region is 22,992 km². The total population of Toscana region is 3,676,285 (2022).

Based on pairwise comparison using the AHP model, final weights are derived. Consistency of judgements was evaluated by calculating CI and CR using eigen vector approach. All thematic data layers have been reclassified by using the final weights derived (Table 5.30). All thematic data layers were finally integrated using weighted layer operation to prepare the landslide susceptibility map for the study area (Figure 5.56).

5.7.1 RESULTS

The AHP-based landslide susceptibility map shown in Figure 5.56 indicates that north-western parts of Toscana region are identified as high to very high susceptible

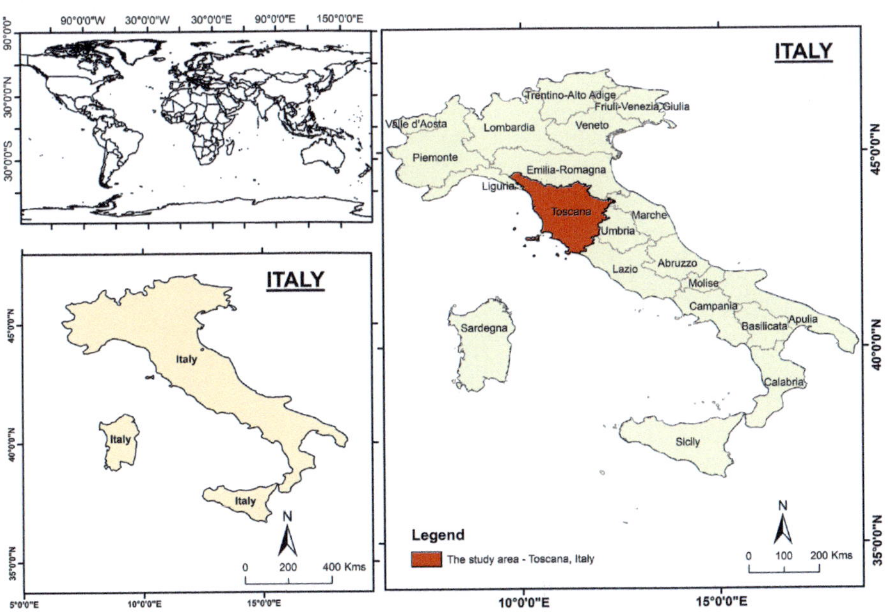

FIGURE 5.50   Location Map of the Toscana Region, Italy.

TABLE 5.30
Final Weights of AHP Pairwise Comparison Matrices

	Slope	Aspect	Elevation	Rainfall	Land Use and Land Cover	Roads	Criteria Weight	Percentage
Slope	0.41	0.51	0.37	0.32	0.28	0.25	**0.36**	36
Aspect	0.21	0.26	0.37	0.32	0.28	0.25	**0.28**	28
Elevation	0.14	0.09	0.12	0.22	0.12	0.14	**0.14**	14
Rainfall	0.14	0.09	0.06	0.11	0.28	0.25	**0.15**	15
Land Use and Land Cover	0.06	0.04	0.04	0.02	0.04	0.08	**0.05**	5
Roads	0.05	0.03	0.02	0.01	0.01	0.03	**0.03**	3
	1	1	1	1	1	1	**1**	100

CI = 0.074   CR = 0.05

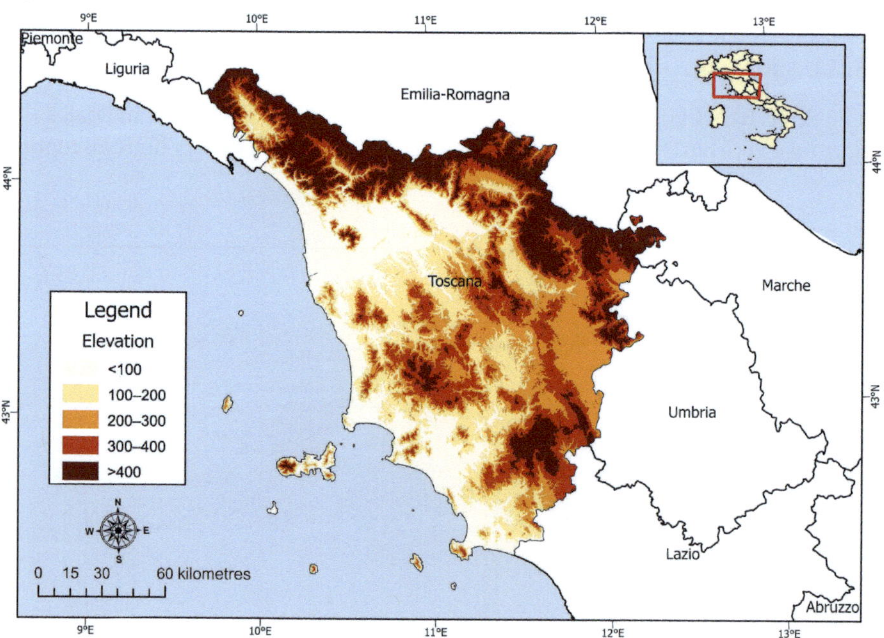

FIGURE 5.51   Elevation Map of the Toscana Region, Italy.

zones as compared to other parts of the region. The northern Apennine region is identified as potentially the most landslide-prone region. Massa-Carrara, Lucca, Pistoia, Prato, Firenze, and Arezzo regions are identified as high to very high landslide-susceptible zones.

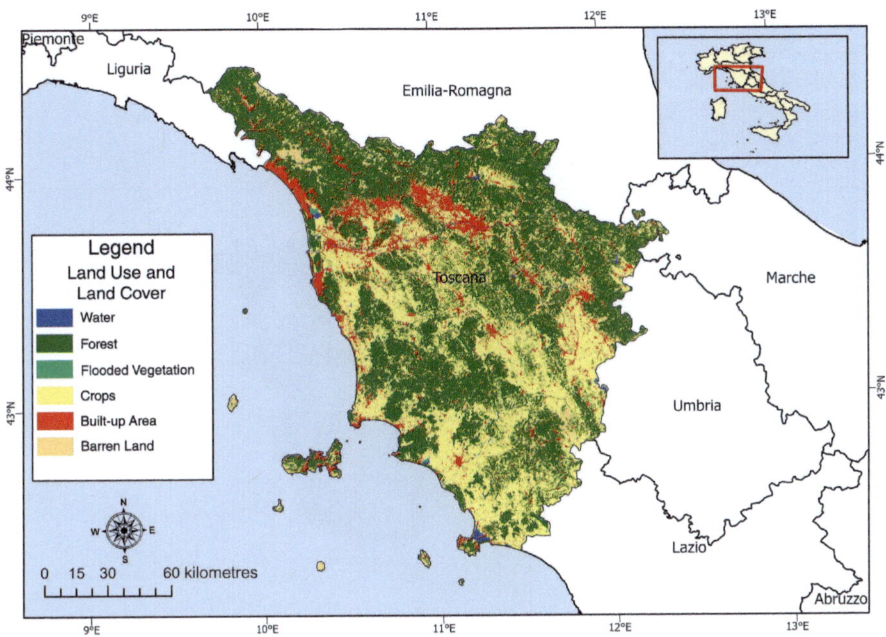

FIGURE 5.52 Land Use and Land Cover Map of the Toscana Region, Italy.

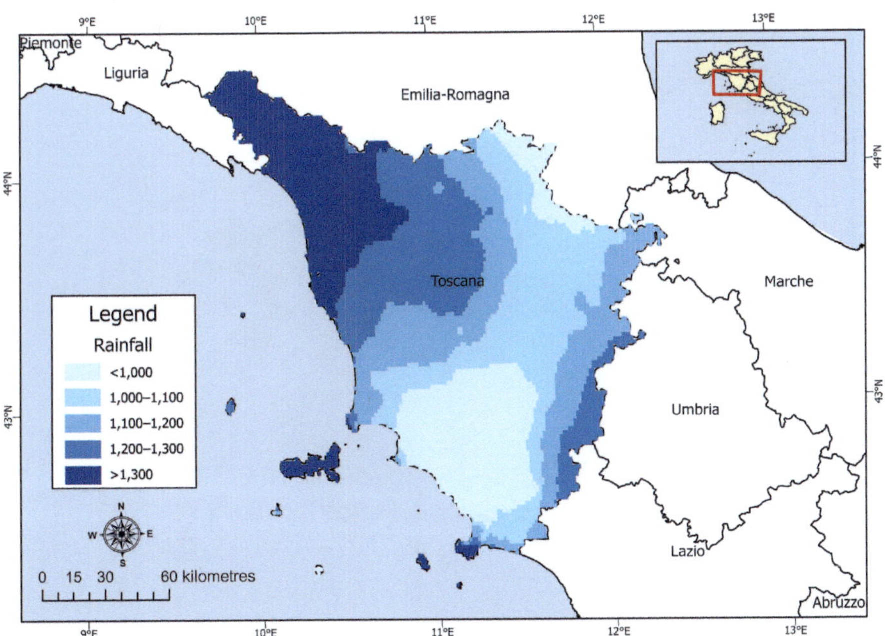

FIGURE 5.53 Rainfall Map of the Toscana Region, Italy.

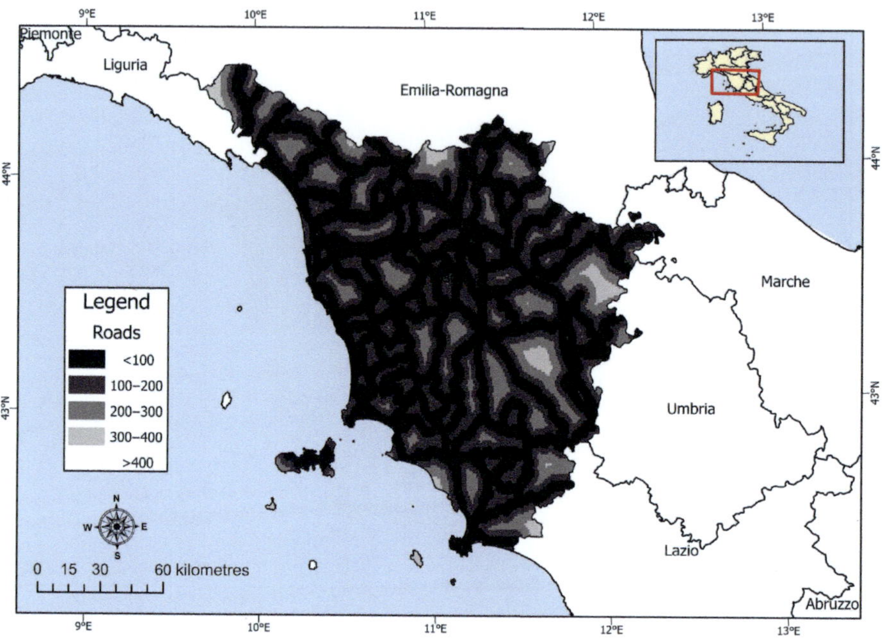

FIGURE 5.54   Road Map of the Toscana Region, Italy.

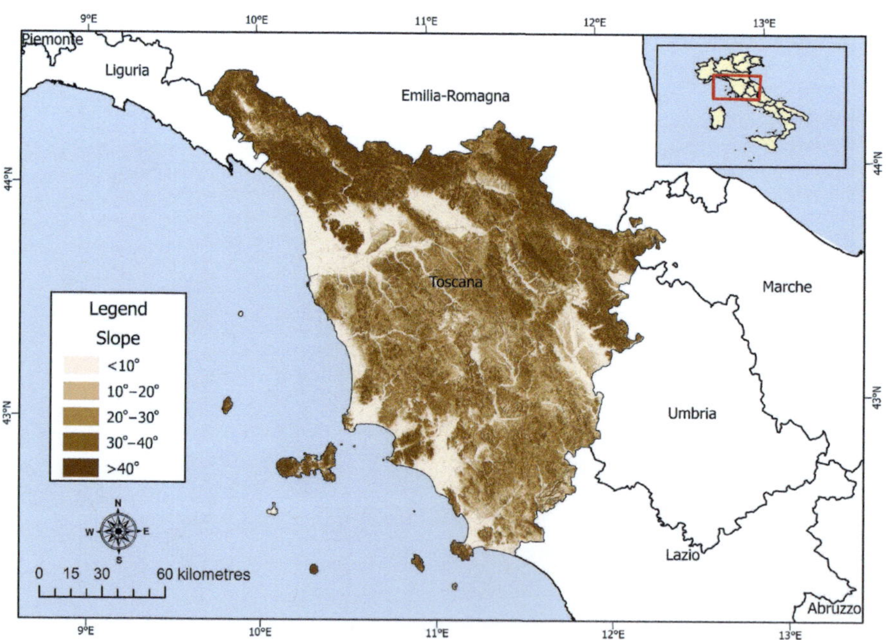

FIGURE 5.55   Slope Map of the Toscana Region, Italy.

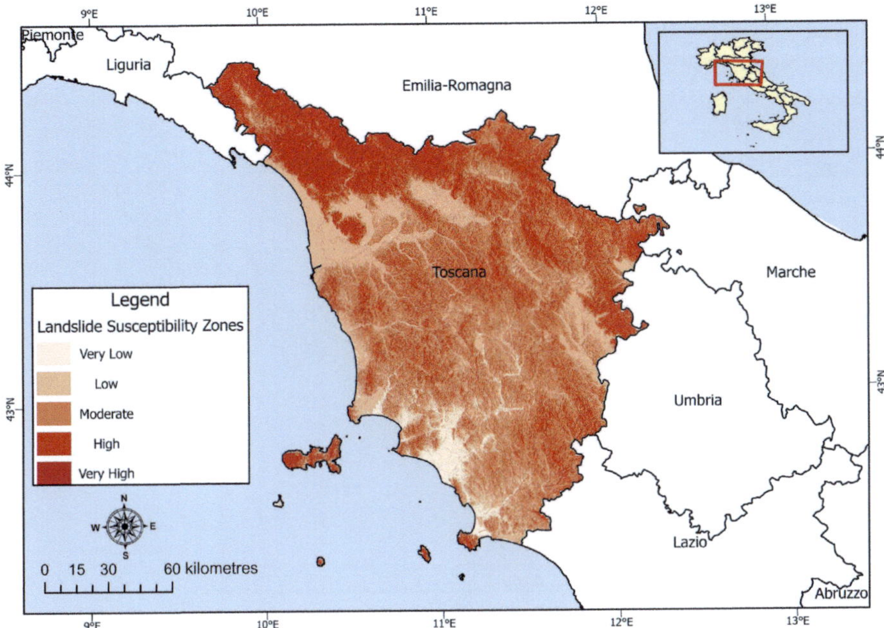

FIGURE 5.56   Landslide Susceptibility Map of the Toscana Region, Italy.

TABLE 5.31

Landslide Susceptibility Zones in Toscana, Italy

S. No.	Landslide Susceptibility Class	Area (Sq. Km)	Percentage Area
1	Very low	721.01	3.17
2	Low	5,478.43	24.06
3	Moderate	9,972.72	43.8
4	High	6,035.04	2.65
5	Very high	559.21	2.46
Total		22,766.41	100

The classification of landslide-susceptible zones in Toscana region of Italy is presented in Table 5.31. The results indicate that 5.09% of the total geographical area of the Toscana region, with an area of 6,594.24 km^2, is identified as having high to very high landslide susceptibility. High elevation, steep slopes, heavy rainfall, and human interventions are important factors affecting landslide occurrence.

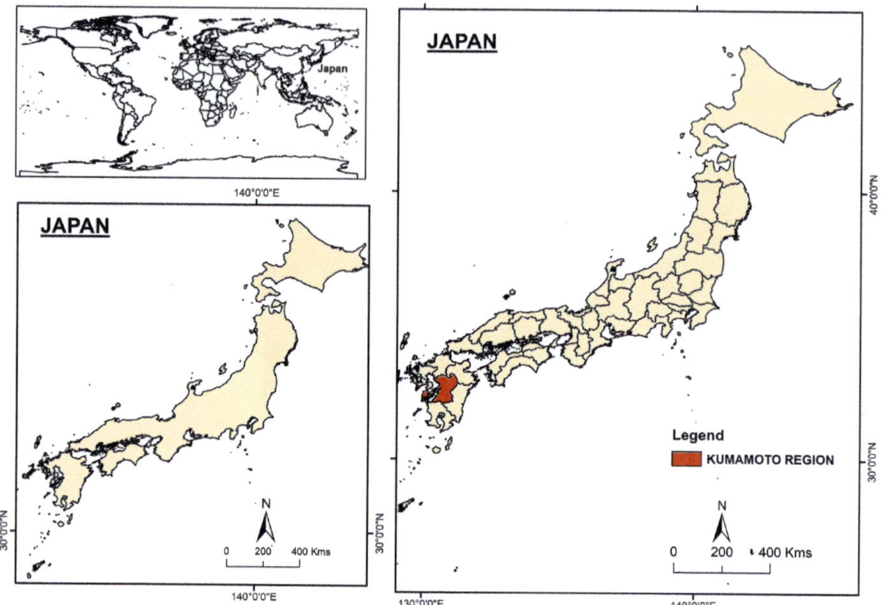

FIGURE 5.57 Location Map of the Kumamoto Region, Japan.

5.8 LANDSLIDE SUSCEPTIBILITY ZONATION USING MULTI-CRITERIA DECISION-MAKING APPROACH: CASE STUDY 7 (A CASE STUDY FROM KUMAMOTO, JAPAN)

The Kumamoto region is located at the centre of Kyushu on the coast of the Ariake Sea and is one of the southernmost four major Japanese islands. The Kumamoto region is in south-west Japan and is dominated by Paleozoic and Mesozoic rocks. Tectonically, this region is active. Several surface ruptures and active faults are observed in this region. The Kumamoto region is bordered by Ariake Sea in the west, Fukuoka in the north, Oita in the north-east, Miyazaki in the east, and Kagoshima in the south (Figure 5.57).

The region is dominated by a tropical climate with humid summers and mild winters. The average annual temperature in the Kumamoto region is 17.2°C, and the average annual rainfall is 200.7 cm.

The Kumamoto region of Japan is one of the important landslide-prone areas in Japan. Besides rainfall, earthquake is also an important landslide triggering factor in this region. Mudslides, shallow debris slides, and debris flow are the common slope failure types in this region.

The parameters used as input data layer for LSZ in this case study include slope, rainfall, elevation, and land use and land cover (Figures 5.58–5.61). The AHP approach has been adopted for LSZ in Kumamoto, Japan. The details of this methodology have been already discussed in Section 5.4.

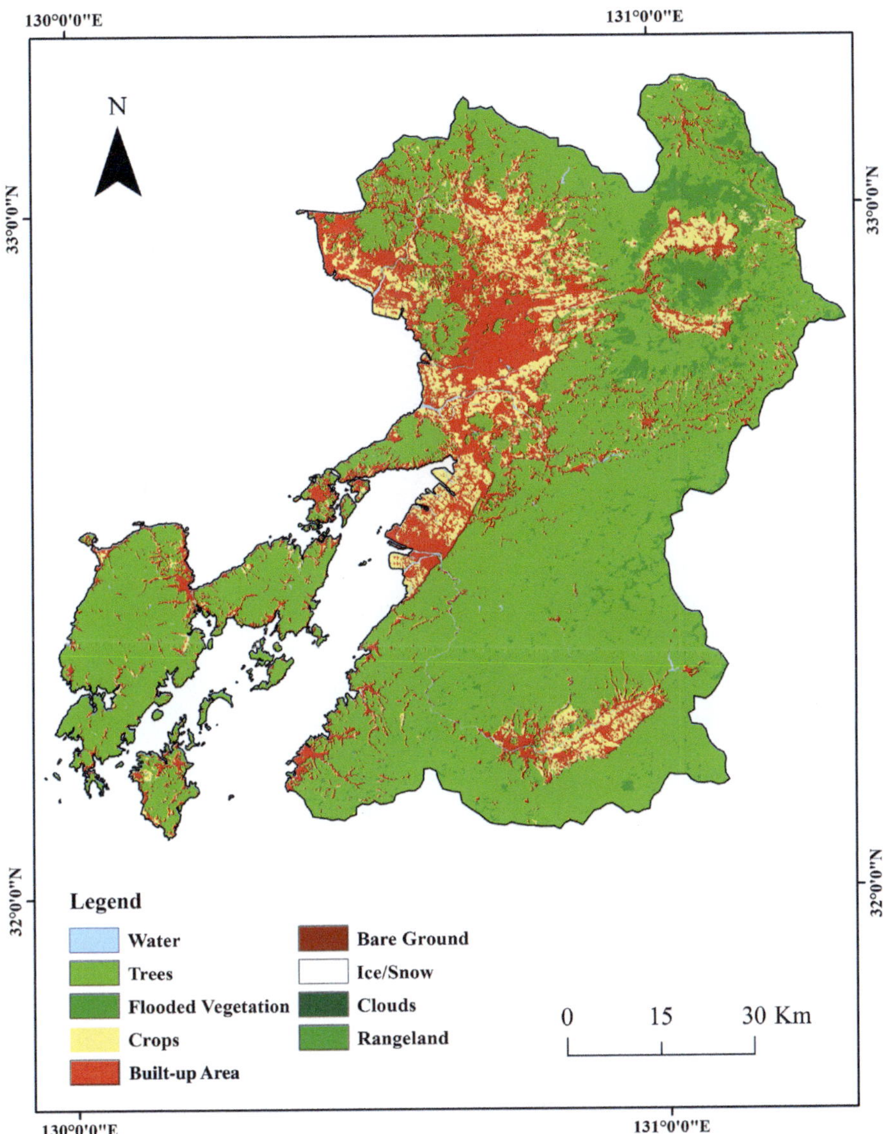

FIGURE 5.58 Land Use and Land Cover Map of Kumamoto Region, Japan.

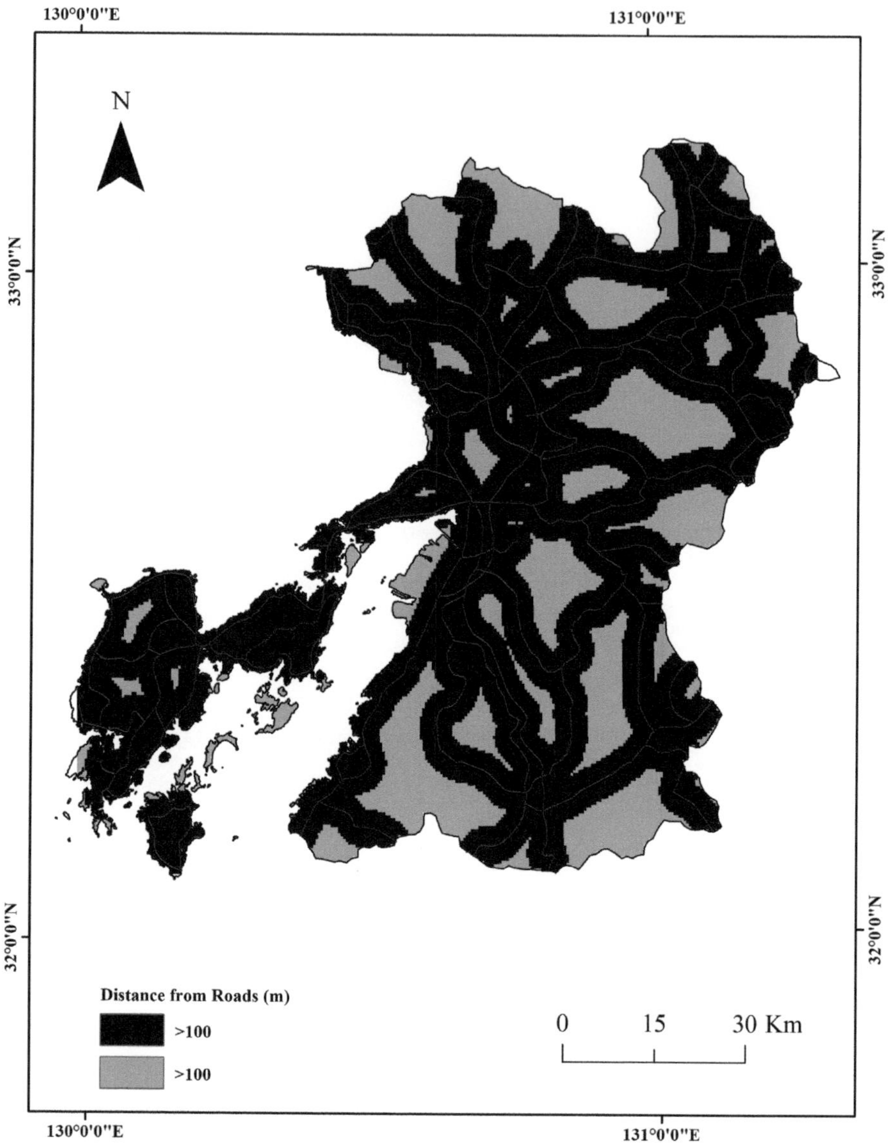

FIGURE 5.59   Road Map of Kumamoto Region, Japan.

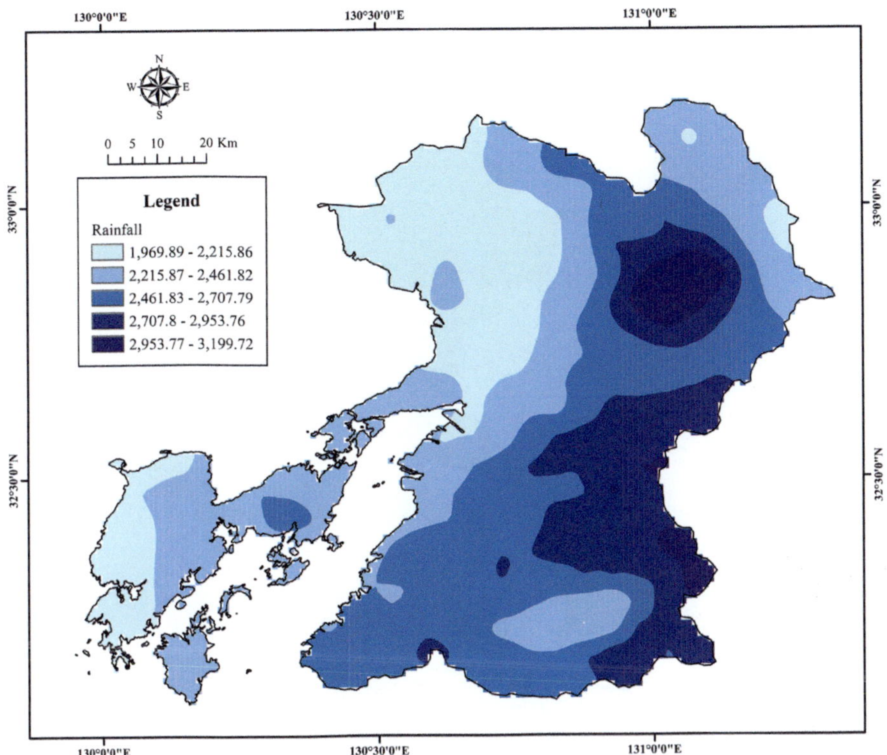

FIGURE 5.60 Rainfall Map of Kumamoto Region, Japan.

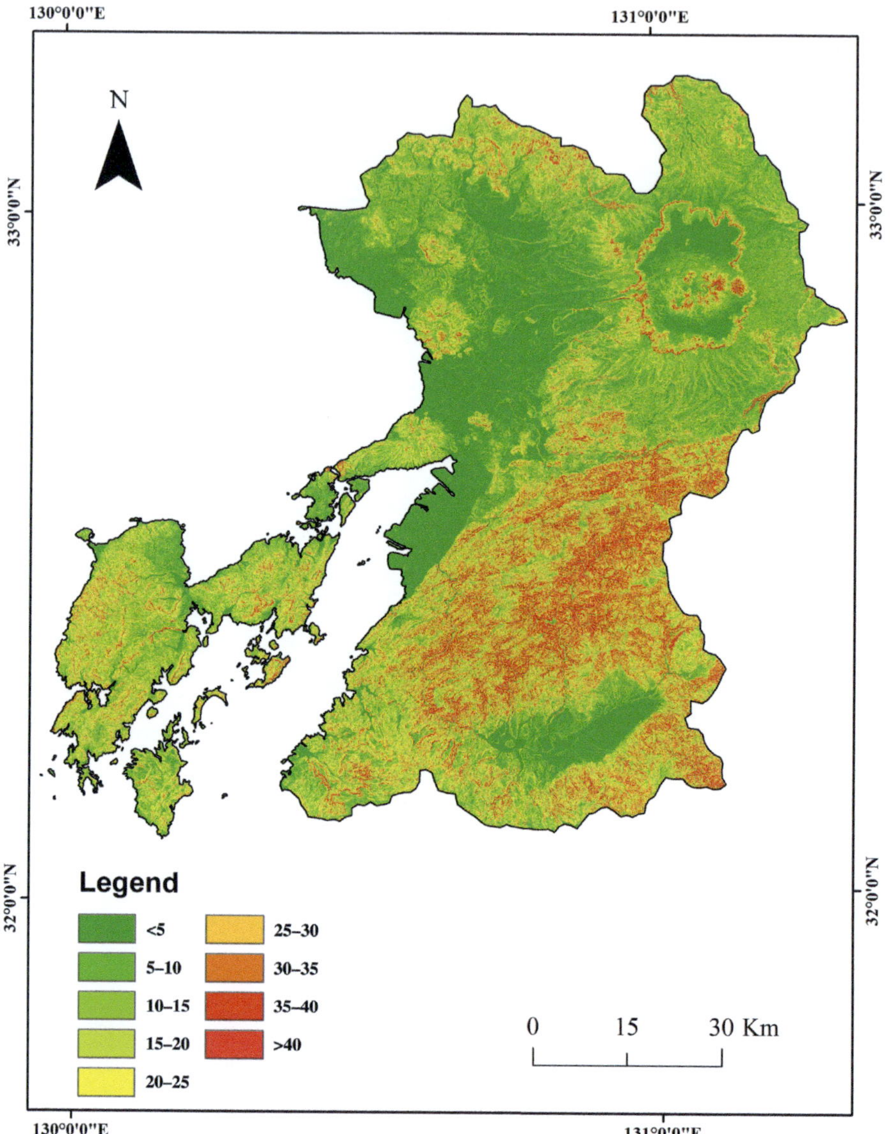

FIGURE 5.61   Slope Map of Kumamoto Region, Japan (slope in degree).

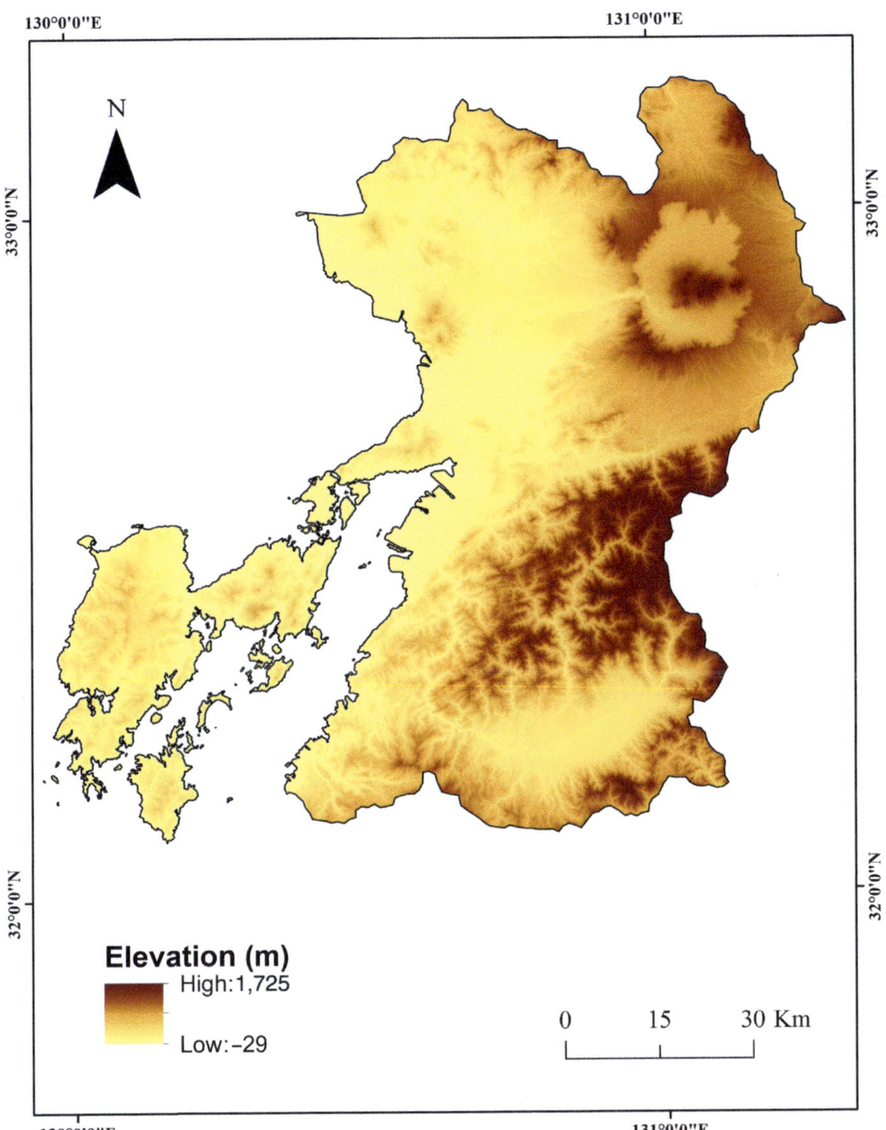

FIGURE 5.62 Elevation Map of Kumamoto Region, Japan.

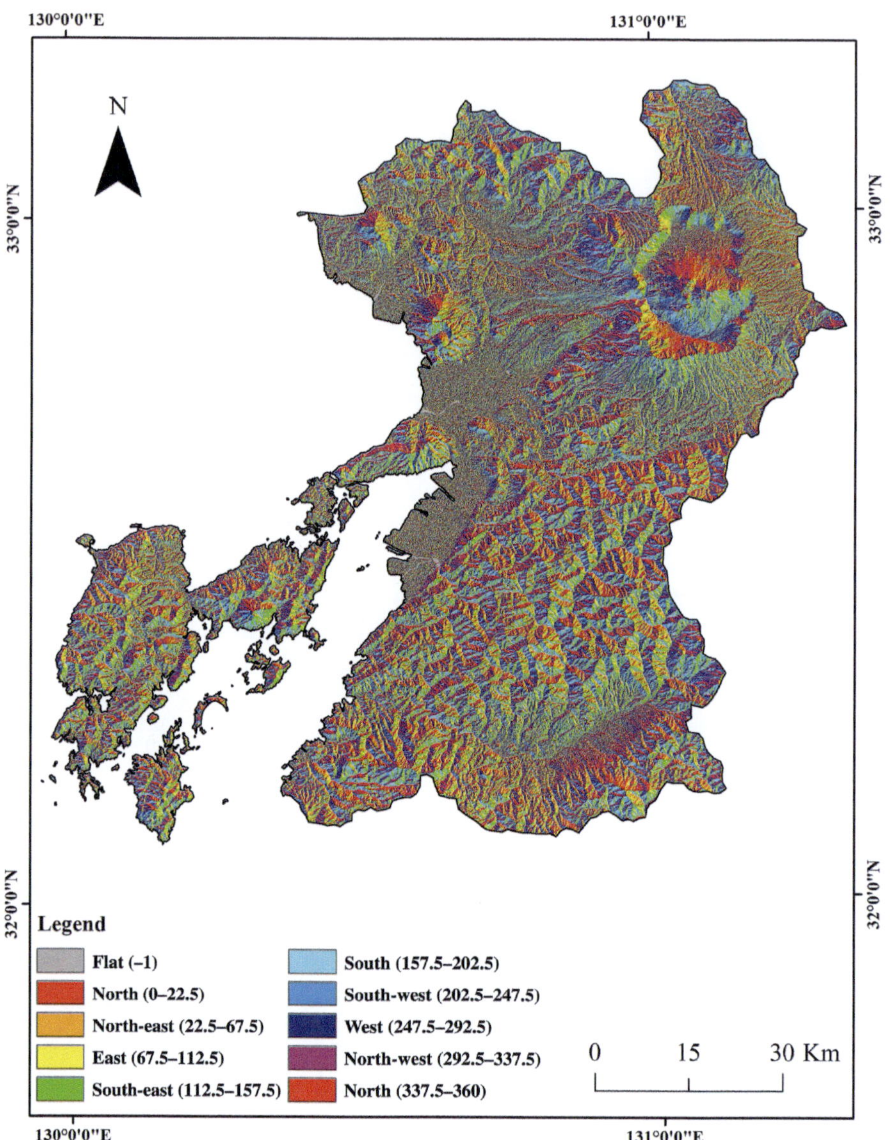

FIGURE 5.63   Slope Aspect Map of Kumamoto Region, Japan.

TABLE 5.32
Final Weights of AHP Pairwise Comparison Matrices

Elevation	Slope	Land Use and Land Cover	Road	Rainfall	Elevation	Slope	Final Weights
7	9	0.56	0.64	0.53	0.42	0.38	**0.50**
5	7	0.19	0.22	0.32	0.30	0.29	**0.26**
3	5	0.11	0.07	0.10	0.18	0.21	**0.14**
1	2	0.08	0.04	0.03	0.06	0.08	**0.06**
0.5	1	0.06	0.03	0.02	0.04	0.04	**0.04**
16.5	**24**	**1**	**1**	**1**	**1**	**1**	**1**

CR = 0.041739

Based on the pairwise comparison using the AHP model, final weights are derived (Table 5.32). Consistency of judgements has been evaluated by calculating CI and CR using eigen vector approach. All thematic data layers have been reclassified by using the final weights derived. All thematic data layers are finally integrated using weighted layer operation to prepare the landslide susceptibility map for the study area (Figure 5.64).

5.8.1 RESULTS

The AHP-based landslide susceptibility map shown in Figure 5.64 indicates that north-eastern parts of the Kumamoto region are identified as high to very high susceptible zones as compared to other parts of the region. The north-eastern hilly terrain is observed to be highly susceptible to landslides.

The classification of landslide-susceptible zones in the Kumamoto region of Japan is presented in Table 5.33. The results indicate that 22.96% of the total geographical area of the Kumamoto region is identified as high to very high landslide susceptibility, with an area of 1,678 km^2. Seismicity, elevation, steep slopes, heavy rainfall, and human interventions are the important factors affecting landslide occurrence.

5.9 LIMITATIONS

A major issue related to LSZ mapping, especially in the Indian context, is the lack of sufficient and reliable data for accurate landslide assessment. Therefore, the application of data-driven methods of LSZ in a data-scarce environment is a major limitation in landslide susceptibility assessment. The present study deals with the application of indirect methods (BIS and AHP) of LSZ mapping in the North Konkan region of Maharashtra. However, the need is to be

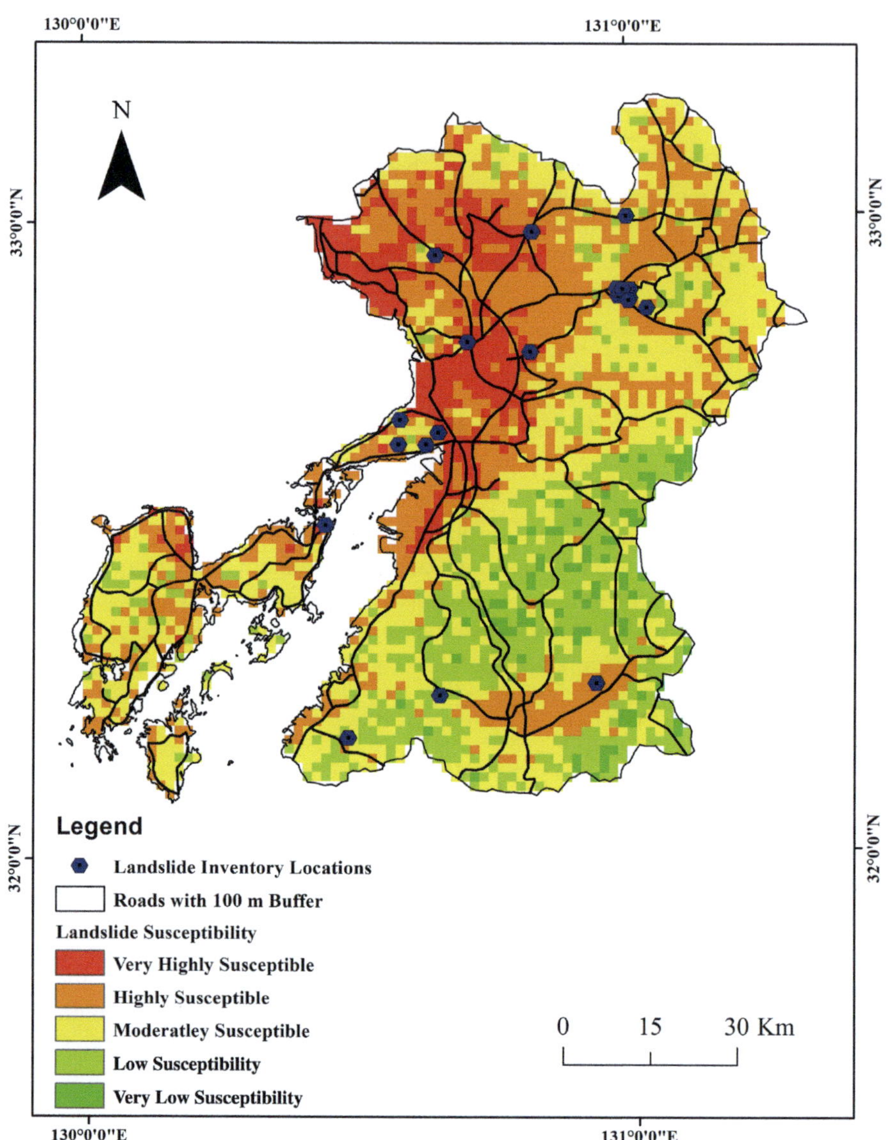

FIGURE 5.64   Landslide Susceptibility Map of Kumamoto Region, Japan.

TABLE 5.33

Landslide Susceptibility Zones in the Kumamoto Region, Japan

S. No.	Landslide Susceptibility Class	Area (Sq. Km)	Percentage Area
1	Very low	776	10.61
2	Low	2,271	31.07
3	Moderate	2,585	35.36
4	High	1,489	20.37
5	Very high	189	2.59
Total		7,310	100

cautious in the selection of the appropriate method in unique geo-environmental conditions.

5.10 CONCLUSION

LSZ mapping is an important step in landslide susceptibility management. The LSZ maps for the North Konkan region, derived from BIS (modified) based LHEF rating scheme and multi-criteria decision-making (AHP) approach, clearly show that the concentration of LH zones are confined to the Western Ghat region of the study area, followed by plateau margins of the dissected Jawhar plateau and hillsides of isolated hills in the central uplands. The distribution of susceptibility zones in the entire study area shows similar areas except for a few locations. Vasai Hill and the hilly area near Palghar town are demarcated as high susceptibility zones in the AHP-based model, whereas it is shown as a low susceptibility zone in the modified BIS-based LSZ map. Plateau margins in the Jawhar area are classified as high to very high in AHP-based LSZ, whereas this area is shown as moderate to high susceptibility zone in the modified BIS-based LSZ map. Major parts of eastern and south-eastern Raigad district fall in the high susceptibility zone per the modified BIS-based LSZ map, whereas these areas indicate moderate landslide susceptibility per the AHP-based model. The results obtained from validation of BIS (modified) and AHP-based LSZ maps show 75.69% and 73.48% of the actual slope failures are represented by high to very high susceptibility classes, respectively. This confirms the validity of LSZ maps prepared for the North Konkan region of Maharashtra.

REFERENCES

Bhatt, B., K. Awasthi, B. Heyojoo, T. Silwal, and G. Kafle. 2013. "Using Geographic Information System and Analytic Hierarchy Process in Landslide Susceptibility Zonation." *Journal of Applied Ecology and Environmental Sciences* 1 (2): 14–22.

Boroumandi, M., M. Khamehchiyan, and M. Nikoudel. 2015. "Using of Analytic Hierarchy Process for Landslide Susceptibility Zonation in Zanjan Province, Iran." *Engineering Geology for Society and Territory* 2: 951–955.

Bureau of Indian Standards. 1998. *Preparation of Landslide Hazard Zonation Maps in Mountainous Terrain – Guidelines (Part2-Macrozonation)*. IS 14496-2, Hill Area Development Engineering (CED 56), New Delhi, pp. 1–19.

Courture, R. 2011. *Landslide Terminology – National Technical Guidelines and Best Practices on Landslides. Open Files 6824, Geological Survey of Canada*. Canada.

Ghosh, S., C. van Westen, E. Carranza, T. Ghoshal, N. Sarkar, and M. Surendranath. 2009. "A Quantitative Approach for Improving BIS (Indian) Method of Medium-Scale Landslide Susceptibility." *Journal Geological Society of India* 74: 625–638.

Habibi, A. 2014. "Landslide Susceptibility Zonation for Determination of Appropriate Region with AHP Model Dry Areas of Iran, Khuzestan." *International Journal of Forest, Soil and Erosion* 4 (1): 16–20.

Kornejady, A., K. Haidary, M. Sarparast, G. Khosravi, and M. Mombeini. 2014. "Performance Assessment of Two 'LNRF' and AHP Area Density Models in Landslide Susceptibility Zonation." *Journal Life Science Biomed* 4 (3): 169–176.

NRDMS (Natural Resources Data Management System). 1984. *Methodology for Landslide Susceptibility Zonation: A Report Prepared under Caoordinated Programme on the Study of Landslides* (pp. 1–32). New Delhi: Department of Science and Technology.

Othman, A., W. Naim, and S. Norani. 2012. "GIS Based Multi-Criteria Decision Making for Landslide Susceptibility Zonation." *Procedia – Social and Behavioral Sciences* 35: 595–602.

Pardeshi, S., S. Pardeshi, and S. Autade. 2013. "Landslide Susceptibility Assessment: Recent Trends and Techniques." *SpringerPlus* 2 (523). https://doi.org/10.1186/2193-1801-2-253.

Phukon, P., D. Chetia, and P. Das. 2012. "Landslide Susceptibility Assessment in the Guwahati City, Assam, Using Analytic Hierarchy Process (AHP) and Geographic Information System." *Internal Journal of Computer Applications in Engineering Sciences* 2 (1): 1–6.

Saadatkhah, N., A. Kassim, and L. Lee. 2014. "Qualitative and Quantitative Landslide SusceptibilityAssessment in Hulu Kelang area, Malaysia." *EJGE* 19: 545–563.

Saaty, T. 2008. "Decision Making with the Analytic Hierarchy Process." *International Journal of Services Sciences* 1 (1): 83–98.

Saaty, T. 2013. "On the Measurement of Intangibles: A Principal Eigenvector Approach to Relative Measurement derived from Paired Comparisons." *Notices of the AMS* 60 (2): 192–208.

Vaidya, D., and S. Kumar. 2006. "Analytic Hierarchy Process: An Overview of Applications." *European Journal of Operational Research* 169: 1–29.

Varnes, D. J. 1984. "Landslide Hazard Zonation: A Review of Principals and Practice." *United Nations International* 1–62.

6 Landslide Risk Assessment

Case Studies

6.1 INTRODUCTION: NEED FOR LANDSLIDE RISK ASSESSMENT

Slope failure is a major geological hazard in the North Konkan region of Maharashtra and poses a risk to both lives and property with associated socioeconomic consequences. Landslide risk assessment is an important step in landslide hazard assessment. Varnes (1984) defines landslide risk as 'the expected number of lives lost, persons injured, property damage and disruptions to economic activities due to landslides in each area and given period'.

Landslide risk (R) is the product of hazard (H) and vulnerability (V) and can be expressed as $R = H \times V$. Identification of landslide risk zones provides the basis for prioritisation of landslide mitigation measures and developing disaster management policy. However, assessment of both direct (lives lost, property damage, repairing cost, number of persons injured, loss of agricultural products, etc.) and indirect risk (disruption in traffic flow, fuel loss, delay, disruption in communication between the settlements along the communication routes, loss of energy, etc.) is a very difficult task. Van Westen et al. (2006) discussed some of the important limitations in carrying out quantitative landslide risk assessment, including the lack of a complete landslide database, insufficient damage details, and difficulty in determining runout distances for different types of slope failures.

Landslide risk mapping involves identification, quantification of elements at risk, and integration of hazard and vulnerability layers to determine landslide risk–prone areas with varying magnitudes. To quantify landslide risk, hazard classes are multiplied by the vulnerability to different types of elements at risk.

There are three major approaches to assessing landslide risk: viz., qualitative, semi-quantitative, and quantitative landslide risk assessment. Keeping in mind the limitations in quantifying elements at risk in a data-scarce environment, qualitative landslide risk assessment, in combination with the heuristic landslide hazard zonation method, can be used effectively (van Westen et al., 2006). For the present work, landslide risk–prone areas in the North Konkan region are identified using qualitative methods.

6.1.1 Qualitative Landslide Risk Assessment

Landslide risk assessment involves the identification of elements at risk, vulnerability assessment, and delineation of landslide risk–prone areas. Qualitative risk assessment involves the identification of landslide risk–prone areas based on spatial patterns of elements at risk.

The input data required for qualitative landslide risk assessment involves a landslide susceptibility zonation map and thematic data layers of elements at risk (e.g., population, settlements, roads, man-made structures, agricultural lands, natural resources). A landslide susceptibility map is combined with each thematic data layer of elements at risk to map landslide vulnerability to each element at risk. Finally, all vulnerability maps for all elements at risk are combined to prepare a landslide risk–prone map.

The qualitative landslide risk assessment approach helps to identify and delineate potential landslide risk–prone zones. This helps planners and administrators to prioritise landslide mitigation measures. One of the most important limitations of qualitative landslide risk assessment is that it fails to predict the exact quantity of potential losses due to landslides. Hence, accurate prediction of losses caused by landslides of a given magnitude becomes difficult.

6.1.2 QUANTITATIVE LANDSLIDE RISK ASSESSMENT

Quantitative risk assessment has proven to be more accurate and useful in predicting potential losses associated with landslide occurrence. In this approach of landslide risk assessment, a landslide susceptibility map is compared with different thematic layers of the vulnerability of elements at risk. Landslide susceptibility classes and subclasses of each thematic layer of elements at risk are assigned numerical values depending on their importance in terms of losses that occurred due to their proximity to areas of landslide occurrence. Both direct and indirect risks are calculated and then combined to get the total landslide risk map. Quantitative landslide risk assessment is a data-driven approach for analysing landslide risk.

6.1.3 ROLE OF RS AND GIS IN LANDSLIDE RISK ASSESSMENT

Over the last two decades, the advancement in Earth observation (EO) techniques (satellite remote sensing and aerial photography) has facilitated effective landslide detection, mapping, monitoring, and hazard and risk assessment (Tofani et al., 2013). In India, satellite data products are used more frequently as compared to aerial photography. Moderate-resolution satellite data such as IRS LISS II, IRS LISS III, and LANDSAT have been frequently used for landslide investigations in India (Thigale and Umrikar 2007; Champatiray et al., 2007; Kanungo et al., 2008; Naithani, 2007; Nagarajan et al., 2000; Das et al., 2012; Prabhu and Ramakrishnan, 2009; Rammohan et al., 2011). High-resolution satellite data (e.g., CARTOSAT 1, RESOURCESAT, SPOT; IRS LISS IV, QuickBird) have recently been used for landslide studies in India (Champatiray, 2009; Chauhan et al., 2010; Das et al., 2011, 2012; Lallianthanga and Lalbiakmawia F., 2013; Saraf et al., 2009; Sharma et al., 2009; Karlekar, 2012; Thigale and Umrikar, 2007; Wagh and Deshpande, 2013).

Digital elevation model (DEM) has wide applications in landslide studies, especially in the extraction of thematic layers such as slope, aspect, curvature, and drainage in pre- and post-event landslide investigations and even in landslide detection. In India, DEM extracted from topographical maps (Topo DEM) and also from SRTM and ASTER GDEM are used in several studies (Deshpande et al., 2009; Sarkar and Kanungo, 2004; Sharma et al., 2009; Saraf et al., 2009; Rawat et al., 2012; Onagh et al., 2012; Chandel et al., 2011; Balsubaramani and Kumaraswamy, 2013). A brief review of the application of RS data in landslide studies in India reveals its importance in accurate and reliable landslide hazard assessment. However, extensive field check is also essential to validate the information extracted from remotely sensed data.

A wide range of spatial information sources is now being put into use for the accurate prediction of landslides. Ferlisi et al. (2021) carried out a quantitative landslide risk assessment using DInSAR images. Nguyen and Kim (2021) have recently attempted to use digital aerial photos and LiDAR images for landslide risk assessment in Mount Umyeon, Korea. Unmanned aerial vehicles are now used widely for landslide investigations across the world for accurate prediction of risk associated with landslides of different intensities (Tempa et al., 2021; Niethammer et al., 2010; Liao et al., 2022; Farina et al., 2017; Prasetyo et al., 2018; Yaprak et al., 2018).

6.2 QUALITATIVE LANDSLIDE RISK ASSESSMENT: A CASE STUDY IN PARTS OF THE WESTERN GHATS, INDIA

6.2.1 Data Sources and Methodology

The present section aims at determining landslide risk–prone areas in the North Konkan region of Maharashtra. Qualitative landslide risk assessment has been carried out using three elements at risk: viz., population, built-up areas, and road traffic density. These thematic data layers have been extracted from Google Earth images, district road development plans (2001–2020), and the Census of India, 2011. All thematic data layers of elements at risk have been reclassified using numerical weights. Reclassified thematic layers are then multiplied with landslide hazards and integrated to produce a landslide risk map. The calculation of total landslide risk is expressed as under (van Westen et al., 2006):

$$R_t = \sum (R_{pop} * LH + R_{built} * LH + R_{roads} * LH),$$

where
    R_t = total landslide risk
    R_{pop} = vulnerability to population
    R_{built} = vulnerability to built-up areas
    R_{roads} = vulnerability to road traffic

The details of the methodology adopted for landslide risk assessment in the North Konkan region of Maharashtra are illustrated in Figure 6.1.

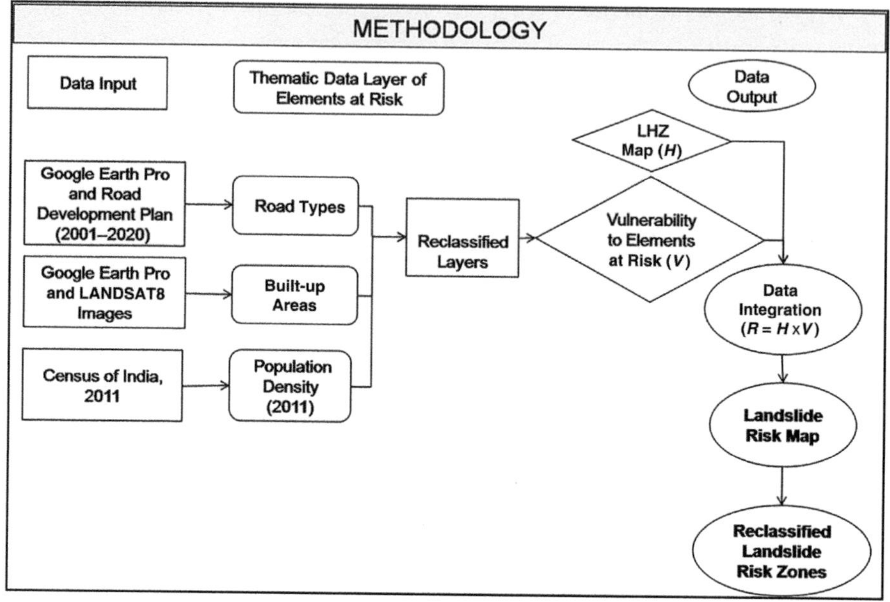

FIGURE 6.1 Methodology Flowchart for Landslide Risk Mapping.

6.2.2 Elements at Risk

To delineate landslide risk–prone areas in the North Konkan region, three elements at risk – viz., population, built-up areas, and traffic density along the roads – have been considered.

6.2.2.1 Population

Risk to human lives caused by slope failure events is an important consideration in landslide risk assessment. To delineate the areas with population concentration, village-wise population data (Census of India, 2011) have been obtained, and population density (arithmetic population density) has been calculated ($Pop_{density}$ = total population/total geographical area). Four population density classes – viz., <1,000, 1,000–10,000, 10,000–20,000 and >20,000 per km^2 area – have been determined. Figure 6.2 illustrates the spatial patterns of population density in the study area. The highest concentration of densely populated areas is associated with urban clusters in the coastal parts of the entire study area, particularly in the Palghar and Thane districts. Some isolated areas with a population density of $10000 - 20000$ are observed at the tehsil headquarters locations. The settlements away from the coastline, central uplands, and dissected Jawhar plateau and foothills of the Western Ghat escarpment of the study area are moderately populated, whereas population density is very low in the hilly tracks of the Western Ghats and in isolated hillocks in the study area.

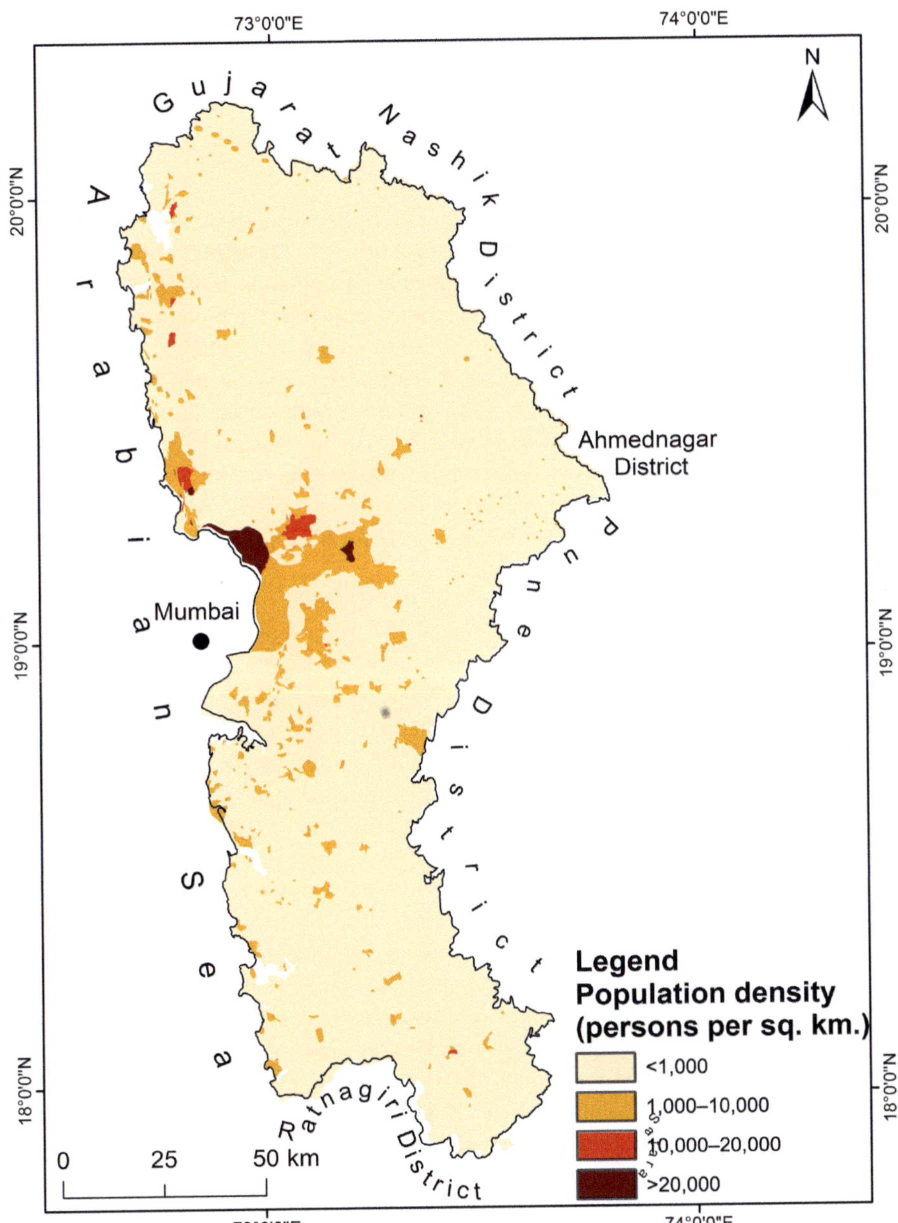

FIGURE 6.2   Village/Town-Wise Population Density Map of the Study Area.

Although population density can be used as an input parameter for landslide risk assessment, it gives a generalised picture of population concentration because the entire geographical area of a village or city is considered for calculating population density. Practically, the population is not distributed over the area, but population concentration is observed in the actual built-up areas (settlements). Therefore, consideration of built-up areas, along with the population density layer, can be more useful in landslide risk assessment.

6.2.2.2 Built-up Areas

Slope failure in inhabited areas also causes damage to man-made structures like houses, buildings, roads, railway tracks, and bridges. Therefore, consideration of spatial patterns of built-up areas is important in landslide risk assessment. The thematic layer of built-up areas has been extracted from the Google Earth image. The distribution of built-up areas in the study area is presented in Figure 6.3. It is concentrated in the highly populated areas in the coastal plains, particularly in the Palghar and Thane districts. Few of these urban settlements located along the foothills such as the Parsik Hill area, Matheran area, and coastal hill range near Palghar are vulnerable to slope failure. Many slope failure events have been reported in Parsik Hill areas, particularly in Mumbra-Kalwa, where a large number of slums are affected as a result. Besides, several small settlements are located in the interior parts of the study area.

6.2.2.3 Roads

Slope failure in North Konkan is mainly associated with road-cut slopes along the road corridors passing through the Western Ghat escarpment. During peak monsoon periods, these roads are frequently affected by slope failures, resulting in disruption to traffic flow, delay in traffic, and sometimes damage to road structures and moving vehicles. Therefore, it is important to consider roads as input parameters for landslide risk mapping. For this, major roads have been digitised using Google Earth images, topographical maps, and district road development plans (2001–2020). All roads have been categorised into four classes: viz., expressway, national highways (NH), state highways (SH), and district roads (DR). Different types of roads have a varying degree of traffic density (Figure 6.4). National highways and expressways experience exceptionally high traffic flow, followed by state highways and district roads. Kasara Ghat (NH 3), Malshej Ghat (NH 222), Mumbai-Pune Expressway, Mumbai-Pune-Bengaluru National Highway (NH 4) along the Bhor Ghat section, and Mumbai-Goa National Highway (NH 17) along the Kashedi Ghat section are major road corridors with high traffic flow density affected by slope failures during the rainy season. The thematic data layer of roads has been reclassified based on traffic density.

6.2.3 Landslide Vulnerability Assessment

After extraction of thematic data layers, it is important to assess the vulnerability of each of the elements at risk caused by slope failures. For this, each thematic data layer is compared with landslide hazard (LH) zones and reclassified using numerical weights (Table 6.1).

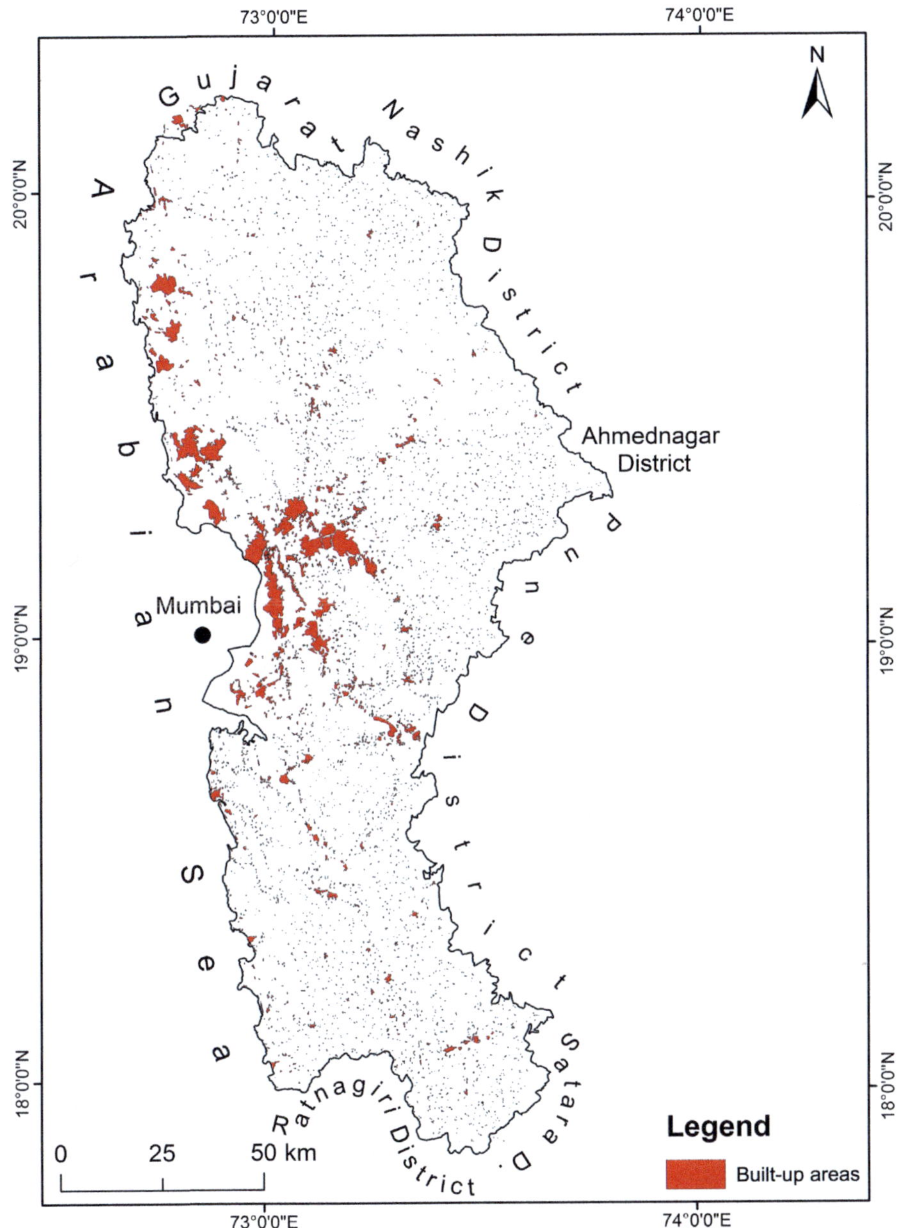

FIGURE 6.3 Distribution of Built-up Areas.

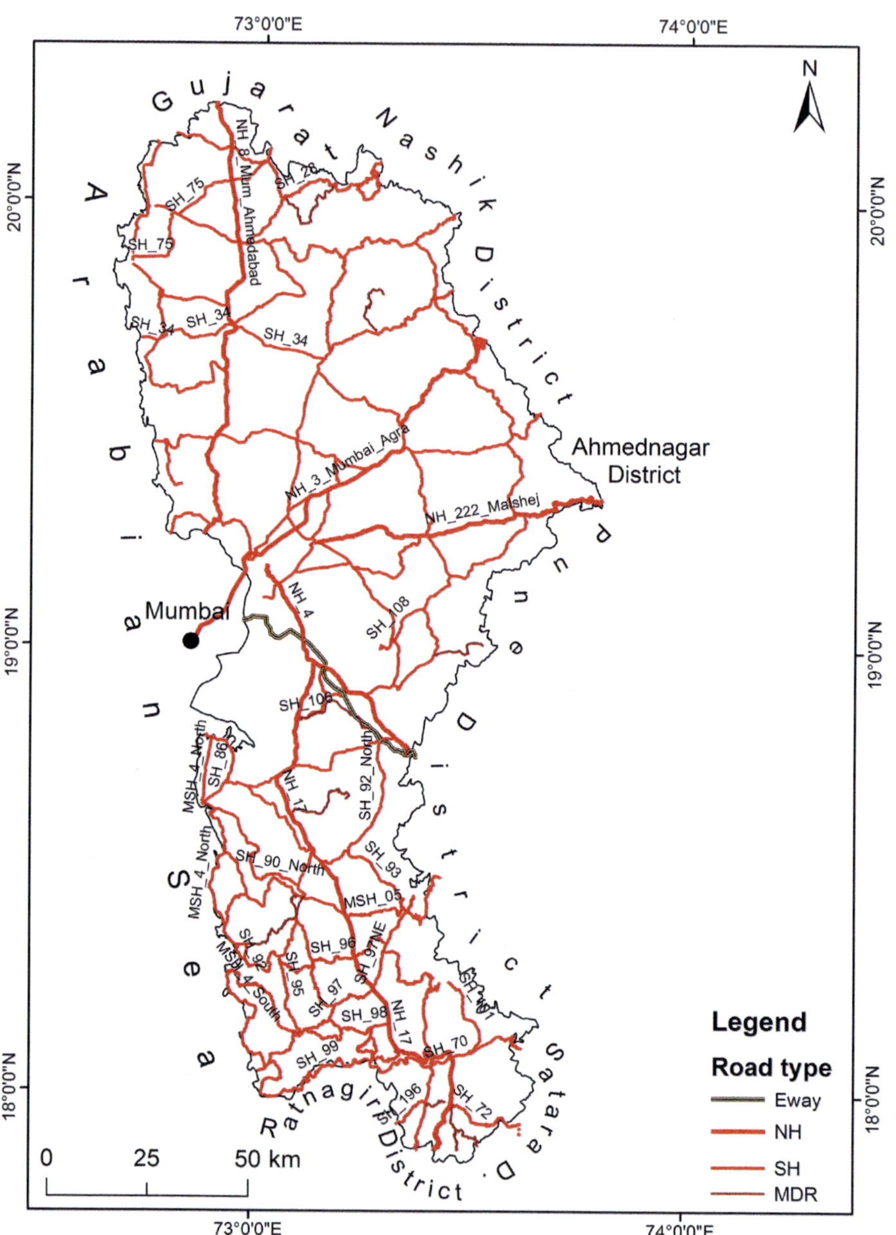

FIGURE 6.4   Road Types.

Note: Eway, Expressway; NH, National highway; SH, State highway; MDR, Main district road.

TABLE 6.1
Rating Scheme Used for Landslide Risk Mapping

S. No.	Thematic Data Layer	Classes	Weights Used for Risk Mapping
1	Population density	> 20,000	4
		10,000–20,000	3
		1,000–10,000	2
		< 1,000	1
2	Built-up areas	Built-up areas	5
		Non-built-up areas	1
3	Roads	Expressway	5
		National highways	3
		State highways	2
		District roads	1
4	Landslide hazard zones	Very high	5
		High	4
		Moderate	3
		Low	2
		Very low	1

(*Source*: Based on the thematic data layers used for risk assessment)

6.2.3.1 Vulnerability to Population

Although historical records show a comparatively low number of fatalities caused by slope failures as compared to other natural disasters, it is important to identify the areas prone to slope failures which may harm people. Even minor slope failures in densely populated areas may pose severe risks to human lives. To determine the zones vulnerable to human lives, the thematic data layer of population density is compared with LH zones. Figure 6.5 depicts population density areas prone to slope failure of different magnitudes. The results indicate that the densely populated settlements in the foothills of the Western Ghat escarpment and isolated hills in the study area are vulnerable to slope instability and thereby pose a risk to the lives of people. Few fatal landslides have been reported in Sainik Nagar, Gholai Nagar (Kalwa), Sainik Nagar (Mumbra), and Shanti Nagar (Mumbra) areas. Besides, the foothills of Matheran Hills, Palghar, and the Wada area are highly vulnerable to the lives of the people in the vicinity. Although landslide susceptibility in the north-eastern corner of the study area is categorised as highly to very highly hazardous, the risk to the lives of people is low, mainly due to low population density. The concentration of landslide vulnerability to people is limited to the highly populated areas in the western coastal belt of the study area and also to the major settlements located near the slopes of isolated hills and the western slopes of the Western Ghat escarpment. Based on population density classes, the thematic data layer of population density is reclassified using

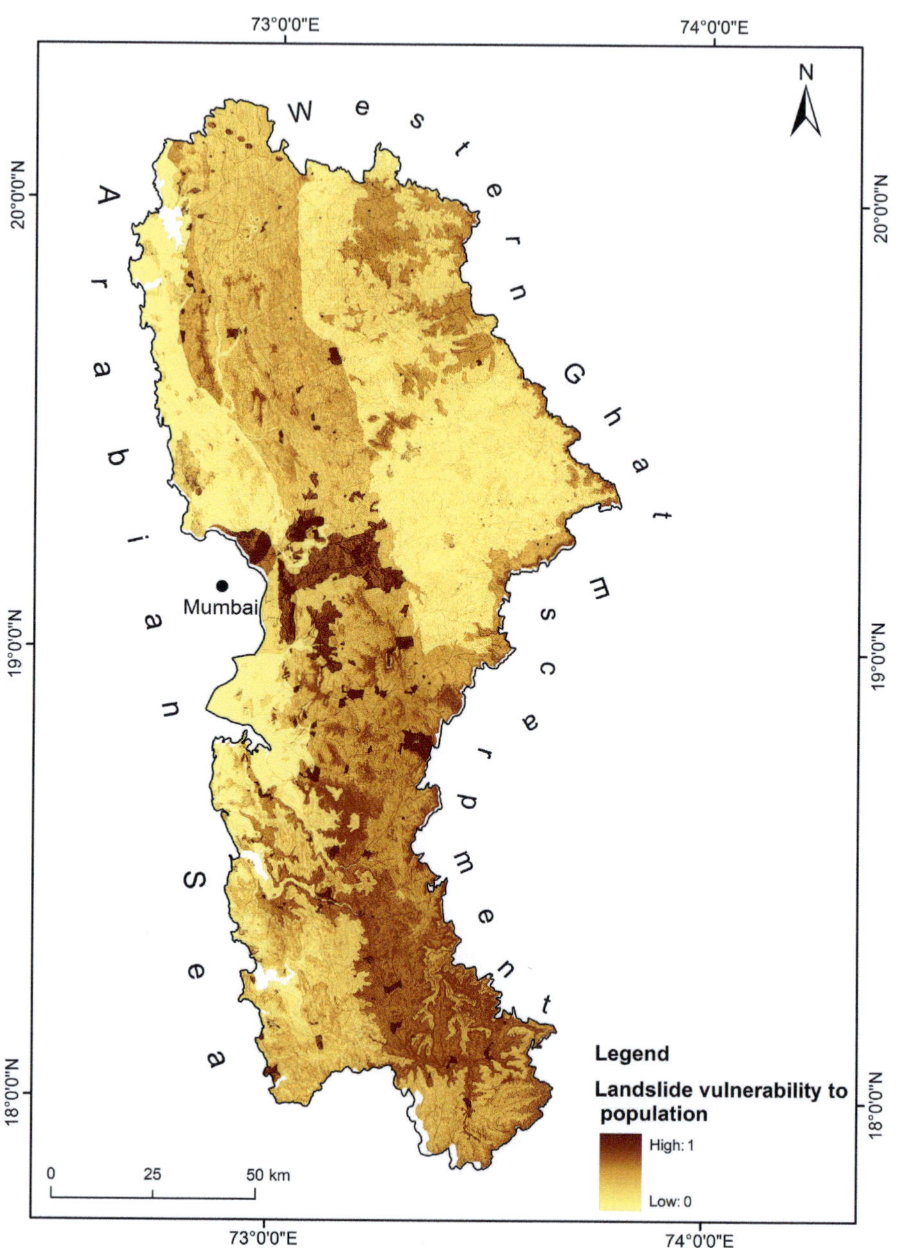

FIGURE 6.5 Landslide Vulnerability to Population.

Note: '0' indicates no loss; '1' indicates total loss of human lives.

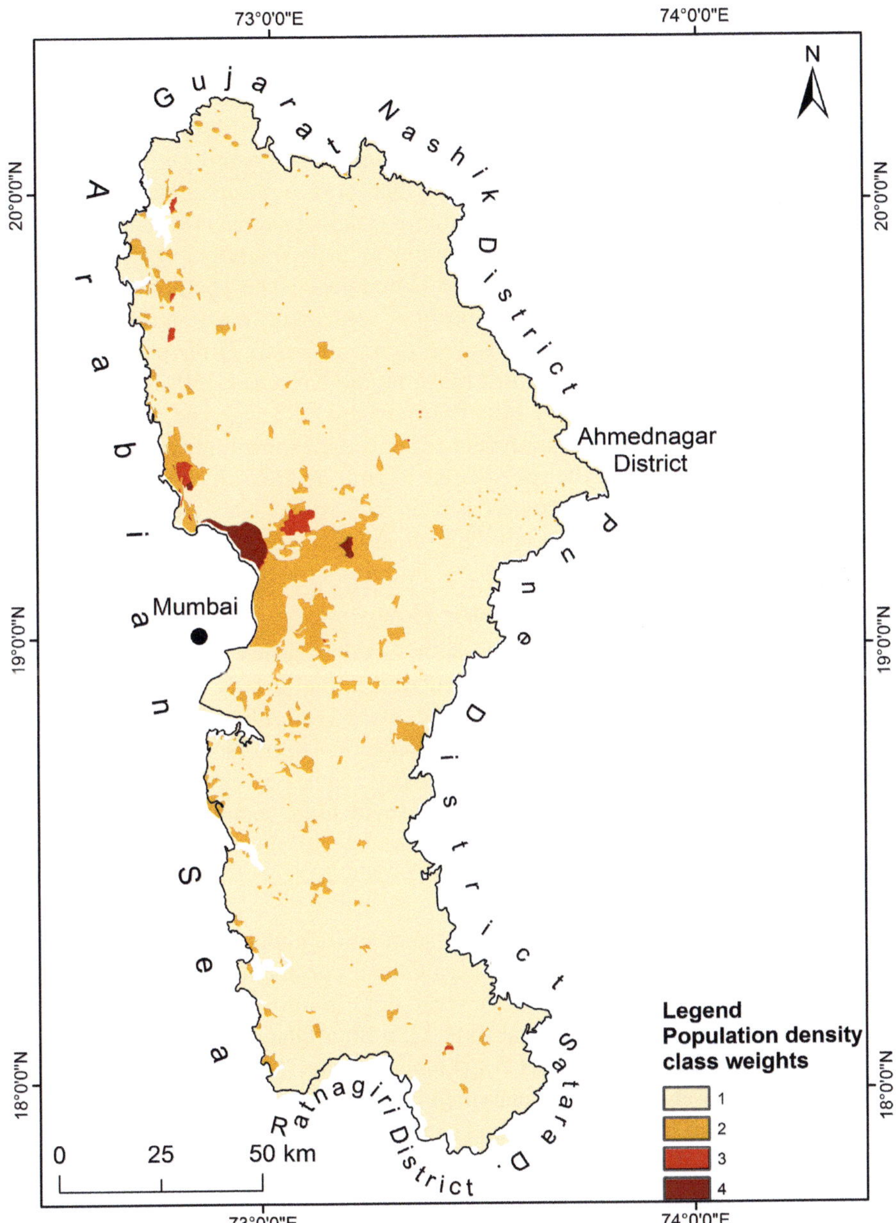

FIGURE 6.6 Reclassified Layer of Population Density.

numerical weights as 4 for a population density of >20,000, 3 for a population density of 10,000–20,000, 2 for a population density of 1000–10,000, and 1 for a population density less than 1000 persons per km^2 (Figure 6.6).

6.2.3.2 Vulnerability to Built-up Areas

Slope failures in inhabited areas often cause damage to man-made structures. The estimation of specific risks to built-up areas caused by slope failures is an important step in landslide hazard mitigation planning. Man-made structures like houses and other infrastructure in built-up areas located on natural slopes are susceptible to slope failures. The distribution of built-up areas in the North Konkan region (Figure 6.7) indicates that the highest concentration of built-up areas is located in the coastal belt of the study area, particularly in Palghar, Vasai, Ulhasnagar, Kalyan, Thane, Panvel, and Uran tehsils. The population growth in these urban centres leads to the expansion of settlements even on the hill slopes of the surrounding areas. Mumbra-Kalwa area near Parsik Hills, parts of Thane, the western slopes of the coastal hill range in Palghar tehsil, hillside settlements in Matheran area, Khalapur, the hilly tracks in south Raigad, and the foothills of the Western Ghat escarpment in the entire study area show high vulnerability for built-up areas. To estimate the specific risk to built-up areas in North Konkan, the thematic data layer of built-up areas has been reclassified using numerical weights as 5 for built-up areas and 1 for non-built-up areas (Figure 6.8).

6.2.3.3 Vulnerability to Road Traffic Flow

Slope failure along roads results in major losses in terms of actual damage to road structure, delayed traffic flow, disruption in the flow of goods and passengers, loss of fuel due to diversion of traffic, mental stress/fatigue, medical emergencies, and loss of energy. All the slope failures in the North Konkan region are associated with slope cutting for roads. Besides geo-environmental conditions, anthropogenic activities such as road widening, removal of upslope vegetation, and blasting along the cut slope also trigger slope failure in the study area. Figure 6.9 illustrates the location of roads with their susceptibility to slope failure.

The distribution of vulnerable road sections in the study area shows that major roads passing through the Western Ghat escarpment of the study area are most vulnerable to slope failures. Several slope failure events are reported in Kasara Ghat (NH 3), Malshej Ghat (NH 222), Bhor Ghat (NH 4), Mumbai-Pune Expressway, Varandh Ghat (MSH 70), Ambenali Ghat (SH 72), and Kashedi Ghat (NH 17) sections. The roads are classified into four categories: viz., expressways, national highways, state highways, and district roads based on their status, importance, and road traffic density. Using these classes, the thematic layer of roads has been reclassified using numerical weights as expressway (5), national highways (3), state highways (2), and district roads (1). The reclassified road classes are presented in Figure 6.10.

6.2.4 Landslide Risk Zones

Reclassified thematic data layers of landslide susceptibility zonation map (Figure 6.11), population density, built-up areas, and road types are integrated

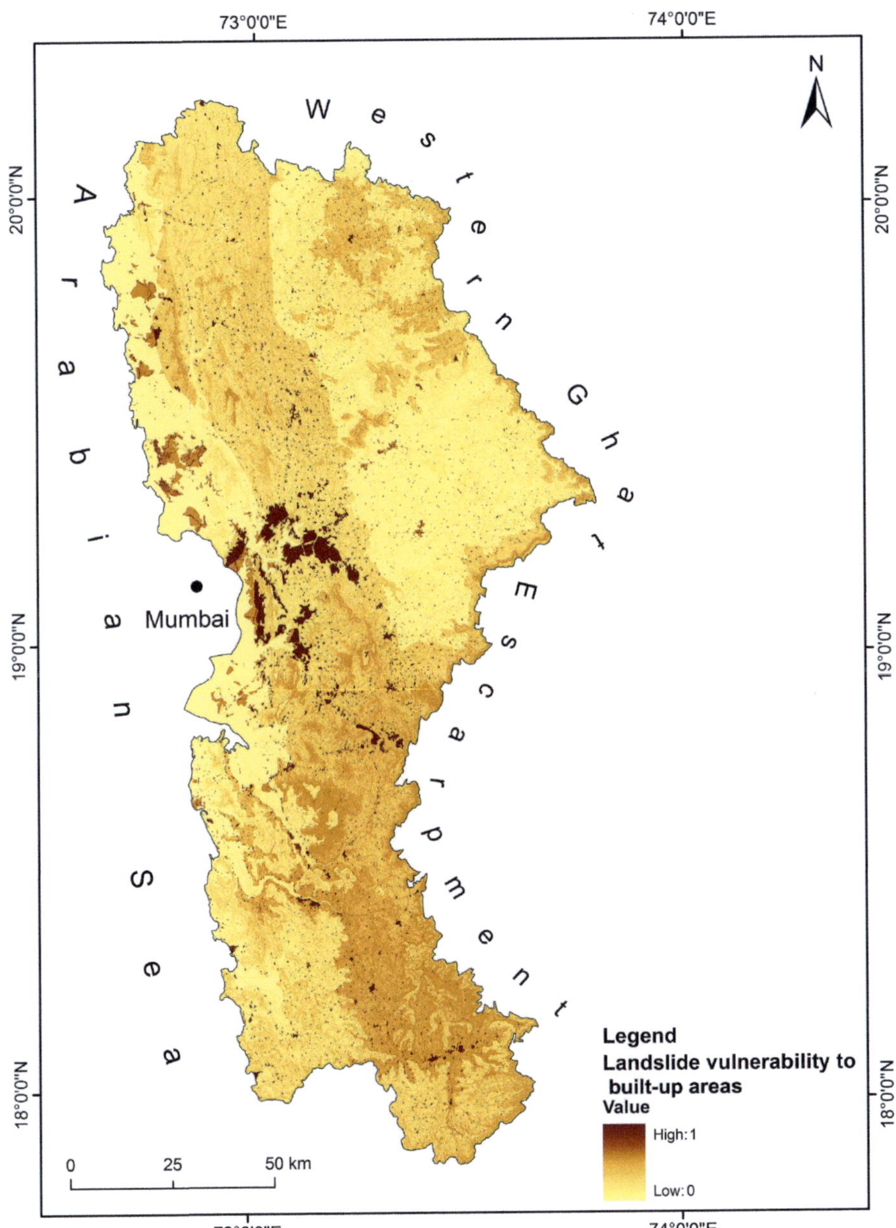

FIGURE 6.7 Landslide Vulnerability to Built-up Areas.

Note: '0' indicate no loss; '1' indicates total loss of built-up structures.

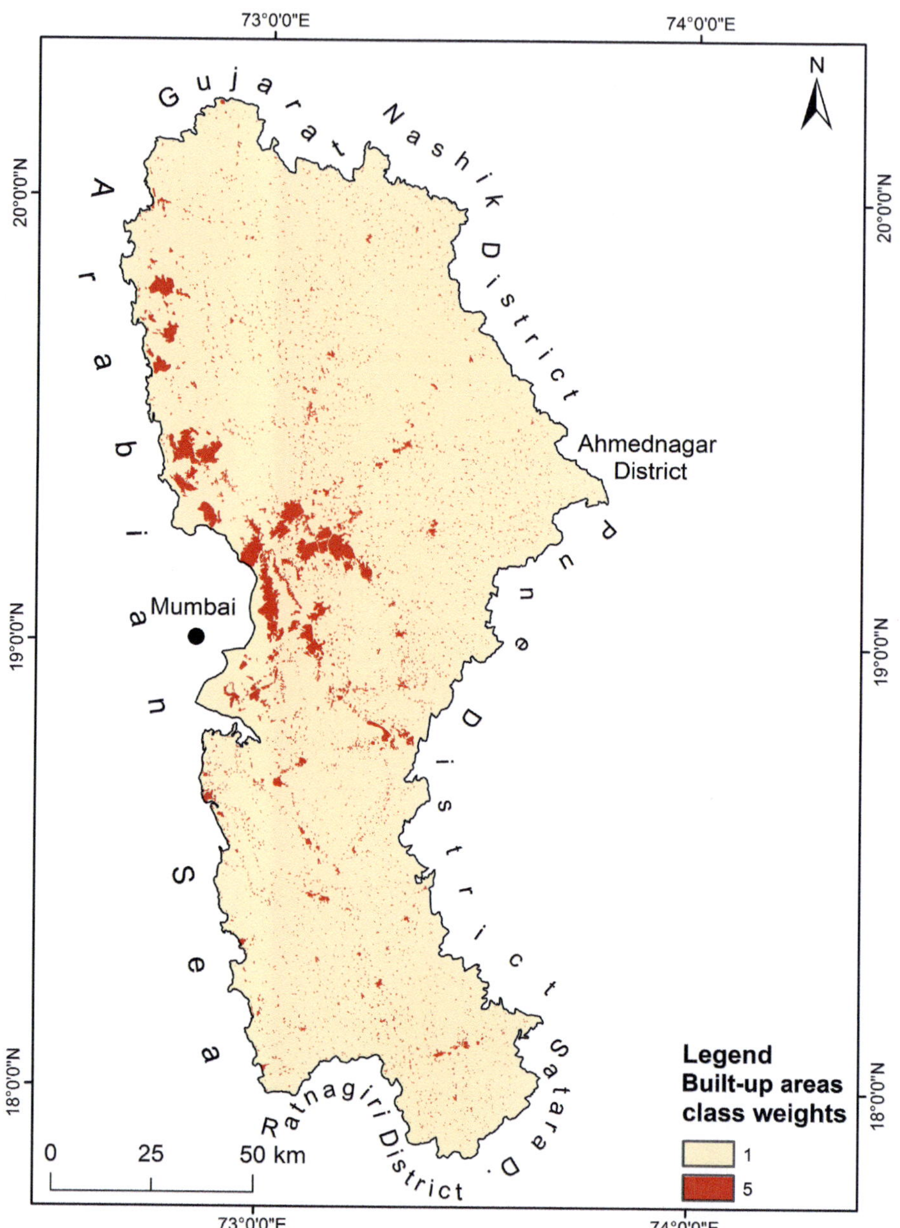

FIGURE 6.8 Reclassified Layer of Built-Up Areas.

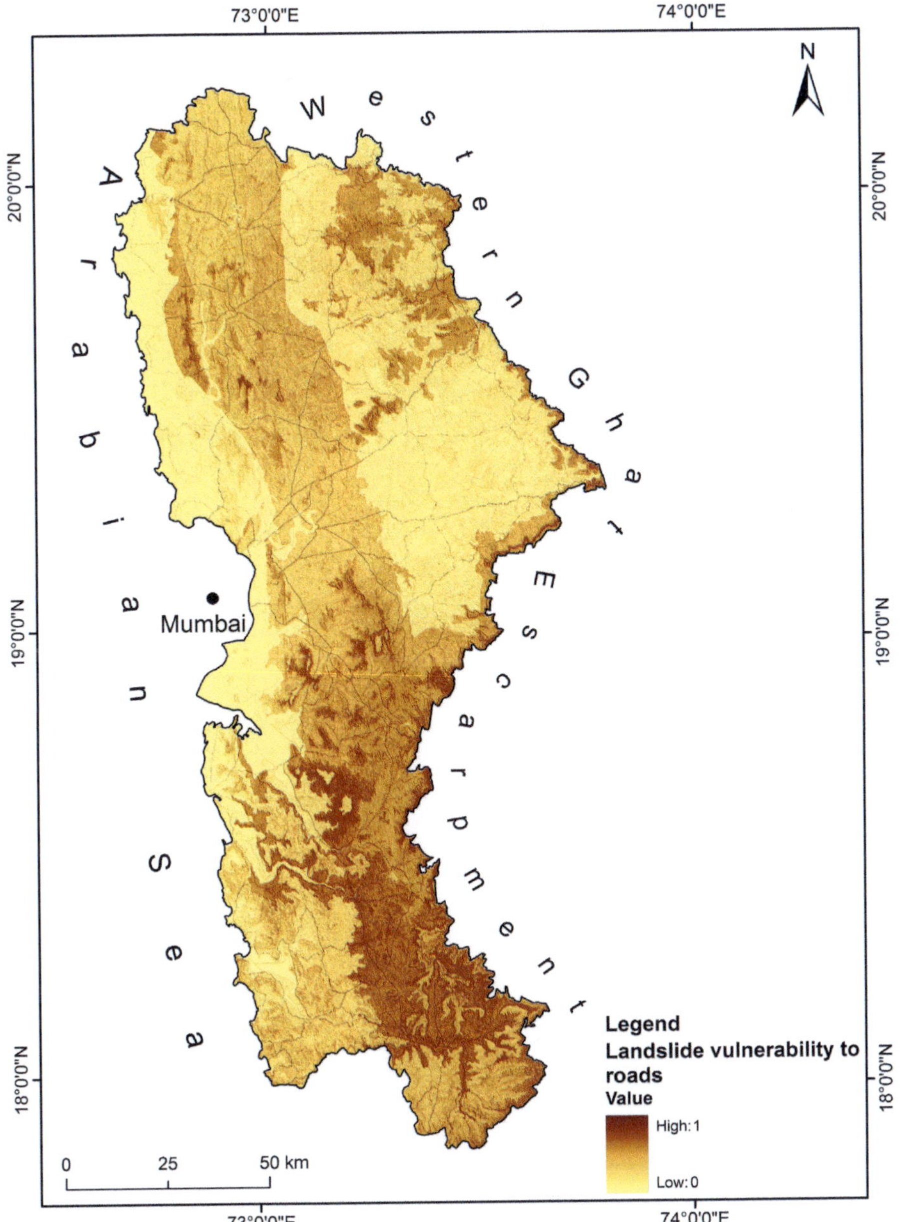

FIGURE 6.9 Landslide Vulnerability to Road Traffic Flow.

Note: '0' indicates no loss; '1' indicates total loss to the road traffic flow.

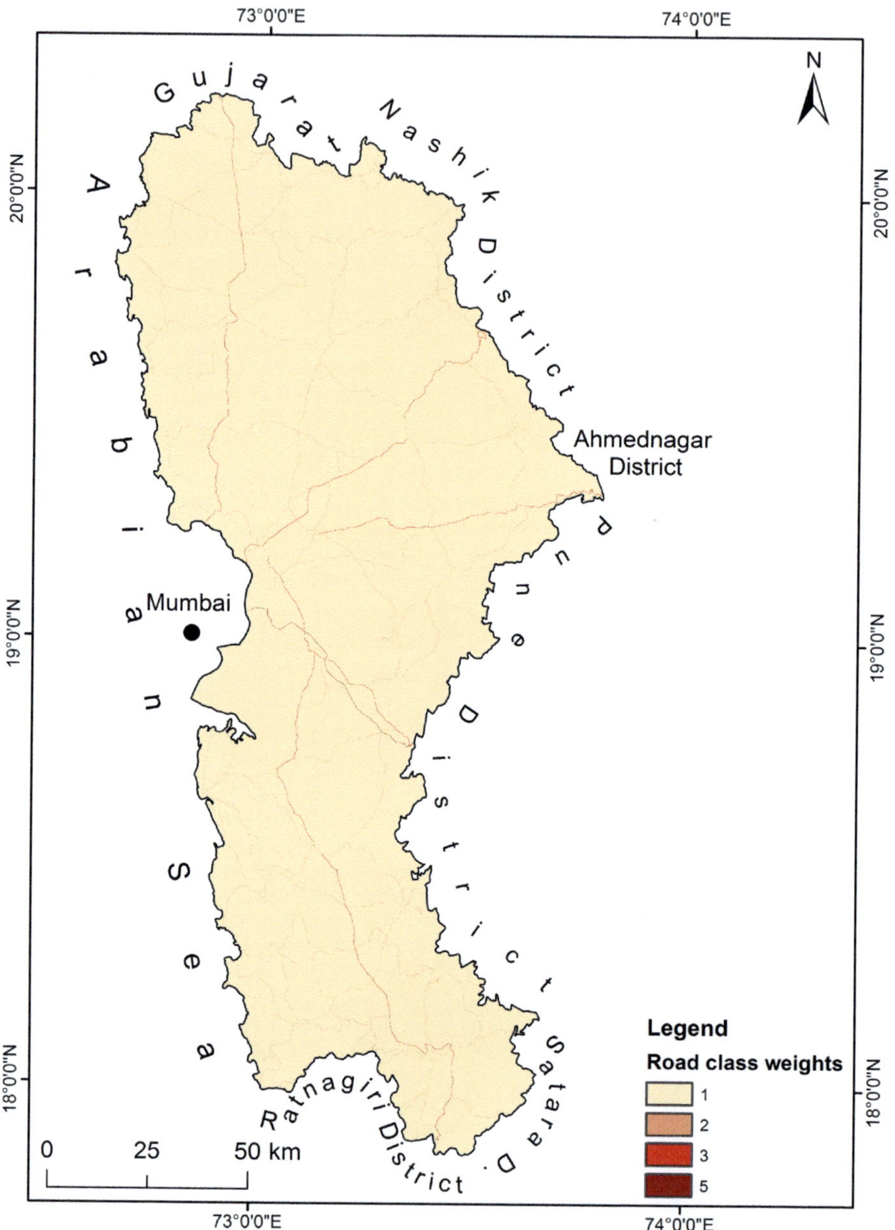

FIGURE 6.10   Reclassified Layer of Road Traffic Density.

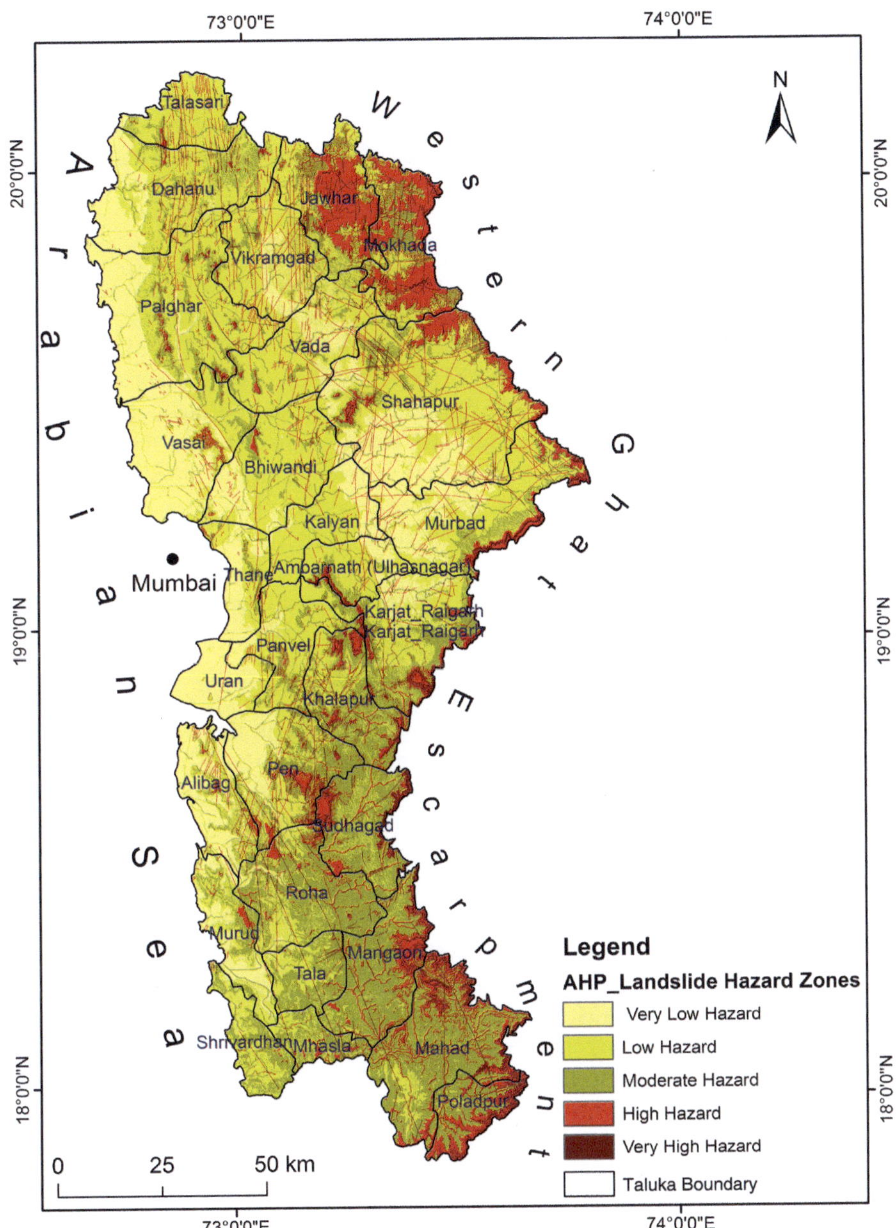

FIGURE 6.11 LHZ Map of the North Konkan Region.

to derive the total landslide risk. Figures 6.12 and 6.13 give details of the distribution of landslide risk classes in the North Konkan region. After integrating all the data layers using the natural breaks method in a GIS environment, the final output is classified into four risk classes: viz., high risk, moderate risk, low risk, and no risk. The spatial patterns of landslide risk areas are shown in Table 6.2 and discussed below.

6.2.4.1   High-Risk Zone

High landslide risk zones in the study area are concentrated in the south-west corner of the Thane district, particularly around the Mumbra-Kalwa slum area, and isolated hill slopes near Kalyan, Panvel, and Ambarnath. The hill slopes of the Matheran area also fall in a high-risk zone. Besides, several other settlements located along the slopes of isolated hills in the rugged south Raigad district show high landslide risk. Few isolated concentrations of high-risk zones are observed along the slopes of small hills around Mahad town. The hill slopes around the Kasara railway station also fall in a high-risk zone. It has been observed that major communication roads passing through the Western Ghat escarpment in the study area fall in a high-risk zone, particularly along the ghat sections. Moreover, the settlements located at the foothills of the Western Ghat escarpment of this region are also in the high-risk zone.

6.2.4.2   Moderate- to Low-Risk Zones

A moderate landslide risk–prone zone is a major zone in the study area, where landslide mitigation measures are to be taken on a priority basis. Of the total geographical area of North Konkan, 12.79% falls under moderate landslide risk zone. The spatial pattern of the moderate-risk zone is uneven. However, the settlements and communication roads in the dissected topography of Raigad district are vulnerable to slope failures. It includes isolated hills near Karnala Fort, the Matheran area, and hilly tracks around Mangaon and Mahad. The inhabited areas along the foothills of the Western Ghat escarpment of Raigad district also fall in this zone.

More than half of the total geographical area of North Konkan region of Maharashtra State, India (54.5%) falls under the low landslide risk zone. It covers a major part of the Palghar and Raigad districts, followed by the Thane district. It covers slopes of small, isolated hills in the coastal lowlands, north-eastern dissected plateau areas, and the north-east and coastal lowlands of Raigad district. Although the plateau margins of Jawhar plateau fall in the high LH zone, it falls under the low-risk zone, probably due to low road traffic density and sparsely populated areas. The spatial distribution of a low-risk zone reveals that the landslide mitigation measures in these areas can be prioritised based on local conditions and the location of settlements with the runout distances of possible slope failures in this zone.

6.2.4.3   No-Risk Zone

Of the total geographical area of North Konkan, 31.24% falls in the no-risk zone. The distribution of safe areas concerning landslide hazards is associated with the coastal lowlands of the study area, particularly in the Palghar and Thane districts,

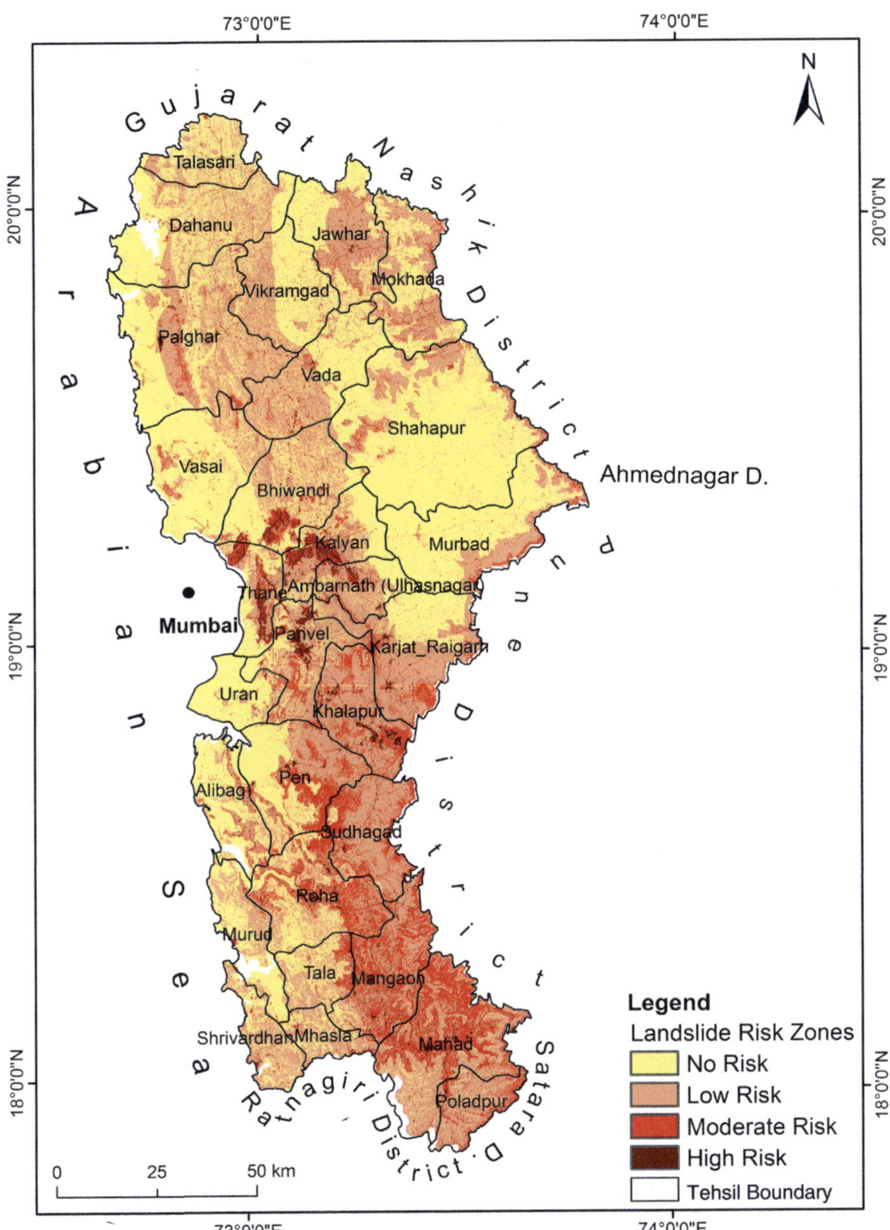

FIGURE 6.12 Landslide Risk Zones.

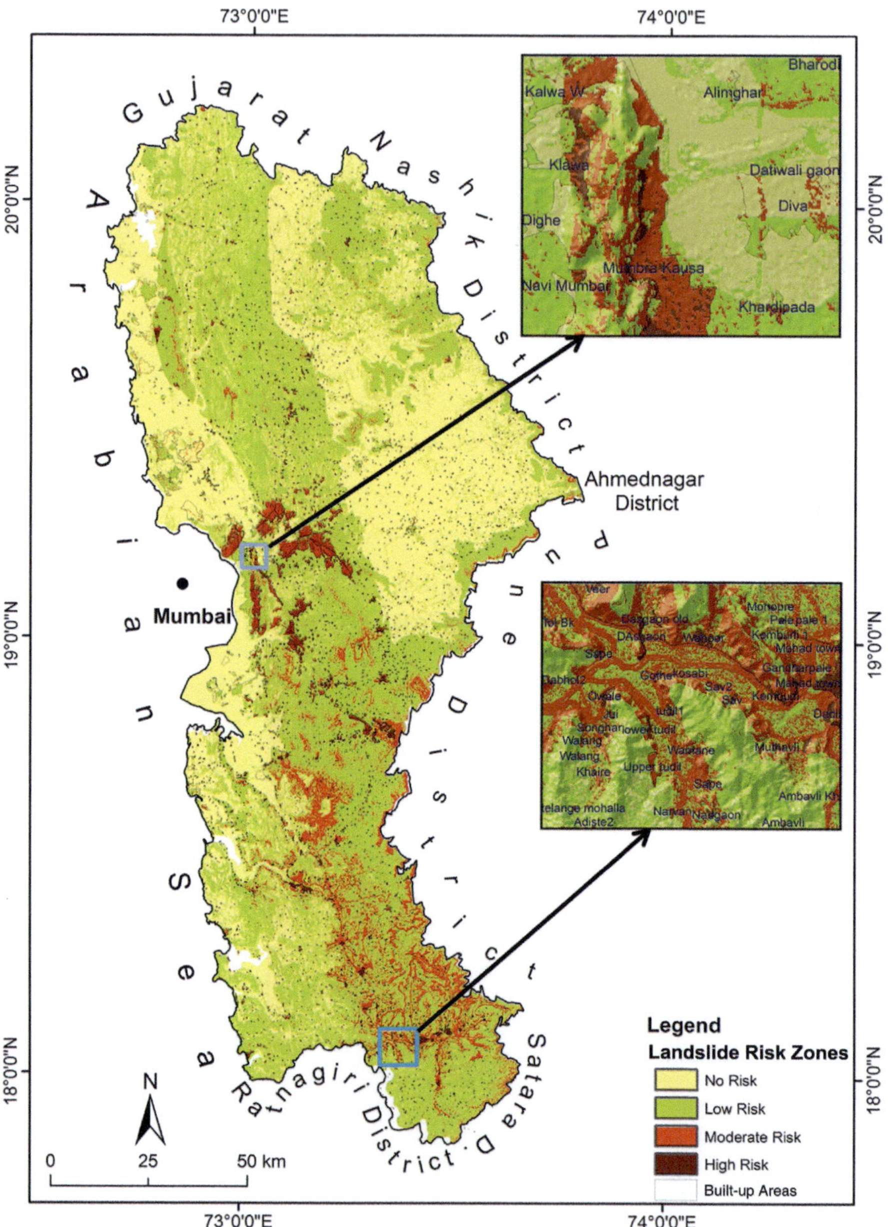

FIGURE 6.13　Landslide Risk Zones: Potentially, Landslide Risk–Prone Settlements.

TABLE 6.2
Landslide Risk Zones in North Konkan

S. No.	Risk Class	Weights (Range)	Area (km²)	Percentage Area
1	High risk	> 21	240.25	1.47
2	Moderate risk	12–21	2,090.34	12.79
3	Low risk	7–12	8,907.23	54.5
4	No risk	< 7	5,105.73	31.24
Total			16,343.55	100

(*Source*: Based on numerical weights used for a population, built-up areas, and roads)

characterised by extensive valley flats with an elevation below 30 m. Besides, the coastal plains in Uran, Panvel, and Alibag tehsils also fall in this zone. The plateau top of the Jawhar plateau also falls in the high risk zone.

The qualitative assessment of landslide risk in the North Konkan region of Maharashtra reveals that high landslide risk–prone areas are associated with the urbanised hilly section in the coastal plains, whereas moderate risk–prone areas are situated in the dissected topography in the south-eastern parts of the study area. All the road corridors passing through the steeply sloping Western Ghats are also observed to be under a moderate- to high-risk zone. The extensive coastal plains in the northern part of the study area and plateau tops in the north-east corner of the study area show no risk associated with slope failures. The risk zones identified using the qualitative method for the North Konkan region can be used for prioritising landslide mitigation measures and for further infrastructural development planning.

6.3 SUMMARY

Landslide risk assessment is one of the most important steps in landslide hazard assessment. Potential losses caused by landslides can be estimated by landslide risk assessment. There are two broad categories of landslide risk assessment: qualitative and quantitative landslide risk assessment. Quantitative landslide risk assessment has proven to be useful in the accurate prediction of landslide risk–prone zones for a given magnitude of landslide. The qualitative landslide risk assessment in the North Konkan region is carried out in this chapter. The spatial patterns of population density, road traffic, and built-up areas, in combination with the LHZ map, have been considered for determining landslide risk–prone areas in the study area. The concentration of both built-up areas and population density are observed in three main pockets: viz., slopes of Parsik Hill near Mumbra-Kalwa, the western slopes of the coastal hill range near Palghar town, and some major settlements in the coastal lowlands, which are highly vulnerable to slope failures.

The landslide risk zones identified using a qualitative approach indicate that 14.26% of the total geographical area in North Konkan falls in a moderate- to high-risk-prone areas, whereas more than half of the North Konkan are (54.5%) is identified as low-risk areas. Densely populated and habited areas in proximity to hill slopes near the Kalwa-Mumbra area and Palghar area show high risk to people and human structures. Besides, small settlements and road sections situated in the Western Ghat region are also depicted as moderate- to high-risk-prone areas. On the other hand, uninhabited areas in the foothills of the Western Ghat escarpment, coastal lowlands, and dissected plateau areas indicate low risk to both people and property. The risk-prone areas identified using qualitative methods may be useful in prioritising landslide mitigation measures and also in policymaking for new infrastructure development.

REFERENCES

Balsubaramani, K., and K. Kumaraswamy. 2013. "Application of Geospatial Technology and Information Value Technique in Landslide Hazard Zonation Mapping: A Case Study of Giri Valley, Himachal Pradesh." *Disaster Advances* 6: 38–47.

Champatiray, P., S. Dimri, R. Lakhera, and S. Sati. 2007. "Fuzzy Based Methods for Landslide Hazard Assessment in Active Seismic Zone of Himalaya." *Landslides, Springer-Verlag* 4: 101–110.

Champatiray, P. 2009. "Perationalization of Cost-Effective Methodology for Landslide Hazard Zonation using RS and GIS: IIRS Initiative." In P. Roy, C. Van Westen, V. Jha, and Lakhera R. Dehradun (Eds.) *Natural Disasters and their Mitigation; Remote Sensing and Geographical Information System Perspectives* (pp. 95–101). India: Indian Institute of Remote Sensing.

Chandel, V., K. Brar, and Y. Chouhan. 2011. "RS and GIS Based Landslide Hazard Zonation of Mountainous Terrain: A Study from Middle Himalayan Kullu District, Himachal Pradesh, India." *International Journal of Geomatics and Geosciences* 2 (1): 121–132.

Chauhan, S., M. Sharma, M. Arrora, and N. Gupta. 2010. "Landslide Susceptibility Zonation through Ratings Derived from Artificial Neural Network." *International Journal of Applied Earth Observation and Geoinformation* 12: 340–350.

Das, I., A. Stein, N. Kerle, and V. Dadhwal. 2011. "Probabilistic Landslide Hazard Assessment Using Homogeneous Susceptible Units (HSU) along a National Highway Corridor in the Northern Himalayas, India." *Landslides* 8: 293–308.

Das, P., P. Phukon, and D. Chetia. 2012. "Landslide Susceptibility Assessment in the Guwahati City, Assam Using Analytic Hierarchy Process (AHP) and Geographic Information System (GIS)." *International Journal of Computer Applications in Engineering Science* II (I): 1–6.

Deshpande, P., J. Patil, D. Nainwal, and M. Kulkarni. 2009. "Landslide Hazard Zonation Mapping in Gopeshwar, Pipalkoti and Nandaprayag Areas of Uttarakhand, GEOTIDE." In *Proceedings of ICC Conference* (pp. 802–812). Guntur, India: Indian Geotechnical Society.

Farina, P., G. Rossi, L. Tanteri, T. Salvatici, T. Gigli, S. Moretti, and N. Casagli. 2017. "The Use of Multi-Copter Drones for Landslide Investigations." *3rd North American Symposium of Landslides* (c): 978–984. https://tinyurl.com/mvv3yvhj.

Ferlisi, Settimio, Antonio Marchese, and Dario Peduto. 2021. "Quantitative Analysis of the Risk to Road Networks Exposed to Slow-Moving Landslides: A Case Study in the Campania Region (Southern Italy)." *Landslides* 18 (1): 303–319. https://doi.org/10.1007/s10346-020-01482-8.

Kanungo, D. P., M. K. Arora, R. P. Gupta, and S. Sarkar. 2008. "Landslide Risk Assessment Using Concepts of Danger Pixels and Fuzzy Set Theory in Darjeeling Himalayas." *Landslides* 5 (4): 407–416. https://doi.org/10.1007/s10346-008-0134-3.

Karlekar, S. 2012. "Landslide Hazard Zonation in Raigad District of Maharashtra: A Multivariate Approach." *Journal of Indian Geomorphology* 1: 75–82.

Lallianthanga, R. K., and F. Lalbiakmawia. 2013. "Micro-Level Landslide Hazard Zonation of Saitual Town, Mizoram, India Using Remote Sensing and GIS Techniques *1,2." *International Journal of Engineering Sciences & Research Technology* 2 (9): 184–194.

Liao, Shifang, Manzhu Ye, Rongcai Yuan, and Wanzhi Ma. 2022. "Unmanned Aerial Vehicle Surveying and Mapping Trajectory Scheduling and Autonomous Control for Landslide Monitoring." *Journal of Robotics*. https://doi.org/10.1155/2022/2365006.

Nagarajan, R., A. Roy, R. Vinodkumar, and M. Khire. 2000. "Landslide Hazard Susceptibility Mapping Based on Terrain and Climatic Factors for Tropical Monsoon Region." *Engineering Geology* 58: 275–287.

Naithani, A. 2007. "Macro Landslide Hazard Zonation Mapping Using Univariate Statistical Analysis in Parts of Garhwal Himalaya." *Journal Geological Society of India* 70: 353–368.

Nguyen, Ba Quang Vinh, and Yun Tae Kim. 2021. "Regional-Scale Landslide Risk Assessment on Mt. Umyeon Using Risk Index Estimation." *Landslides* 18 (7): 2547–2564. https://doi.org/10.1007/s10346-021-01622-8.

Niethammer, U., S. Rothmund, M. R. James, J. Travelletti, M. Joswig, and Lancaster Environment Centre. 2010. "Uav-Based Remote Sensing of Landslides." In *International Archives of Photogrammetry, Remote Sensing and Spatial Information Sciences, Vol. XXXVIII, Part 5 Commission V Symposium*. Newcastle upon Tyne, UK.

Onagh, Mohammad, V. K. Kumra, and Praveen Kumar Rai. 2012. "Landslide Susceptibility Mapping in a Part of Uttarkashi District (India) By Multiple Linear Regression Method." *International Journal of Geology, Earth and Environmental Sciences* 2 (2): 102–120.

Prabhu, S., and S. Ramakrishnan. 2009. "Combined Use of Socio-Economic Analysis: Remote Sensing and GIS Data for Landslide Hazard Mapping Using Artificial Neural Network." *Journal of the Indian Society of Remote Sensing* 37 (3): 409–421.

Prasetyo, Yudo, Nurhadi Bashit, and Reyhan Azeriansyah. 2018. "Analysis of Landslide Disaster Impact Identification Using Unmanned Aerial Vehicle (UAV) and Geographic Information System (GIS) (Case Study: Ngesrep Sub District, Semarang City)." *MATEC Web of Conferences* 159: 0–5. https://doi.org/10.1051/matecconf/2018159001041.

Rammohan, V., A. Jayaseelan, T. Naveenraj, T. Narmatha, and M. Jayprakash. 2011. "Landslide Susceptibility Mapping Using Frequency Ratio Method and GIS in South Eastern Part of Nilgiri District, Tamil Nadu, India." *International Journal of Geomatics and Geosciences* 1 (4): 951–961.

Rawat, M., B. Rawat, V. Joshi, and M. Kimothi. 2012. "Statistical Analysis of Landslide in South District, Sikkim, India: Using RS and GIS." *IOSR Journal of Environmental Science, Toxicology and Food Technology* 2 (3): 47–61.

Saraf, A., J. Das, and V. Rawat. 2009. "Satellite Based Detection of Early Occurring of Co-Seismic Landslides." *Journal of South Asia Disaster Studies (Journal of SAARC Disaster Management Centre)* 2 (1): 47–55.

Sarkar, S., and D. Kanungo. 2004. "An Integrated Approach for Landslide Susceptibility Mapping Using RS and GIS." *Photogrammetric Engineering and Remote Sensing* 70 (5): 617–625.

Sharma, L., N. Patel, M. Ghosh, and P. Debnath. 2009. "Geographical Information System Based Landslide Probabilistic Model with Trivariate Approach – A Case Study in Sikkim Himalaya." In *Eighteenth United Nations Regional Cartographic Conference for Asia and the Pacific.* Bangkok: Economic and Social Council, United Nations.

Tempa, Karma, Kinley Peljor, Sangay Wangdi, Rupesh Ghalley, Kelzang Jamtsho, Samir Ghalley, and Pratima Pradhan. 2021. "UAV Technique to Localize Landslide Susceptibility and Mitigation Proposal: A Case of Rinchending Goenpa Landslide in Bhutan." *Natural Hazards Research* 1 (4): 171–186. https://doi.org/10.1016/j.nhres.2021.09.001.

Thigale, S., and B. Umrikar. 2007. "Disastrous Landslide Episode of July 2005 in the Konkan Plains of Maharashtra, India with Special Reference to Tectonic Control and Hydrothermal Anomaly." *Current Science* 92 (3): 383–385.

Tofani, V., S. Segoni, A. Agostini, F. Catani, and N. Casagli. 2013. "Use of Remote Sensing for Landslide Studies in Europe." *Natural Hazards and Earth System Sciences* 13: 299–309.

van Westen, C., T. Van Asch, and R. van Sester. 2006. "Landslide Hazard and Risk Zonation – Why is It Still Difficult?" *Bulletin of Engineering Geology and the Environment* 65: 167–184.

Varnes, D. I. 1984. *Landslide Hazard Zonation: A Review of Principles and Practice.* Paris: United Nations Scientific and Cultural Organization.

Wagh, K., and P. Deshpande. 2013. "Landslide Locations and Vulnerability Assessment in RS/GIS Environment." In *Proceedings of Indian Geotechnical Conference* (pp. 1–13). Institute of Indian Technology, Roorkee, India.

Yaprak, Servet, Omer Yildirim, Tekin Susam, Samed Inyurt, and Irfan Oguz. 2018. "The Role of Unmanned Aerial Vehicles in Monitoring Rapidly Occurring Landslides." *Geodetski List* 72 (2): 113–132.

7 Landslide Hazard Management/ Mitigation Measures
Recent Developments

7.1 INTRODUCTION

Landslide hazard mitigation is the process of reducing the impact of landslides on people, property, economic activities, and natural features, including vegetation and the habitats of plants and animals. Landslides are the reason for losses in various ways. Amongst these losses, economic and social losses can be controlled or reduced through effective planning and management (Dai et al., 2002). To reduce or control the risk of landslide hazards, it is very necessary to carry out mitigation. Proper execution of mitigation work can reduce losses in terms of life and economy. Mitigation of landslide hazards is also essential to reduce social and economic stress, for the rehabilitation of the people, and to avoid unnecessary expenses during rescue operations. The Sendai Framework of 2015–2030 indicated targets and priorities for reducing or preventing disaster risks (UNISDR, 2015).

7.1.1 Need for Landslide Hazard Mitigation

Mitigation is the process that avoids or reduces the impact of the hazard. In the mitigation process, different factors play a vital role. It can be implemented before and after the hazard events. In the pre-hazard phase, it is mainly in the form of preparedness, awareness of the local people, and releasing warnings of hazards. The nature of mitigation in the post-hazard phase is rescue and rehabilitation. Before designing and planning mitigation measures, it is necessary to understand the region's physical environment. Besides this, the social environment of the region should be well understood for effective mitigation measures. The occurrence of landslide events is directly related to the surface slope. Surface slopes also determine human habitation and economic activities; therefore, there is always a risk of landslides in areas with unstable slopes or areas of landslide history. The areas with very high population density and intensive economic activities are more susceptible to landslide hazards.

DOI: 10.1201/9781003384205-7

The areas become a 'high landslide risk zone' not because of natural processes or human activities but because of the presence of human habitation and other human-economic interests such as mining, agriculture, and infrastructure development. In developed and developing countries, such risk is multiplied due to intense economic activities, great demand, and pressure on land resources. In densely populated regions of the world, a burden on land is observed. Therefore, in landslide hazard mitigation, it is very obvious to prioritise such areas.

It is possible to reduce the impact of landslide hazards through various measures. These are practised worldwide. All these measures are meant to reduce the impact of landslides, but in some cases, the results are not seen even after mitigation measures. Some of these are very successful methods, and many of these could not reduce the risk of losses due to landslides.

7.2 MAJOR LANDSLIDE MITIGATION STRATEGIES

Landslide mitigation strategies depend on the types of landslides that occur in the prone areas. There are different methods adopted by different international, national, and regional authorities.

In landslide mitigation strategies, the most important part is preparedness for the expected hazard event. In the preparedness part, it is essential to involve the scientific community. Data collection about landslides, prone areas, and other human habitation and economic details is also an important step towards the development of landslide inventory.

While executing mitigation plans, one should have information related to landslides such as the location of the event, the type of landslide, landslide inventory, and the population of the region and their economic activities.

The rescue team should know about the lithology, slope, and drainage characteristics related to the physical environment. It is also essential to have a landslide inventory of the landslide-affected areas. For effective implementation of the mitigation, a landslide susceptibility map should be available to know the landslide-susceptible zones. Information such as the region's infrastructural development will also help better to manage resources. Human activities like the construction of roads and settlements and infrastructural development should be planned as per the landslide-prone areas. Based on such information, a mitigation plan can be designed and successfully implemented to avoid or minimise losses.

Mitigation of landslides can be planned in such a way that the losses would be none or very less. Mitigation measures should be designed and planned as per the requirement of the locals. While designing mitigation measures, it is required to consider the parameters related to the environmental, scientific, technological, economic, social, and political factors.

There are various landslide mitigation strategies:

1 Impose limits on development in landslide-affected areas.
2 Activities that alter the land surface, including landscaping, excavation, and construction, should be limited.

3 Engineering structures and physical measures, such as a change in slope geometry and diversion of drainage, should be used to prevent or control landslides.
4 Awareness of local people related to landslide triggering factors.
5 Mock drills and use of warning systems.

While selecting appropriate measures to reduce landslide risks, it is very much necessary to understand the root cause(s) of the slope failure process. AGS (2007) have described various parameters and causes of slope failure or landslides. For the mitigation purpose, it is essential to understand the causes and triggering factors of landslides in those areas.

7.2.1 GENERALISED STEPS OF HAZARD MITIGATION

As per FEMA (2022), the mitigation process varies from region to region. The mitigation process involves various factors. Besides government bodies, the involvement of experts and local people is very much essential. Following are the generalised steps required to mitigate the hazards.

There are four core steps in any type of mitigation (FEMA, 2022). This includes the following:

1 Resource organisation and planning
2 Landslide hazard risk assessment
3 Developing hazard mitigation plans
4 Adoption of mitigation plans and their implementation

These core steps are adopted in every type of hazard assessment and practice throughout the world. In the past few decades, the availability of geospatial data has made these steps very effective.

The resources involved in the landslide mitigation process can be divided into the following components:

1 People
2 Government and nongovernment agencies
3 Field executives

7.2.2 ROLE OF PEOPLE IN HAZARD MITIGATION

Mitigating any hazard is a collective effort. It is believed by many scholars and many other scientists that people's participation is very much necessary. It is also observed that in many parts of the world, the mitigation measure is considered the duty of the government and administration. In such cases, people are the least bothered.

It is observed that in many countries, people do not take part in mitigation work. The participation of people plays an important role in mitigation work. Local people are aware of many aspects related to the local physical, social, and cultural conditions.

In landslide hazard mitigation, people can assist at various stages of mitigation. People can provide information on historical events of landslides, which is useful in the preparation of landslide inventory. People can also help in a hazard risk awareness campaign.

Finally, it is very important to take the help of local people in rescue and rehabilitation.

In the case of mitigation planning and management, involving local people will make it effective. Before and after the event, if the appropriate information is collected from and with the help of locals, the execution of the plans and mitigation measures will become relatively easy for the administrators. Before the event, it is the preparedness phase in which locals can help in the identification and confirmation of potential sites of hazard, sites for temporary shelters, and also rehabilitation.

After the event, locals can help in the rescue operation (Figure 7.1).

7.2.3 Government and Nongovernment Organisations

The second important part is government and nongovernment organisations. Local, provincial, and federal government authorities are equally responsible for mitigation planning and management at all three stages of hazard mitigation.

It is also expected that the government authority revise the information based on the landslide inventories updated every year. For instance, the National Disaster Management Authority (NDMA) is an apex government organisation that is involved in making policies and planning disaster management. The National Disaster Response Force (NDRF) is a trained force that specialises in rescue and relief operations. NDMA (2019) has also suggested various steps to

FIGURE 7.1   Human Resources in Hazard Mitigation.

reduce landslide risk and measures to mitigate landslide hazard in 'The National Disaster Management Plan 2019'.

Nongovernment organisations are also involved in mitigation measures. Particularly in India, many nongovernment organisations volunteer for rescue operations. The people also volunteer in collaboration with government officials and NDRF personnel.

7.2.4 Field Executives

Field executives play an important part in the mitigation of any hazards. They are involved in the execution of the mitigation plan, rescue operations, and relief and rehabilitation. Government organisations have different bodies working for different purposes, such as preparedness, planning, and execution of relief and rescue operations.

Field executives are government employees or belong to nongovernment organisations. Besides, the locals also volunteer in rescue and relief operations as executives.

Giving training to field executives is an integral part of mitigation. Government and nongovernment personnel, in disaster rescue forces, are trained. It is very essential to give training to the local people who are from landslide-prone areas.

7.3 LANDSLIDE MITIGATION METHODS

Mitigation methods have evolved with advancements in technology over time. These methods and structures are developed mainly to reduce the impacts of landslides. They can be broadly divided into conventional and modern methods. Modern methods mainly emphasise the monitoring of slope movements and deformities of slopes. These include the application of geospatial technology.

7.3.1 Conventional Mitigation Measures

A landslide is triggered by various factors. These natural and man-made factors can be controlled partially or completely through different actions. Among them are physical structures and bioengineering structures, through which the level of landslide hazard can be reduced (Prasad, 1995).

The physical structures are engineering structures designed and constructed to stop or reduce landslides and the effects of landslides. Conventional structures have their advantages and disadvantages. The main problem with these structures is that they change the entire slope system. The construction cost is very high. These structures cannot be sustained over a long period unless they are massive. Big-size structures are costly and require more area for construction. On transport-limited slopes, the construction of such structures is recommended when the slope length is more and the pressure exerted is relatively high.

There are different structures given as under.

7.3.2 Retention Structures

Various structures are designed to retain the landslides (Popescu and Sasahara, 2015):

- Retaining walls
- Gabion walls
- Block walls
- Passive piles for slope stabilisation
- Passive piers for deep slides
- Reinforced concrete walls
- Reinforced metallic or polymer earth–retaining structures
- Bolting and anchoring of boulders to prevent rockfall
- Retention steel nets to trap the free fall of boulders
- Protective cement surfaces
- Preventive boulders or concrete blocks

7.3.3 Slope Modification Includes the Following Processes

For the prevention of slope failure, a change in slope geometry can be a long-lasting solution, but these changes also have disadvantages. Due to the removal of a large volume of earth material, the slope geometry modification causes loss of weathered mantle and plants present on the surface. The costs involved in the removal and dumping of material are also considerable. Dumping of material at new places also causes damage to new areas. Keiko and Satoru (2006) suggested a slope management approach to mitigate landslide hazards in urban areas:

- Removing of material through slope cutting
- Dumping of material in a downslope direction as 'fill' or 'waste'
- Cutting of slope to reduce slope angle
- Terracing the slope to control the downslope movement of material

7.3.4 Bioengineering Methods

Bioengineering methods are used to prevent slope movement as well as to reduce the moisture present in the weathered mantle at the potential landslide site.

7.3.5 Surface Drainage Diversion

Surface water in the form of surface and subsurface flows is a triggering factor for landslides. The porosity and permeability of the slope material allow the movement of water through the mass. It leads to an increase in the mass of the material and further leads to downwards movement. To avoid such types of movement and slope failure, different techniques of surface flow diversion are used.

Surface drains, diverted from the slide area, are used to drain the water out of the slide area via deep and shallow trenches. These are lined trenches so that the water moves quickly without percolating into the slide areas. In structures like retention walls, 'weep holes' are prepared through which subsurface water drains out. Other structures are also used; these are mainly to trap subsurface water flow and to empty it at an appropriate location. These structures include vertical boreholes, which are self-draining, pumping vertical wells of large diameters. The sole purpose is to create drainage in different forms to remove excess water from potential landslide sites. Tree plantation is also an option to reduce moisture from the weathered mantle. Trees and grasses also hold the material to a certain extent.

7.3.6 Other Techniques Used to Reduce Potential Slope Failure and Debris Flow

Land surface treatment is done to reduce surface erosion. Debris flow mitigation is done through the reduction of soil erosion, plantation, mitigating wildfire, creating debris flow retention walls, and debris flow basins (Highland and Bobrowsky, 2008).

7.3.7 Modern Techniques

Besides mitigation measures in the form of conventional structures, new, emerging technology is predominantly used to monitor landslides. Pre-hazard conditions are monitored to reduce the impact and losses caused by landslides.

Modern techniques are highly dependent on instrumentation. Geospatial technology is one such method deployed to monitor changes in the land surface. Following are the geospatial methods used to prepare different thematic maps. These maps are directly or indirectly used in the identification of landslides, prone areas, and potential areas, in mitigation plans, and also in development plans of landslide-affected areas.

Remote sensing, GIS, and location positioning techniques such as GPS are essential techniques in mapping landslides.

Following are the techniques and types of maps used in landslide hazard zonation, risk maps, and mitigation planning maps:

- Use of digital elevation models (DEM) and topographic maps to show slopes, slope aspects, and type of terrain.
- Remote sensing techniques like digital aerial photogrammetry, InSAR and LiDAR imaging, and use of unmanned aerial vehicles (UAV) for mapping and monitoring slope deformation.
- Slope deformation can be monitored through the use of GNSS, terrestrial laser scanning, using a network of wireless sensors (WSN), and ground instrumentation.

- Surface drainage, network and pattern, and drainage density can also be generated using DEMs.
- Geological maps show the type of rocks, material present, and geomorphology of the surface. It is also possible to extract the mapping of the weathered mantle from the geomorphic map of the area.
- Map of vegetation cover to show surface vegetation.
- Soil maps show the distribution of particular soils with their characteristics.
- Assessment of such maps also provide information on landslide triggering factors.
- Data related to human settlements, infrastructure, elements of risk, areas at risk, and risk zonation maps are generated using geospatial techniques.

Other techniques such as the use of a piezometer and extensometer, though not related to geospatial technology, also play important roles in the understanding of landslide hazards.

7.4 LANDSLIDE HAZARD MANAGEMENT SYSTEM IN MINIMISING LOSSES CAUSED

The landslide hazard management system is a part of landslide mitigation. The steps involved in the hazard management system are discussed in Section 7.1. For the management of landslide hazards, the steps to be taken are in the form of preparation of a database related to landslides before the events, rescue and relief just after the events, and the third part is rehabilitation. These important steps are discussed in earlier parts of the chapter. Hazard management systems must consider the technical systems required to get the details of the landslide as a process, landslide causative factors, and their monitoring, making people aware of all aspects of landslides. People living in landslide-prone areas should be trained with respect to the identification of potential landslide sites or zones, responding to the warning system, and monitoring and reporting slope deformation in the habited areas. The adoption of various hazard mitigation measures also depends on the geographical location and the type of landslide occurrence. People living in landslide-prone areas must be made aware of the entire procedure of landslide mitigation. The administrative authorities must involve local people and seek their opinion in certain cases.

REFERENCES

AGS. 2007. "Australian Geoguide (LR2)." *Geomechanics* 42 (1): 162–163.
Dai, F. C., C. F. Lee, and Y. Y. Ngai. 2002. "Landslide Risk Assessment and Management: An Overview." *Engineering Geology* 64 (1): 65–87.
FEMA. 2022. *Hazard Mitigation Planning Process.* www.fema.gov/emergency-managers/risk-management/hazard-mitigation-planning/create-hazard-plan/process. Accessed in December, 2022.

Highland, L. M., and P. Bobrowsky. 2008. *The Landslide Handbook-a Guide to Understanding Landslides* (p. 129). Reston, Virginia: U.S. Geological Survey Circular 1325.

Keiko, I., and S. Satoru. 2006. "Slope Management Planning for the Mitigation of Landslide Disaster in Urban Areas." *Journal of Asian Architecture and Building Engineering* 5 (1): 183–190. https://doi.org/10.3130/jaabe.5.183.

NDMA. 2019. *National Disaster Management Plan-2019. A Publication of the National Disaster Management Authority*. New Delhi: Government of India.

Popescu, M., and K. Sasahara. 2015. *Engineering Measures for Landslide Disaster Mitigation*. https://damfailures.org/wp-content/uploads/2015/07/082_Engineering-Measures-for-Landslide-Disaster-Mitigation.pdf.

Prasad, N. B. 1995. *Landslides-Causes and Mitigation*. www.researchgate.net/publication/317328970_Landslides-Causes_Mitigation.

UNISDR. 2015, March 14–18. "Sendai Framework for Disaster Risk Reduction 2015–2030." In *UN World Conference on Disaster Risk Reduction*. Sendai, Japan; Geneva: United Nations Office for Disaster Risk Reduction. www.unisdr.org/files/43291_sendaiframe workfordrren.pdf.

Author Index

Subject Index